S0-BWV-534

OUR ENERGY:
REGAINING CONTROL

A STRATEGY FOR ECONOMIC REVIVAL THROUGH REDESIGN IN ENERGY USE

MARC H. ROSS
ROBERT H. WILLIAMS

McGRAW-HILL BOOK COMPANY

New York St. Louis San Francisco
London Paris Tokyo Toronto

To our families and especially to our children
Chris, Ann, and Peter
and
Wesley and Ruth
in the hope that theirs will become a more livable world.

Thomas H. Quinn and Michael Hennelly were the editors of this book.
Christine Aulicino was the designer. Thomas G. Kowalczyk supervised the production. It was set in Electra by Rocappi, Inc.
Printed and bound by R. R. Donnelley and Sons, Inc.

Library of Congress Cataloging in Publication Data

Ross, Marc H.
Our energy, regaining control.

Bibliography: p.
Includes index.
1. Energy policy—United States. 2. Energy conservation—United States.
I. Williams, Robert H., joint author. II. Title.
HD9502.U52W56 333.79'0973 80-23612
ISBN 0-07-053894-8
123456789RRDRRD8987654321

CONTENTS

PART III: REINVENTING AMERICA

PREFACE

IN THE LATE 1960s both of us were working in the mainstreams of our physics profession. But roughly at the time when we first met in 1970, concerns about environmental and other adverse impacts of technology on modern society were beginning to capture our professional attention. We came to feel that if society is to cope with these problems there would have to be a redirection of the course of technological change. Since the evolution of technology is strongly shaped by political decisions, we came to believe that it would be necessary for some scientists and engineers to ply their special skills to help clarify technological alternatives, in order to improve this decision-making process.

By 1972 we had both made decisions to work professionally on problems of technology and were particularly attracted to energy issues (in part simply because the topic is relatively close to our physics background). In 1972 R. W. became staff scientist to the Ford Foundation's Energy Policy Project and in 1974 M. R. was co-director of the American Physical Society's summer study on efficient use of energy.

At the end of these efforts we decided to collaborate on a book that would explore a long-term energy strategy for the U.S. emphasizing the opportunities for utilizing energy more efficiently. The writing proceeded slowly because of our continuing involvement in various energy projects. But these projects and our participation in the public debates on energy policy during the late 1970s gave us valuable insights, many of which have been incorporated into this book. Both in terms of our own understanding and in terms of general knowledge about energy demand, a book like this could not have been written in 1974.

As the writing proceeded, critical choices were made defining its scope and emphases. A few are worth mentioning here. First, the book presents a domestic rather than a global strategy. We demonstrate in this book that actions at the national and local levels could resolve most of the nation's energy problems and would be the first big steps toward resolving the remaining inherently international issues. The greatest potential for decisively improving the energy situation rests within our own country because the U.S. is still the world's leader in technology and is the largest producer and consumer of energy.

Another major choice was to concentrate on futures for this country in which established trends in the average individual's lifestyle persist. Many analysts believe that radical changes in lifestyle will be dictated by increasingly troublesome problems of energy and natural resource management. While some radical changes in lifestyle may be desirable, we demonstrate in this book that such changes are not necessary. Thus in order to keep an already complex topic from getting out of hand, we adopted recent trends for the society as a whole as our guide in formulating a description of lifestyles in the future.

We also chose to emphasize cost analysis in the book. We did not make this choice with the thought that if energy option A appears to be a few percent cheaper than B, then A is preferable. Indeed we emphasize that prices in the marketplace are flawed as guides because energy costs are, on the one hand, influenced by important subsidies and, on the other hand, do not adequately reflect the side-effects of energy activities (such as adverse environmental and national security impacts). Moreover, no cost analysis provides the final word on the value of an activity, because many values cannot be so quantified. Nevertheless energy is a pervasively used, everyday commodity which already absorbs more than 10% of all economic effort, so that energy choices are constrained by simple costs. An energy option for which direct costs are relatively high would imply major sacrifices. Thus simple cost analysis is a key component in good decision making about energy.

One minor choice we made was to use a mixture of units for physical quantities. There is no unique set of units for energy and power, even in the standardized metric system. In our judgment, any advantages of global uniformity from use of metric units are completely outweighed by an increase in intuitive understanding by our readers which would be facilitated by use of familiar units such as gallons of gasoline and square feet of floor space. This choice reflects the need to involve in decisions relating energy-efficiency investments not just the experts but millions of people—in our homes, in our plants and offices, and in our communities.

We would like to acknowledge contributions of many people and several organizations during the six long years this book was in preparation, but will explicitly cite only a few. The book was started in the summer of 1974 during a delightful interlude as guests of Walt and Barbara Carnahan in the wine country of Hammondsport, N.Y. A major boost in 1977 was two weeks of secluded writing we enjoyed in the wilderness cabin made available to us by the von Hippel family. Especially important to the writing of this book was the hospitable atmosphere provided by Princeton University's Center for Energy and Environmental Studies, where undertakings such as ours, which do not fall into any neat academic niche, are not only tolerated but nurtured.

We are grateful for various sources of support for this undertaking. An advance from the Levinson Foundation enabled us to work on the book together at Princeton University during 1977–78. Also grants from the Levinson Foundation, the Rockefeller Brothers Fund, and the Ford Foundation supported research efforts related to this book by one of us (R. W.).

Among our colleagues, Claire Adler and Arthur Schwartz of the University of Michigan and Harold Feiveson, Robert Gilpin, Bert Smiley, Robert Socolow, and Frank von Hippel of Princeton University provided us with constructive criticism and advice regarding parts or all of the manuscript as it was being developed. One of us (M. R.) would like to thank Roger Sant and his colleagues at the Mellon Institute's Energy Productivity Center for valuable discussions. Various students also provided us with helpful insights. We wish in particular to acknowledge the assistance of Howard Geller, who performed detailed district heating calculations for our chapter on a solar community energy system.

Among other individuals who provided us with invaluable assistance in the preparation of this book, we are especially grateful to Sidney Shapiro of the Levinson Foundation for his continuing encouragement; to the late Daniel Lapedes for bringing our manuscript to the attention of McGraw-Hill; to Tom Quinn and Mike Hennelly at McGraw-Hill for their superb help in editing the manuscript; and to Alice Carroll, Joann Poli, and Barbara Pinkham for their patient typing and retyping of the many drafts of the manuscript. Finally we appreciate the enduring support of our wives, Joan Ross and Elinor Williams.

August, 1980

CHAPTER 1

INTRODUCTION

THE UNITED STATES HAS LOST CONTROL over its energy system. With the decline in domestic production of oil and gas and the great increase in the world oil price accompanying the Arab oil embargo of 1973, the United States should have made dramatic advances in bringing forth alternative energy supplies. This has not happened. Attempts to build coal gasifiers on a commercial scale have been frustrated by repeated delays; efforts to extract shale oil have met critical technical and environmental setbacks; the coal industry is in the doldrums; twice as many nuclear power plants were canceled as were ordered between 1975 and 1979; all major new construction projects to expand our supply of energy have been plagued with runaway costs; and oil imports increased nearly twofold in volume and sixteenfold in total cost between 1972 and 1979.

These failings cannot be blamed on governmental neglect. On the contrary, there has been a crescendo of incentives to expand energy supply, government demonstration projects, and new regulations and bureaucracies. A policy emphasizing the expansion of energy supply requires this move toward central planning because supplies are far more difficult to create today than in the bygone era of cheap oil and gas. For many projects, the uncertainties associated with technical performance, future energy prices, and environmental regulation create financial risks which the private sector is unwilling to assume.

We do need new sources of energy, but the present emphasis on developing high-cost energy sources to support growing energy demands is fundamentally ill-conceived. If present federal programs should be *successful* in meeting their objectives, the result would be that the energy system would divert economic resources away from needed non-energy activities, that environmental problems would become increasingly unmanageable, and that global political stability would be threatened.

An even more disturbing, perhaps more realistic outlook is that present energy programs will not meet their goals. If these programs continue to be plagued with delays and other setbacks, the resulting *unexpected* shortfalls in production will lead to repeated crises requiring rationing, the closing of industrial plants, and other dislocations caused by emergency fuel-allocation directives.

This book describes how American society can regain control of its energy system. Fundamental changes in energy policy are needed to break the present impasse. Our present policy, which is preoccupied with energy supplies, does not take into account the fact that supply expansion is only one way, and an increasingly ineffective way, of meeting energy needs. There are also vast opportunities for meeting these needs, at a lower cost to society, by saving energy through energy efficiency improvements. Some examples: we can reduce the fuel requirements for space heating from a major to a minor fraction of our housing-energy budget by plugging heat leaks; we can cut back gasoline consumption substantially by tripling or even quadrupling the fuel economy of our automobiles; we can generate electricity with only half the fuel required today by exploiting the advantages of industrial cogeneration, in which electricity is produced at industrial sites as a byproduct of the generation of heat for industrial processes.

But to exploit these opportunities and raise saved energy to the status of energy supply, the nation and its energy users must stop conceiving of the energy crisis in terms of energy supply and consider instead the services that energy provides. The homeowner, for example, should think not of buying heating oil, but of buying warmth and comfort; he should think not of buying electricity, but of buying illumination and the other services provided by electricity. And energy-supply industries should reorganize themselves to provide these services instead of simply providing energy supplies. This reorganization would provide a framework conducive to improving energy efficiency. At the same time, the focus of government energy policy should be shifted away from promoting a handful of energy-supply technologies toward fostering myriad approaches to improving the efficiency of energy use. This shift in priorities would require a fundamental change in the use of policy tools—away from the present trend to central planning and toward a greater reliance on market mechanisms for shaping the course of energy-related development. This approach is necessary because detailed central management of the diverse uses of energy is wholly impractical. Because the present energy market functions so badly, the challenges to public policy would be to foster competitive conditions for providing energy services and to create an economic climate in which consumer decisions would take into account the total costs of the services—including social costs not normally reflected in market prices.

Once policies were adopted to achieve these goals, the task of providing new supplies would be far easier than at present because *the opportunities for improving energy efficiency are so great that total energy demand could actually decrease over several decades without adversely affecting the economy.* Indeed this fuel-conservation strategy would provide the means for Americans to realize their economic goals in the face of more costly energy. In shifting capital resources from energy-supply expansion to corresponding energy-efficiency improvements, there would usually be net capital savings, so that capital resources would be freed up for the much-needed modernization of American industry. Thus a fuel-conservation strategy would foster improvement in the total productivity of the economy.

By pursuing this course, environmental and global-security risks would be reduced—first because these risks tend to increase with the level of energy use, and second because it would be possible to avoid or minimize use of the most risky technologies.

Perhaps the most important benefit of this alternative strategy, however, is that because its logic is so compelling, the prospects are good for achieving a broad consensus to pursue it among consumers, producers, and environmentalists—factions which today are embroiled in highly polarized debates on the nation's energy course.

Our exposition of this strategy is divided into three parts. Part I describes the present energy impasse. *Readers who have closely followed energy developments may want to skim through or skip certain chapters* (e.g., Chapters 4 and 5) in this part of the book and refer back to them only as necessary. Part II focuses on the use of energy throughout the economy and describes opportunities for innovation and cost reduction through energy-efficiency improvements. Part III integrates the specific proposals from Part II into a coherent policy proposal for the management of natural resources and technological innovation relating to energy in the United States.

PART I

TREND CANNOT BE DESTINY

CHAPTER 2

MATERIAL GROWTH AT A TURNING POINT

MATERIAL GROWTH IN THE PAST

During the past century the advances of science and technology have given rise to an expansion in production of goods to levels the imagination can hardly grasp. The resulting levels of consumption can be brought down to earth by expressing them in per capita terms. Figure 2.1 shows that *every day the average American consumes his weight in the most important basic materials.* Can one fail to be astonished that use of natural resources is so high?

This staggering level of resource consumption is not devoted mainly to the growth of industry as such or to the very rich or to the military. It is true that nearly all raw materials are consumed by industry and that, in the case of fuels, industry and commerce account for roughly ⅔ while individual households directly account for only ⅓ of total consumption. However, most industrial consumption is for the immediate production of consumer goods. Only about 15% of all basic materials are used for purposes which do not serve immediate consumer needs. The United States is truly a consumer society.

It is also a society where even the poor consume materials at a high rate. A poor American consumes more basic materials than income would suggest—for example, more than the *average* person in Denmark, and more than the typical rich person of a century ago. (In fact a

Fig. 2.1

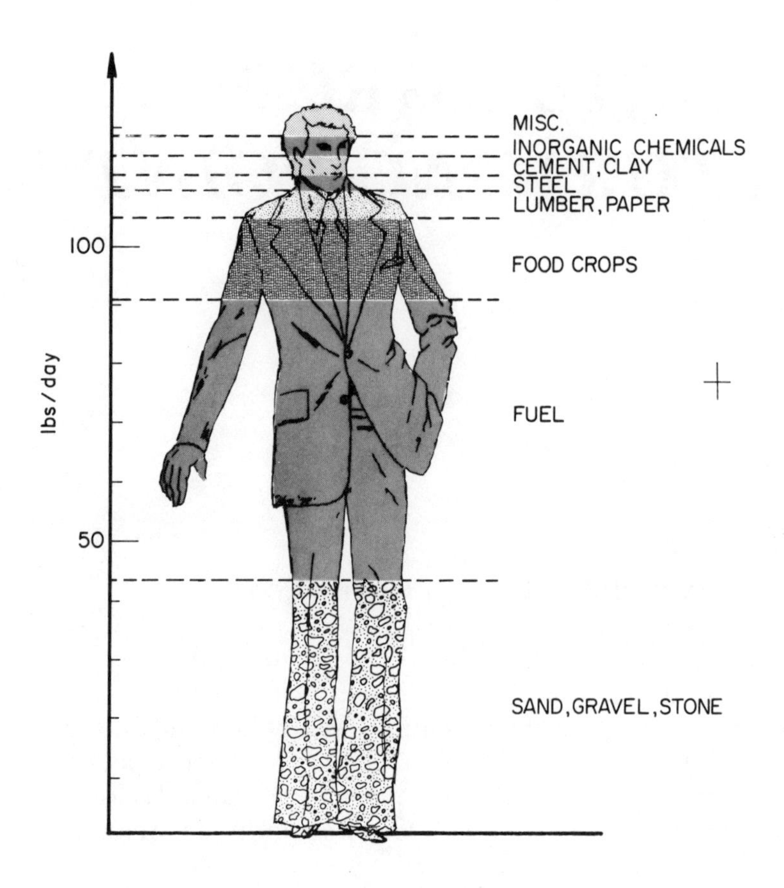

PER CAPITA CONSUMPTION OF BASIC MATERIALS IN THE U.S.

high level of consumption of materials is not a good indicator of well-being. This issue is examined in the context of household energy consumption in Chapter 8.)

Much of the growth that led to these levels of material consumption took place in this century. From the '20s to the '70s GNP (gross national product) grew almost fivefold in real, or constant-dollar, terms, i.e., after correcting for inflation. While not all basic materials experienced such growth (e.g., lumber production did not increase), steel, cement, paper, key inorganic chemicals, and fuels experienced phenomenal growth of production almost equaling and in some cases exceeding the overall growth in the economy. Also, a number of "new" materials such as organic chemicals (plastics, synthetic fibers, resins), ammonia (for fertilizer), and aluminum have experienced especially rapid growth of production in recent decades.

How has this country been able to achieve these levels of use of natural resources? Can and should this growth be sustained? As the use of any finite natural resource increases, other things being equal, further use becomes more difficult as the quality of the remaining resource declines. It may, however, be possible to mitigate these difficulties by means of new technology. John Stuart Mill stated these relationships over a century ago:

> . . . all natural agents which are limited in quantity, are not only limited in their ultimate productive power, but long before that power is stretched to the utmost, they yield to any additional demands on progressively harder terms. This law may, however, be suspended, or temporarily controlled, by whatever adds to the general power of mankind over nature; and especially by any extension of their knowledge, and their consequent command, of the properties and powers of natural agents.[1]

Some 19th-century economists and policy makers predicted that scarcity of natural resources would soon become a dominant force in industrial society. Yet in 1963 Barnett and Morse[2] could show that since 1870, the earliest date for which they had significant statistics, the costs of various foods and minerals, including fuels, had declined or remained steady. They examined the question in two ways. They showed that the labor involved in producing food and minerals (including both the direct labor of production and indirect labor to provide the needed equipment and structures) had declined severalfold per unit of physical product (per pound of copper, bushel of wheat, etc.). This means that the average person could avail himself of more and more food and minerals. Barnett and Morse also showed that the prices of food and minerals were fairly steady with respect to the price of all other products. Thus as Figure 2.2 shows, for a typical product the fraction of the cost devoted to natural resources remained fairly steady. These results are hardly in accord with the notion of growing scarcity of natural resources. In the past, while use of natural resources grew to astonishingly high levels, the resulting decline in the quality of resources was well compensated for by the advance of technology.*

* Barnett and Morse found, however, that one other category of natural resource, forestry products, did increase in cost (by both measures). It is no coincidence that growth of consumption of lumber, the dominant forest product in this period, was not great.

Fig. 2.2

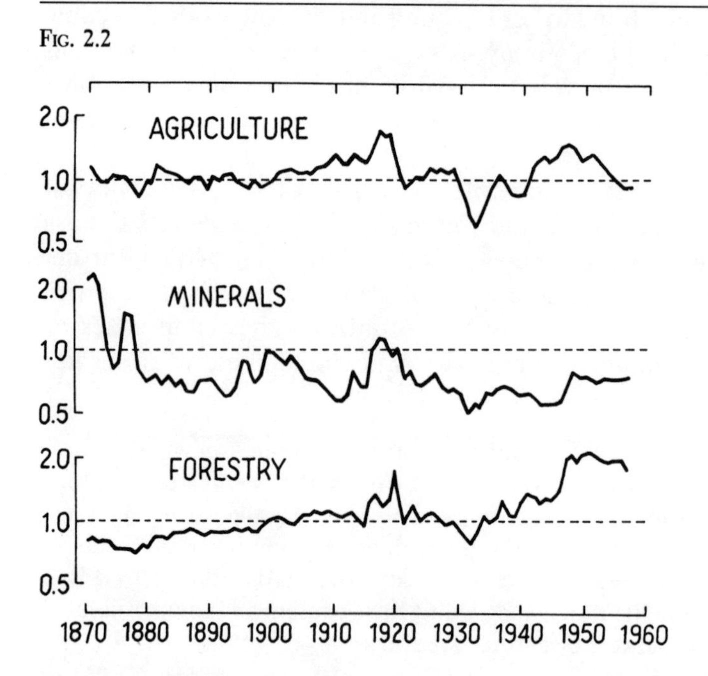

Trends in unit prices of extractive products relative to non-extractive products in the United States, 1870–1957

SOURCE: H.J. Barnett and C. Morse, *Scarcity and Growth: The Economics of Natural Resource Availability*, ref. (2).

The role of technology in keeping down prices is dramatically illustrated by examples from the energy sector of the economy. It would be quite a feat to retail Gulf of Mexico water in the upper Midwest for 30¢ a gallon—including substantial taxes and services. Yet before the 1973 price increases that was roughly the retail price of gasoline.*

The bounty of nature made possible very low production costs for petroleum and thus allowed the giant oil corporations to continually expand their markets and maximize long-term profits. Investments in large-scale refineries, pipelines, and later in tankers and a huge, growing sales volume kept costs down, with the result that the price of gasoline fell in real terms until 1973. And even after the price hike brought on by the oil embargo, the real price of gasoline in 1978 was less than in 1960, as shown in Figure 2.3.

* The production and transportation costs were actually much less. In 1971 the retail price of gasoline in Detroit was 32¢ per gallon, of which 13¢ was for production and transportation, 7¢ was for distribution services, and 12¢ was for taxes. See ref. 3.

Fig. 2.3

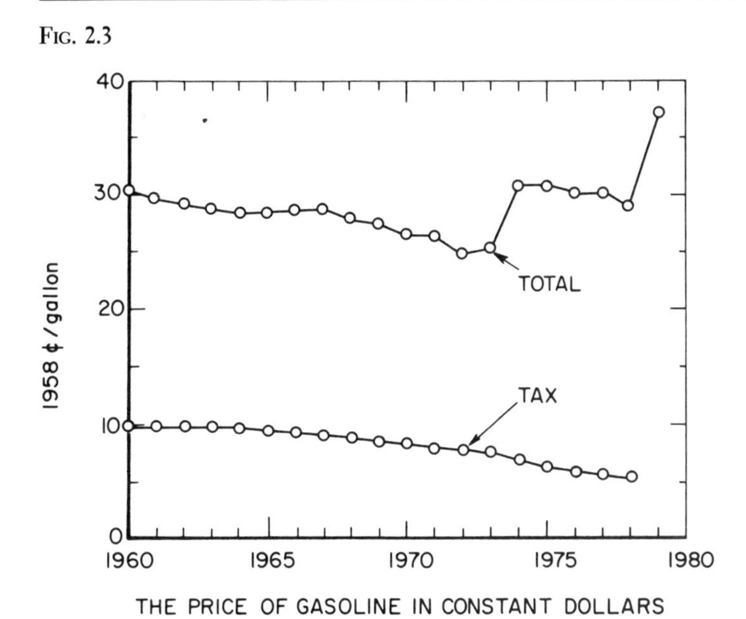

NOTE: Here the price of gasoline is converted to constant dollars using as a deflator the consumer price index.

Natural gas has been even cheaper than oil. The price of natural gas rose slightly during the '50s and '60s but remained substantially below the price of oil as measured in energy-equivalent units. When natural gas was introduced to most parts of the United States after World War II, its price at the wellhead was so low that for 15¢ one could buy the energy equivalent of the physical work a man could do in one year!

The price of electricity also fell rapidly up to 1970. As Figure 2.4 shows, the price drop was particularly dramatic for residential customers, who paid only ⅕ as much for electricity (in real terms) in 1970 as in 1940. Rapidly improving generation and transmission equipment, larger facilities, and expanding markets all made lower prices possible.

Thus in the first ¾ of the 20th century the prices of most natural resources were steady or falling and consumption was rapidly increasing. Energy became abundant and astonishingly inexpensive during this period. Is this to be regarded as the normal state of affairs, or was this a rare moment in human history? What's in store for the final quarter century and beyond?

FIG. 2.4

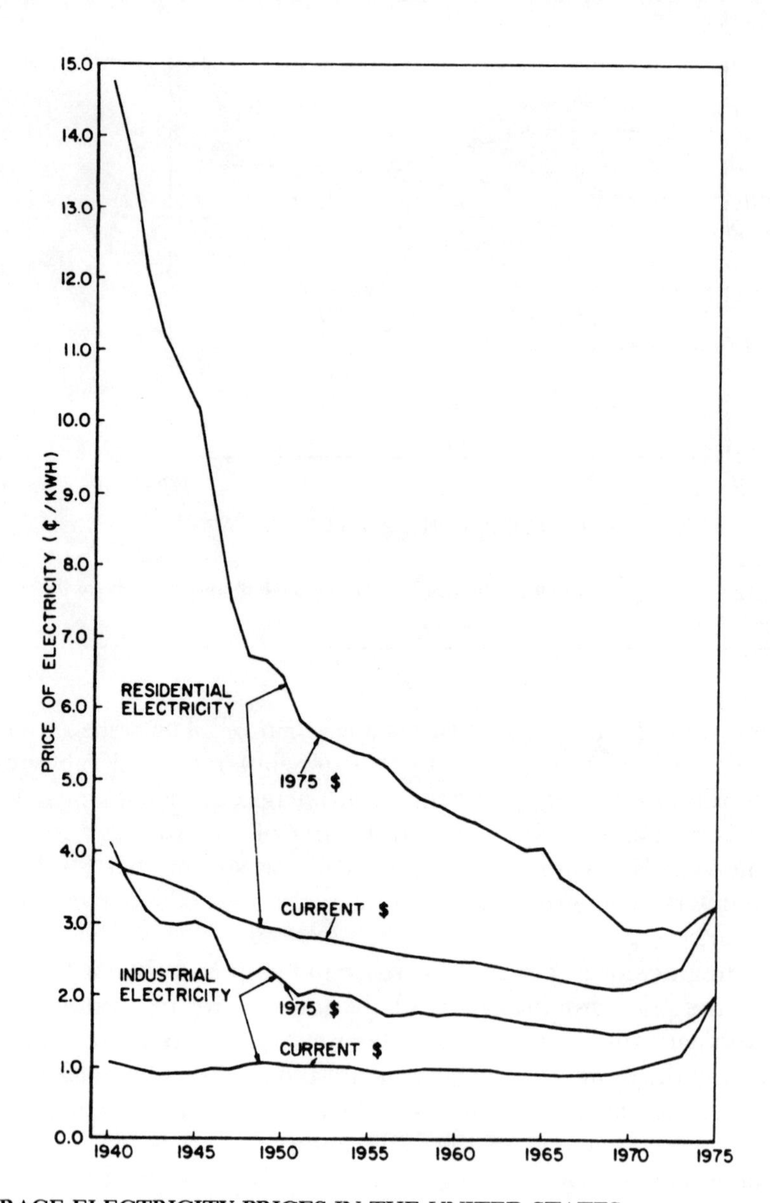

AVERAGE ELECTRICITY PRICES IN THE UNITED STATES

The price of residential electricity is converted to constant 1975 dollars using as a deflator the consumer price index. For industrial electricity, the deflator used is the wholesale price index for industrial commodities.

A NEW ERA FOR INDUSTRIAL SOCIETY

While the *long-term* historical record suggests that material growth in the future might continue apace, *recent* trends strongly indicate that industrial society has reached a turning point relating to material consumption. There are four major factors working together that point to a fundamental change: the diminished capacity of technology to keep down the prices of raw materials (including energy), the declining importance of material inputs in the production of economic product, the need to establish stringent controls to limit the adverse side effects of technology on society, and demographic changes that will inevitably slow down economic growth in the future.

PRICES

For many basic materials prices have begun to rise in real terms. This reversal of trend began rather early for copper. The grade of copper ores has fallen dramatically over time. By modern times the grade of copper ores had fallen to a few percent of the grade of copper-bearing materials available in the Bronze Age and in pre-Columbian America, when copper was found in certain areas in nugget form. During the first half of this century the quality of copper-bearing materials in the United States declined continuously from ores containing 4% copper in 1900 to ores with 1% copper in 1950, as shown in Figure 2.5. But in this same period technology advanced so as to keep prices from rising. Over the last quarter century the quality of copper ores was again cut in half, but technological change in this period was inadequate to keep prices in check. Figure 2.5 shows that as a result, copper consumption has ceased to grow.

Everyone is familiar with the more recent trend of rising fuel prices. While the sharp rise in fuel prices after 1973 was in response to price fixing by the OPEC (Organization of Petroleum Exporting Countries) cartel, this dramatic political action was possible because the principal *new* energy resources of the world other than OPEC oil appear to be much more costly than energy resources have been historically. New oil and gas resources in the United States are going to be much more costly because they must be sought in environments that are more difficult to exploit, such as above the Arctic Circle on the Alaskan North Slope, and in the ocean using platforms serviced by helicopter. In addition, new oil reserves onshore from known reservoirs often will require the use of costly "tertiary" recovery technology, which, in contrast with oil recovery from gushers, involves pumping hot water or chemicals

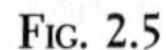
Fig. 2.5

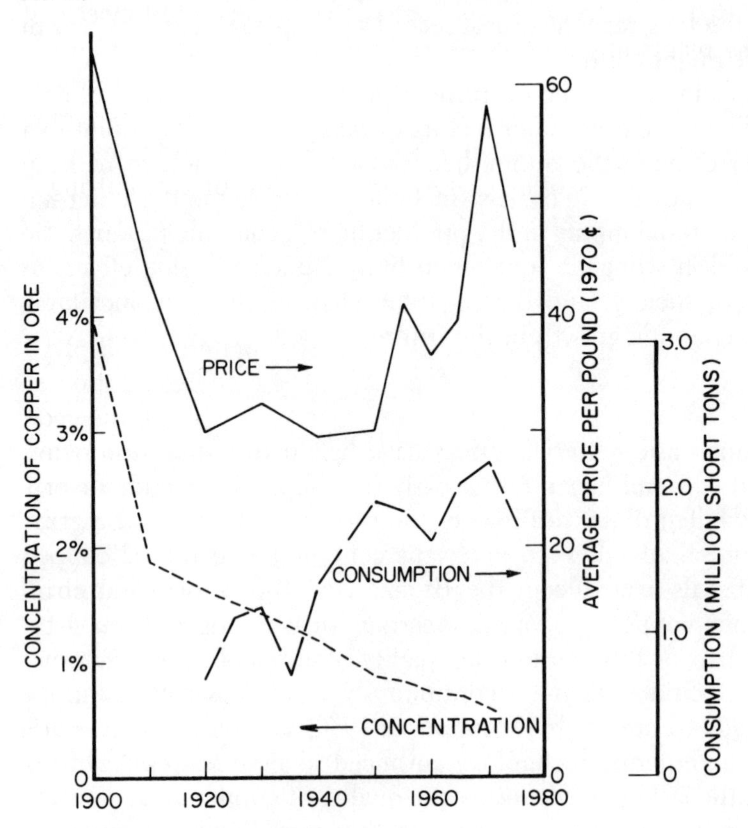

THE HISTORY OF THE ECONOMICS OF COPPER

SOURCE: Ref. (4)

down into wells to coax some more of the oil out. The cost of this "new" oil, or, as economists call it, the replacement cost of oil, is much greater than the cost of "old" oil.

Electricity prices are also rising much faster than general inflation. In part, the recent reversal of the long-term downward price trend for electricity came about because of the hike in world oil prices and its repercussions in markets for natural gas, coal, and uranium. However, close scrutiny of Figure 2.4 shows that the price trend for electricity actually reversed before the oil embargo of 1973, reflecting the increasing costs for electric power plants. The cost of electricity from new plants, the replacement cost, is much greater than the cost of electricity from existing plants for several reasons.* New requirements for air-pollution-abatement technologies are driving up the price of electricity based on coal. Quality-control problems and toughening safety regulations are driving up the cost of nuclear electricity. In addition, as we show in Chapter 4, the technological opportunities for cost cutting in central-station power generation are drying up. Thus the prospects are that the price of electricity will continue to rise for decades.†

MATERIALS SATURATION

As a natural adjustment to rising prices, basic materials will play a smaller role in economic growth. There is evidence that saturation effects are also leading in a declining role for basic materials in the economy. While in the second quarter of this century consumption of important basic materials generally grew about as fast as the overall economy, Figure 2.6 shows that in the third quarter there was a marked shift away from major basic-materials-intensive activities. This new trend was particularly pronounced in the steel and cement industries. Even in the chemical industry, often viewed as a major growth industry, the consumption of basic materials has grown only about as fast over the last 10 years as total industrial output. This is shown in Figure 2.6 by the curves for feedstocks and three inorganics, which have been almost level since the late 1960s. It is noteworthy that this relative decline in basic materials was already becoming established before the 1970s in the period of declining prices for basic materials, showing that saturation is leading the United States away from materials-intensive activities.

That saturation is occurring should hardly be surprising. In a society so consumption oriented that each person consumes daily his weight in "stuff," consuming even more stuff becomes less desirable and less

* When a new plant is added to an electric utility's system, the cost of electricity from this plant is averaged in with the cost from all existing plants. As a result of this averaging, the price of electricity rises only slowly toward the replacement cost.

† We have cited factors relating to *electricity supply* that are driving up replacement costs. There is also an important *electricity-demand* factor at work driving up replacement costs: the ongoing trend to use electricity more and more for "temperature-sensitive loads" (air conditioning and space heating). For these loads most consumers tend to demand large quantities of electricity simultaneously. The utility is thus faced with a very sharp peak demand for electricity on the very hottest and the very coldest days. While these peaks are very infrequent, the utility must be prepared to meet them. Because the needed reserve generation, transmission, and distribution capacity is idle much of the time, these peak demands drive up costs. For details see Chapter 8, pp. 113–114.

FIG. 2.6

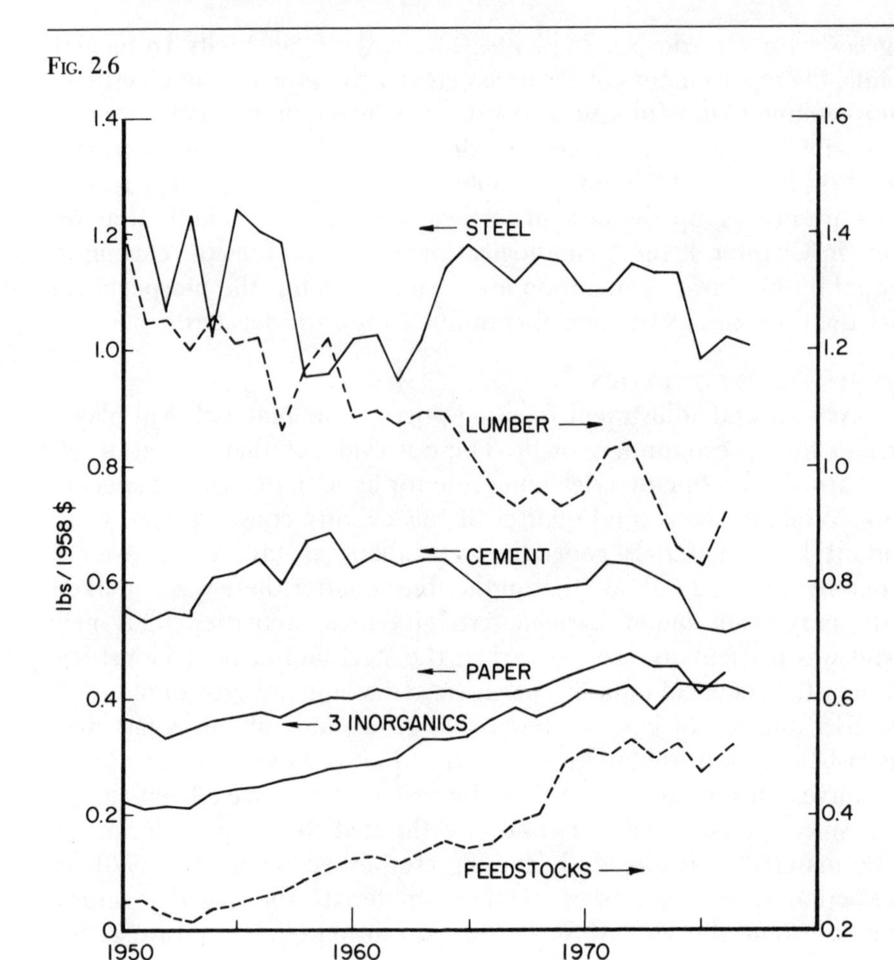

CONSUMPTION OF BASIC MATERIALS PER DOLLAR OF INDUSTRIAL VALUE ADDED

SOURCE: See note (5)

manageable. Although the United States is still a consumer society, services, which are less materials-intensive than goods, are accounting for a growing share of consumption. Moreover, the goods consumers buy are tending to contain more "value added" and fewer raw materials. Such shifts away from materials-intensive products throughout the economy are reflected in the industrial sector by a trend away from processing basic materials toward a greater emphasis on finishing and fabrication. In the steel industry, for example, where heavy steel rails represented a major component of production in the past, steel output today is made up more of thin-rolled steel of special qualities. Thus the recent decline in steel production relative to total industrial output does not mean that the United States is doing without steel. Far from it. Rather it means that in today's products steel is used more judiciously and more delicately.

Are we short-sighted in making this case for materials saturation? What about new consumer products? We find there are no product innovations on the horizon which might create major requirements for materials and fuels so as to sustain energy and materials consumption relative to economic activity. Fresh applications of petrochemicals (aside from substitutions for other highly processed materials), major energy-intensive household appliances such as swimming pools and hot tubs, and private boats and airplanes, are not becoming widespread enough to have a major impact; and many new products are not, in fact, very intensive in their consumption of materials and energy relative to their cost. Most of the innovative products which *are* taking off—light equipment for home and recreational use, information technologies, biomedical technologies, and other services—have low material and fuel requirements relative to their cost.

SOCIAL REGULATION

The adverse side effects posed by modern technologies are becoming so serious that the trend will be toward stronger social regulation in areas involving public and occupational health and safety, national security, and environmental protection. This trend will inevitably lead to less growth in very hazardous activities, which often tend to be materials-intensive activities. Such regulatory activity does not imply economic stagnation, however, since there are substantial opportunities for economic growth in areas involving less hazardous, less materials-intensive activities.

DEMOGRAPHY

Finally, economic growth itself can be expected to slow down dramatically in the coming decades because of slowed population growth, or more specifically slower growth in the labor force. The "labor-force boom" that followed the post-World War II "baby boom" and has been a major factor propelling economic growth will have run its course by the mid-1980s.

* * *

THIS BRIEF DISCUSSION has surveyed the major factors that lead us to conclude that even if the well-being of the American people continues to improve in the long term, the era of phenomenal growth in material consumption is coming to an end. At this point we turn from a general discussion of material growth to a more focused discussion of future trends in energy growth.

CHAPTER 3

OUR PRESENT ENERGY COURSE

It is very difficult to make an accurate prediction, especially about the future.

NIELS BOHR

WE WILL SHOW IN THIS BOOK that the level of future energy demand is more a matter of societal choice than of prediction. It is nevertheless useful to project how energy demand might grow in the future in a surprise-free world where demographic, economic, and energy-price trends developed more or less as now expected, and where no significant new energy policies were enacted that might change the situation. Knowledge of energy-demand trends helps provide a quantitative basis for understanding the problems of energy-supply expansion.

We describe such a projection in this chapter and call it a "Business-as-Usual" projection.* We shall show that under Business-as-Usual conditions energy use would grow much more slowly in the future than in the past, owing to the collective impact of factors discussed in the preceding chapter.

While our Business-as-Usual projection involves much slower growth than conventional projections made recently by government and industry groups, it still represents substantial growth. As we shall show in Chapters 4, 5, and 6, this Business-as-Usual growth in demand still poses formidable problems relating to energy supplies.

* The meaning of Business-as-Usual has, in a sense, been changing. Our perspective at the time of writing includes 6 years of experience since the events of 1973.

FALSE PROPHETS

We first discuss how one might derive the conventional high projections. Most energy projections involve viewing the future as an extension of the past. One way to make a projection is to observe that since the end of World War II energy consumption has increased almost as fast as the overall rate of production of goods and services in the economy, as measured by GNP, corrected for inflation. Energy consumption has grown sometimes more rapidly than GNP and sometimes more slowly. On the average it grew 0.5% per year more slowly than GNP between 1950 and 1975. Perhaps this trend will continue. If so, then one might estimate energy growth, naively, as follows:

$$\begin{pmatrix}\text{annual energy}\\ \text{growth rate}\end{pmatrix} = \begin{pmatrix}\text{annual GNP}\\ \text{growth rate}\end{pmatrix} - 0.5\%/\text{year} \qquad (3.1)$$

Since in the third quarter of this century GNP grew at an average annual rate of 3.5% per year, energy demand might be expected to grow in the future at an average annual rate of

$$3.5 - 0.5 = 3.0\%/\text{year} \qquad (3.2)$$

If this growth rate were sustained, then energy demand would grow over the next quarter century, from 71 *quads** in 1975 to 149 quads in 2000—more than double the 1975 level.

Many governmental and industrial organizations have spent a great deal of time and effort in recent years estimating future energy demand. As shown in Table 3.1, many of the recent projections are remarkably close to our naive projection given above.[1] These projections are important because they are targets for energy policy making: they influence the course of multi-billion-dollar investments by industry, they influence the shaping of government subsidy and taxation policy, and they influence the shaping of the government's energy research and development program.

But how good are these projections? Let us consider the projection shown in Table 3.1 by the Edison Electric Institute (EEI), the trade association for the electric utility industry. It was set forth in a weighty hardback volume entitled *Economic Growth in the Future.* The institute retained the services of a very prestigious economic consulting firm to make the projection using a complex and sophisticated econometric model linking energy, economic, and demographic quantities.

* One quad (quadrillion, or 10^{15} Btu) per year is equivalent to half a million barrels of oil per day, or the fuel-consumption rate for 16 large new electric power plants.

Table 3.1 Some Recent Government/Industry Energy Projections

DATE OF PROJECTION	STUDY	PROJECTION FOR 2000 (QUADS)
1975	U.S. Bureau of Mines[a]	163
1975	ERDA-48[b]	
	Intensive Electrification Scenario	160
	Combination of All Technologies Scenario	131
1976	Edison Electric Institute (Moderate Growth)[c]	161
1977	Stanford Research Institute (Base Case)[d]	143
1979	Electric Power Research Institute[e]	150

NOTES:

[a] W. G. Dupree and J. S. Corsentino, *United States Energy Through the Year 2000 (Revised)*, Bureau of Mines, U.S. Department of Interior, December 1975.

[b] See ref. 13

[c] See ref. 7

[d] *Fuel and Energy Price Forecasts*, volumes I and II, 1977, Report EPRI EA-433, prepared for the Electric Power Research Institute, Menlo Park, California.

[e] Chauncey Starr and Stanford Field, "'Economic Growth, Employment and Energy," *Energy Policy*, March 1979, p. 2.

A major problem with this and many similar projections is that it is based on very dubious assumptions. Consider first GNP. The EEI projection involves an average real-GNP growth rate for the period 1975–2000 of 3.7% per year, slightly faster than the average for the preceding 25 years. This, we believe, is highly unrealistic. To understand why we expect a slower growth rate, it's useful to express the growth rate for GNP as follows:

$$\begin{pmatrix}\text{annual GNP}\\ \text{growth rate}\end{pmatrix} = \begin{pmatrix}\text{annual growth rate}\\ \text{for employment}\end{pmatrix} + \begin{pmatrix}\text{annual growth rate for}\\ \text{labor productivity}\end{pmatrix} \tag{3.3}$$

where the term *labor productivity* means the average GNP per employee. Fundamental changes are taking place in both employment and labor productivity that will have a profound effect on future GNP growth.

The level of employment depends on population factors (e.g., birth rates, death rates, immigration, age distribution), on the participation rate (i.e., the fraction of the working-age population that wants to work), and on the capacity of the economy to employ those who want to work. We can estimate with a fair degree of confidence how fast employment would grow if the nation returned to and maintained a "full employment" economy (4–5% unemployment). This can be done because the population factors bearing on employment till the year 2000 are fairly well established (the growth in the labor force at any time is determined by the birth rate 16 years earlier)[2] and because it appears that the labor-force participation rate will not increase much in the future.[3]

Because of a reduced fertility rate employment growth is going to be much slower in the future than in the past. As Figure 3.1 shows, there has been a long-term decline in the U.S. fertility rate, interrupted by the postwar baby boom. After 1960 the fertility rate dropped sharply until it stabilized at about 1.8 children per woman during the 6-year period ending in 1979.* This reduced fertility rate means that in the 1980s and beyond the labor force and therefore employment will be growing relatively slowly. Specifically, if the U.S. economy returns to "full employment" by 1985, employment will grow only about ½ as fast in the period 1985–2000 as it did in the period 1950–1975.

We can also expect slower GNP growth in the future because of a trend toward slower growth in labor productivity, the total output per worker.[5,6] Labor productivity tends to increase with technical change and capital investment. Goods—food and appliances, for instance—have been produced by fewer and fewer workers as time has passed. Indeed roughly 10 times as many goods are produced by a worker today as a century ago.[12] Mechanization had led to a large reduction in the number of farm workers and has kept down growth in factory employment. But at the same time there has been a large increase in service-sector employment—in supervision, sales, promotion, real estate, insurance adjustment, government services, etc. Such activities are less easily mechanized than the production of goods. While it is possible to automate mass services, employees providing personal services are roughly no more productive now than they ever were; e.g., a teacher today teaches as many pupils about as well as a century ago. If the trend toward service employment persists, then the growth rate for productivity of the economy as a whole will be somewhat less in the future than in the past.†[4]

* For the population to replace itself in the absence of immigration, the fertility rate in the long run should be about 2.1 children per woman. However, the U.S. population will continue to increase for many years even if the fertility rate is below 2.0, because of immigration and because the present population has such a large cohort of women of childbearing age. The reader should keep an eye on the fertility rate as an important indicator for America's future.

† Growth rates for productivity have been very low over the last decade. See Chapter 14 for a discussion of the reasons for this slowdown. Our projection is based on productivity trends for 1950–1975 and so is not dominated by the last decade. See note 4.

FIG. 3.1

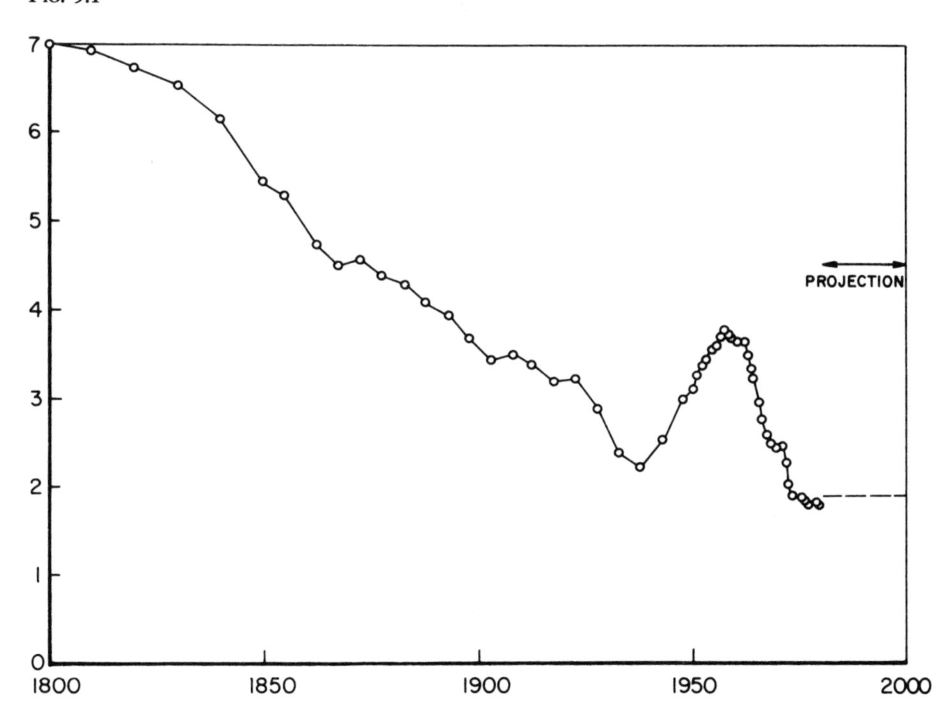

U.S. FERTILITY RATE

For this figure, the births in each year have been converted to a "fertility rate," the effective number of births over the lifetime of an average woman. At left the figure shows the remarkable decline from about seven children per woman in the early decades of the nation, to about 3 children per woman in the early 1900's. A sharp temporary drop occurred in the Great Depression. The subsequent baby boom began in the war years and peaked in the late 1950's. There has been a spectacular drop to half the peak level since that time. The present fertility rate is slightly below the "replacement level" of about 2.1 children per woman. Even at this level the U.S. population will grow for several decades because of the large cohort of women of childbearing age in the present population (in addition to growth from immigration).

SOURCE: Data for 1971-1978 were obtained from "Population Profile of the United States: 1978," *Current Population Reports: Population Characteristics*, Bureau of the Census Series p. 20, No. 336, April 1979. Data for 1940-1970 were obtained from *Historical Statistics of the United States: Colonial Times to 1970*, Part I, Bureau of the Census, 1975. Data for 1800-1940 are for the white population only and were obtained from A.J. Coale and M. Zelnik, *New Estimates of Fertility and Population in the United States*, Princeton University Press, Princeton, 1963.

The combined effect of these employment and productivity trends is that even if we should return to and maintain a "full employment" economy, GNP would grow no faster between 1975 and 2000 than about 3.0% per year on the average, instead of the 3.7%-per-year growth rate used in the EEI projection. A reduction in the GNP growth rate by 0.7% per year may not seem very significant. But small changes in exponential growth rates over periods like 25 years translate into enormous differences in consumption levels in economies where material consumption is initially high. In this case the level of energy use in the year 2000 would be reduced 26 quads from the EEI projection. The difference is equivalent to the fuel consumed annually by 450 large nuclear power plants—enough nuclear plants to produce more electricity each year than all electric power plants in the United States today!

Another flaw underlying the EEI projection involves energy prices. As we pointed out in Chapter 2, there is strong evidence that real energy prices in the United States will be much higher in 2000 than today. The EEI projection is based on the assumption that the price of gas in the year 2000 will be only about 50% higher than in 1975,[7] a gross underestimate because the cost of gas from new wells is already much higher.[8] In the case of electricity EEI errs even more outrageously. The price of electricity is projected to be 37% *less* in 2000 than in 1975.[10] This is pure fantasy, because electricity prices are now rising and will continue to rise. We show in Chapter 8, for example, that the cost of providing electricity to an all-electric home from new electricity sources (the replacement cost of electricity) is about twice the average residential electricity price in 1975.[11]

Let us examine the EEI projection from another viewpoint. What would the energy and in particular the electricity be used for in 2000? The econometric model used by EEI can't provide this kind of information. Engineering models have, however, been developed at Brookhaven National Laboratory (BNL) and elsewhere that suggest how energy might be used in detail. For example, BNL modelers developed the federal government's ERDA-48 scenarios shown in Table 3.1.[13] One of these projections, the Intensive Electrification Scenario, is very similar to the EEI projection. The overall levels of energy use are about the same, as are the projected degrees of electrification.[*,14] This model therefore can be used to gain insights into electricity use in the EEI

Perhaps the most remarkable feature of the Intensive Electrification Scenario is that the BNL modelers were hard pressed to find

* The degree of electrification is the percent of all fuel used to generate electricity. At present electrification is 29%; EEI projects it will rise to about 50% by 2000.

enough demand for electricity. There is so much electricity available that they were forced to "get rid of" excess electricity by channeling more than *all* the electricity produced in the United States in 1975 into resistance heaters to produce heat needed for industrial processes.[15] It makes no sense to use resistance heaters to generate heat for industrial process-heat applications, except in a few specialized operations. The direct use of fuels is almost always more economical, especially for low-temperature process-steam applications (e.g., in canning foods and converting petroleum into plastics), which account for most industrial process heat.[16] Electric resistive heating is so unsuitable for such applications that it wasn't even included among the eight alternatives for generating process steam in the future evaluated in a recent study by Dow Chemical Co.[17]

A general critique of the ERDA/BNL energy scenarios of 1975 has been given by von Hippel and Williams.[18] They found that even the projections which nominally emphasize conservation involve waste which is not credible in the light of present energy price trends. They also pointed out important areas where saturation effects were ignored.

Perhaps the most flagrant neglect of demand saturation in these scenarios relates to automobiles. The ERDA/BNL scenarios project that the use of the automobile will double between 1975 and 2000.[19] That this level of automobile use is simply incredible can be understood in terms of the time people spend in cars. While at present the average person spends about an hour a day in a car,[20] the ERDA/BNL projections imply that auto usage is growing so fast that in 2000 the average person will spend an hour and 40 minutes in a car each day,[20] which is preposterous. To look at saturation another way, at present there are 0.75 autos per adult, defined as a person of age 16 or over. The fraction of fully utilized autos certainly cannot be greater than 1.00 and cannot become much bigger than 0.75 because there are young, infirm, and very elderly people who will remain without cars and because there will always be one-car households. Yet the ERDA/BNL projections imply that there will be 1.2 fully utilized cars per adult in 2000.[21]

These examples of exaggerated future growth in output, underestimation of future energy prices, and neglect of important saturation effects are fundamental errors characteristic of all high-energy-growth projections. Furthermore, the complex econometric models often used to develop these projections are inherently unsuitable for making projections into periods when experience can be expected to be very different from that of the past. Thus the EEI projection has no more truth value than the naive projection set forth at the beginning of this discussion. But the use of this complex model is certainly worth every penny to the utility industry. The industry bought not just results, but the authority of a prestigious consulting firm which used the most advanced "software" technology. The EEI, with this projection in hand, can plead with public policy makers for more tax relief and other subsidies to help the industry get on with the business of meeting the projected exponential growth in demand. While policy makers may be unable to understand how the results were generated, how can they fail to be awed?

A BUSINESS-AS-USUAL PROJECTION

It would be grossly unfair to say that government and industry energy forecasters are oblivious to the problems inherent in most high-energy-demand projections. While most of these analysts have a strong bias to "guess high," they cannot work for too long in the world of fantasy regarding future growth, because the powerful forces that will slow growth are already having an impact. For example, the trade publication *Electrical World* has substantially reduced its electricity-demand projection every year since 1973.[22] If we compare its projections for demand in 1990 relative to actual use in 1975, we find that in 1973 it projected an increase of 190% but 5 years later projected an increase of 80%. The reduction in estimated use of electricity in 1990 corresponds to the elimination of generating capacity greater than the total in operation in 1975!

In discussing its 1978 electricity forecast *Electrical World* analysts expressed astonishment that electricity demand grew so slowly in 1978, a year of booming economic activity:

> We are, by consensus, in the fourth year of a steady economic recovery. So the component of peak demand attributable to economic activity should be normal. We had a "normal" summer—that is, cool weather in some regions of the country—notably the midwestern area—and extreme heat in others. Common wisdom tells us, therefore, that with all other things being equal, price should be the determining factor in peak growth this year. From what we know of price elasticity, price should not have a severe and abrupt depressive effect on peak. And yet, noncoincident peak demand for the country rose a feeble 1.3%.[23]

It is to be expected that *Electrical World* projections and other industry and government projections will continue to drop in the years ahead.

In this section we shall set forth a Business-as-Usual projection, one based on the assumption that there will be no policy changes that would significantly alter the present course. We believe this projection is close to what most industry and government projections would converge to over the next few years if recent trends were adequately taken into account.

We project energy demand from 1975 to 2010. We chose 1975 as the base year because relatively good data exist for patterns of energy consumption disaggregated by end use for that year. We focus on the year 2010 so as to facilitate a comparison with the alternative "Fuel Conservation" projection made in Chapter 13. The purpose of the Fuel Conservation projection is to illustrate the demand and supply implications of an energy policy aimed at exploiting the opportunities for using energy more efficiently throughout the economy. We believe that 30 years or more may be required to carry out a transition to such an energy-efficient economy—to bring about policy changes in the political process, to implement the new policies, and to replace the existing energy-consuming capital stock with more energy efficient equipment.

The Business-as-Usual energy economy will be determined by demographic trends, increasing incomes, the changing product mix of the economy, and rising energy prices. To take into account these factors, we have found it useful to express the growth rate for energy use as follows:

$$\begin{pmatrix}\text{annual energy}\\ \text{growth rate}\end{pmatrix} = \begin{pmatrix}\text{annual growth rate}\\ \text{in functional demand}\end{pmatrix} - \begin{pmatrix}\text{annual rate of decline}\\ \text{of energy intensity}\end{pmatrix} \quad (3.4)$$

The functional demand in equation 3.4 is an appropriate measure of economic activity. In the aggregate it can be the total economic product (GNP), expressed in real terms. However, over the period of time of this projection, GNP may become a less and less adequate measure of activity; for example, household production activities such as carpentry and gardening, which are not counted in GNP, may grow substantially,[24] and many growing service activities are difficult to quantify in real terms. In order to reduce reliance on GNP in making our projections, we use more specific functional demands where feasible. We define below the functional demand, or demand for energy functions, in a way most appropriate for each sector of energy use; for example, for automobiles it is miles driven.

The energy intensity in equation 3.4 is the fuel consumption per average unit of functional demand; for example, for automobiles it is fuel use per mile. Changes in energy intensity occur because the efficiency of fuel-consuming equipment changes and/or because the mix of activities and products changes. At this point we estimate changes in the energy intensity simply as an average response to higher energy prices* in all sectors of energy use except automotive transportation (discussed below). As we have already pointed out, present replacement costs for energy are much higher than the average prices consumers now pay. Replacement costs will probably be even higher for some fuels 10 to 15 years from now. In this analysis, however, we conservatively assume, as many other analyses anticipate,[25] that real energy prices in the United States will only double between 1975 and 2010, corresponding to a growth rate of 2.0% per annum in price. We also assume that the economy will respond to each 1% increase in energy price with a ½% reduction in energy consumption per unit of product after consumers have had ample time to adjust to the higher prices. This means that the annual rate of decline in the energy intensity in equation 3.4 is 1% per year. Recent evidence suggests that the response to price increases over the long run may be higher than this.[26]

Demography will be an especially important factor affecting the functional demand in equation 3.4, because the period to the year 2010 will be one of rapid demographic change. While the total population will grow slowly, the age distribution of the population will change significantly in this period. At present there are an unusually large number of young adults in the U.S. population. A graph of the number of people at each age shows a large bulge centered at 21 years of age. This bulge exists because the post-World War II baby boom interrupted the long downward trend in the fertility rate in the United States (see Figure 3.1). Over time this bulge will propagate to higher and higher age groups, like "a watermelon passing through a boa constrictor," according to Harvard University demographer William Alonso,[28] until the bulge group dies of old age. This shifting age distribution (see Figure 3.2) has far-reaching implications for the patterns of future energy demand.

RESIDENTIAL

The functional demand for energy in the residential sector can be analyzed in terms of the number of households and income per household. Residential energy consumption of course increases with the number of households. As the income of a household rises, energy use in the

* When we derive an alternative energy projection for the year 2010 in Chapter 13, we look much more closely at energy-intensity changes.

household can also increase, because household floor space may increase, more energy-consuming equipment may be purchased, and existing equipment may be used more intensively.

The major contributor to increasing demand for energy in housing is the increasing number of households. While the total population will grow at an average rate of only 0.6% per year (1975–2010), the number of households will increase more than twice as fast because of the rap-

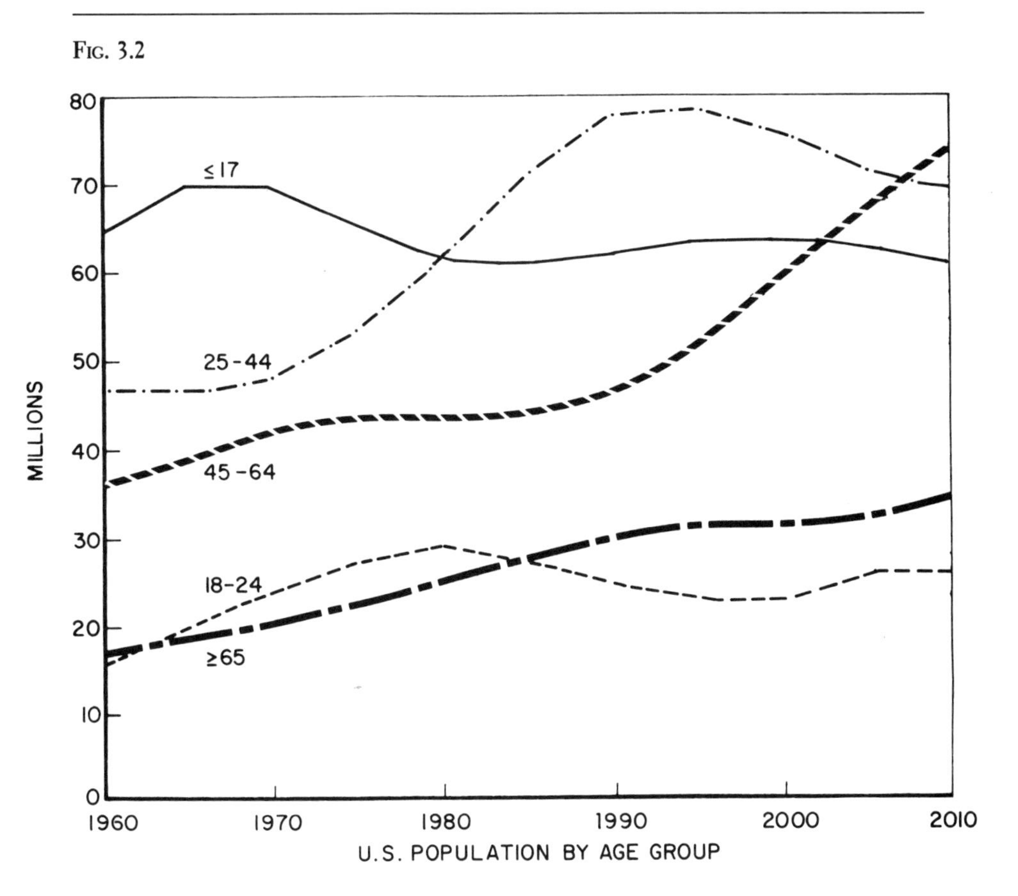

The strong variations in the age distribution, due to the variations in the fertility rate shown in Fig. 3.1, go a long way to explaining the preoccupation with juveniles and young adults in the '60s and '70s and suggest a focus on adult concerns in the '80s and '90s. In this period, for example, the work force will increase more and more slowly; meanwhile there will be strong growth in housing construction.

SOURCE: See Ref. 41.

idly increasing adult population and the trend toward fewer people per household. Most of this growth in housing will occur over the first decade or so while the first-home-buying population is still growing rapidly. The housing boom will have run its course by 1990 as the 25-to-44-year-old age group reaches its peak and begins to decline (see Figure 3.2).

In contrast to household formation, income is becoming rather unimportant here. We estimate that the functional demand for energy due to rising affluence will increase only 0.4% per year in the period 1975–2010, less than ⅓ as fast as household income.[29] This weak dependence on income reflects saturation for major energy-consuming activities. Extra income is being spent mainly on gadgetry that is not very energy intensive (the snow blower, the electric hedge clipper, the extra TV, hi-fi, hand-held calculators, food processors, etc.) or on activities outside the home. Nearly all houses have lighting, central heating, water heaters, refrigerators, and stoves—which accounted for 83% of residential energy use in 1975. Another 7% of 1975 residential energy use was for air conditioning, which is now present in half of all housing units.[30] Many of the houses that don't have air conditioning (in areas such as the Pacific Northwest and the mountain states) will never need it. Food freezers and dishwashers are about the most significant energy-using applicances that have not come close to saturation.[31] However, greater penetration of freezers, which accounted for only 2% of residential energy use in 1975,[30] cannot change the household energy budget much. Nor will more widespread use of dishwashers; a University of Illinois group has shown that dishwashers consume no more energy than washing dishes by hand.[32]

Finally, increased incomes are not likely to lead to bigger houses *on the average*, because the number of persons per household is expected to decrease from 3.0 in 1975 to about 2.4 in 2010.[33] Single-person households accounted for 21% of all households in 1978, compared to 17% in 1970.[28] These considerations are reflected in a recent National Academy of Sciences report[33] which projects that with this trend toward smaller household units there will be a significant shift from single-family to multifamily housing units, which not only tend to be much smaller but also use less energy per unit area.*

The net effect of price, income, and demography is summarized in Table 3.2, which shows that the decreasing energy intensity of residences that is a response to rising prices compensates for most of the

* Less space-heating energy is needed because there is less outside wall area per unit floor area and thus less heat loss. Energy consumption for lighting, refrigeration, and hot water tends to be less because there are fewer people per household.

growth in the number of households. The result is that residential energy use in the period 1975–2010 grows under Business-as-Usual conditions at a rate of only 0.6% per year, compared to 3.5% per year for the period 1950–1975.[34] Thus Business-as-Usual growth is far less than historical growth.

COMMERCIAL

Commercial buildings are our schools, hospitals, stores, hotels, office buildings, warehouses, and garages. The functional demand for energy in commercial buildings can be analyzed in terms of income and demography. However, as in the case of residential buildings, demography is much more important than income. Over the last decade building space has increased in proportion to service-sector employment, a trend which we assume will continue.[36] But a brief survey of how commercial buildings use energy shows that rising incomes will not have a significant effect on future commercial energy use per unit of floor space. Saturation effects are apparent in all three of the energy-intensive activities that account for 87% of commercial energy use—space heating (43%), lighting (23%), and air conditioning (21%).[35]

Commercial buildings are in fact grossly overlighted. Each 100 square feet of commercial space (the size of a small room) has about 200 watts of lighting capacity.[35] This is mainly from fluorescent bulbs which provide about as much light as six 100-watt incandescent light bulbs. This is the *average* lighting level for all commercial space, which includes not just offices, stores, hospitals, and schools, but warehouses and garages as well. This level of lighting arose from standards set by the industries promoting lighting products. It is simply not credible that this intensity would increase in the future; the trend is now toward lower lighting levels.

Air-conditioning penetration also cannot be expected to increase significantly. In 1975 about 63% of commercial space was air conditioned.[35] Much of the remaining commercial space either is in regions that don't need air conditioning or involves building types such as warehouses, only a small fraction of which will ever be air conditioned. It may come as a surprise, moreover, to learn that the heat from lighting is responsible for up to 60% of the air-conditioning load in modern commercial buildings. More sensible lighting practices will thus give rise to a reduced demand for air conditioning.

We project that the net effect of rising energy prices and increasing service-sector employment under Business-as-Usual conditions is that commercial energy use grows at an average rate of 1.4% per year, only ⅓ as fast as in the 1950–1975 period (see Table 3.2).

Table 3.2 A Business-as-Usual Projection of Energy Consumption in 2010

	1975 ENERGY DEMAND (QUADRILLION BTU)	DEMAND GROWTH		AVERAGE ANNUAL RATE OF DECLINE OF ENERGY INTENSITY (%)	AVERAGE ANNUAL GROWTH RATE OF ENERGY DEMAND (%) (1975–2010)	2010 ENERGY DEMAND (QUADRILLION BTU)
		PARAMETER	AVE. ANNUAL GROWTH RATE (%)			
Residential Buildings	16.2	Households, Household-Income Effect	1.3, 0.4[a]	1.0[f]	+0.6	20.2
Commercial Buildings	9.9	Service-Sector Employment	2.5[b]	1.0[f]	+1.4	16.3
Transportation						
Autos and Light Trucks	10.2	Adult Population	0.9[c]	1.7[g]	−0.8	7.6
Other	8.2	GNP	2.7[d]	1.0[f]	+1.7	14.6
Industry	26.1	Goods Production	2.5[e]	1.0[f]	+1.5	43.8
	70.6				+1.1	102.5

NOTES:
[a] See text, note 29.
[b] See text, note 36.
[c] See text, note 2.
[d] See text, note 4.
[e] See text, note 4.
[f] See text.
[g] See text, note 37.

AUTOMOBILES AND LIGHT TRUCKS

As pointed out above, the average person spends about an hour a day in a car, and today there are about 0.75 autos in the United States per adult. Both of these considerations indicate that the use of automobiles has essentially become saturated. People are not likely to spend much more time in a car on the average, and there are some people who will remain without cars. Thus we project automobile use (vehicle miles driven per year) to increase simply in proportion to the adult population, those 16 years of age and older.

For automobiles we assume that energy intensity will decrease mainly as a consequence of mandated fuel-economy standards. We assume that present standards will not be tightened and that the average fuel economy in 2010 will be 25 mpg.[37] This is less than the federally mandated 27.5 mpg value for new cars in 1985 to allow for two effects: (1) the actual average fuel economy will probably be somewhat less than the laboratory measurements of fuel economy for new cars; (2) the fuel economy for light trucks will be less than that for automobiles.

The combined effect of the fuel-economy standards and the relatively slow growth in automobile use is that in the period 1975–2010 energy use by automobiles declines at an average rate of 0.8% per year to a level in 2010 only ¾ of that in 1975, as shown in Table 3.2.

OTHER TRANSPORTATION

About 85% of energy use for other transportation is for freight transport, and most of the rest is for air passenger transport. The volume of freight increased in the past about ⅔ as rapidly as GNP. While in the 1950s and 1960s air-passenger-transport demand grew nearly 4 times as rapidly as GNP, the growth in recent years, as the airline industry has "matured," has been much closer to, but still more rapid than, that of GNP itself.[38] We shall assume that the demand for "other transportation" overall increases as GNP. As shown in Table 3.2, this means that we expect transportation other than autos and light trucks to be the fastest-growing energy-consuming sector.

INDUSTRIAL

Industry is made up of manufacturing, mining, construction, and agriculture. Industrial output can be expected to grow at about the same rate as goods production, or slightly more slowly than GNP.*

* We define goods production as agriculture, forestry, and fisheries; mining; contract construction; manufacturing; transportation; electric, gas, and sanitary services; and wholesale and retail trade—as these industries are defined by the Office of Business Economics, U.S. Department of Commerce.

Industry has established a remarkable trend toward energy efficiency in the 1970s. While the energy required per dollar of industrial output declined at an average rate of 0.5% per year from 1950 to 1975, it fell at an average rate of 1.6% per year from 1973 to 1978.[39] We shall assume that this trend slows down and that, as in other sectors, industrial energy intensity declines, as a response to rising energy prices, at an average annual rate of 1.0% per year.

The net effect of rising output and rising prices is that industrial energy use grows at about the same rate as commercial energy use and energy use for "other transportation" (see Table 3.2).

TOTAL ENERGY DEMAND

When the energy demands for the five sectors considered here are aggregated, the result, shown in Table 3.2, is that under Business-as-Usual conditions energy demand between 1975 and 2010 grows at a rate of about 1.1% per year, or only about ⅓ as fast as in the period 1950–1975. Total energy consumption in 2010 would be 103 quads, which is 46 quads less than our naive projection for 2000.

DEMAND FOR ENERGY CARRIERS

Because energy demand must be matched by supply, it is important to know the distribution of energy carriers, the forms in which energy is used by consumers—as electricity, as fluid fuels (liquids and gases), and as solid fuels.* The carrier concept is developed in Appendix A. The energy-carrier mix has been changing markedly over time. Between 1950 and 1975 the solid-fuel fraction declined dramatically from 32 to 5%, the fluids share increased from 54 to 66%, and the electrical share doubled, going from 14 to 29% (see Fig. 3.3).

We can expect the mix of energy carriers to continue changing. Because cheap oil and gas supplies are running out, fluid fuels will very likely play less of a role in the energy economy in the future. The conventional view is that electricity will substitute for oil and gas, and that nuclear fission will become the dominant source of electricity. But while the trend toward electrification of the energy economy is well established, there are clear signs that practical opportunities for further electrification are becoming more and more limited.

Since, as we have seen, many of the important conventional uses for electricity are becoming saturated in residences and in commercial

* We refer here to the direct consumption of solid fuels, i.e., for purposes such as the generation of steam for industrial process applications (as distinct from the use of solid fuels for the generation of electricity). Direct use of heat from solar or geothermal sources is included here in the solid-fuel category.

Fig. 3.3

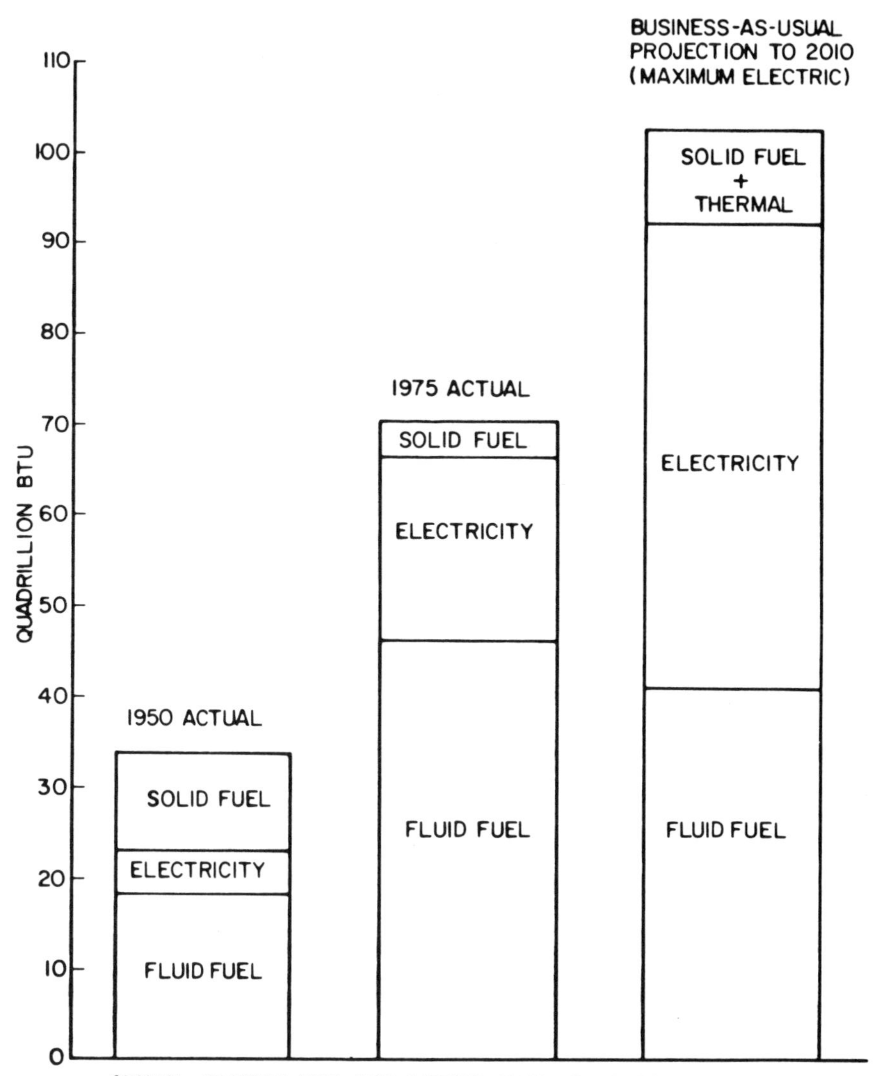

NOTE: ENERGY CONSUMPTION FOR ELECTRICITY REPRESENTS FUEL CONSUMED IN ELECTRICITY GENERATION buildings, electric utilities are seeking new markets for electricity. Two of the most widely discussed potential new markets are space heating and electric cars; indeed if electricity replaced all the oil and gas now used for space heating and if all today's cars were converted to electricity, the demand for electricity would double.

The problem with such speculation is that electricity is a poor match for these energy uses. Consider first space heating. One reason we can expect that electricity will not come to dominate the space-heating market is price. At 1978 prices electricity cost 3 times as much as gas and twice as much as oil for space-heating purposes. Fluid fuels can be expected to maintain a price advantage over electricity well into the future (see Chapter 8), so that as long as consumers have a choice, they will heat with oil or gas instead of electricity. Of course public policy could force a shift to electricity for space heating by limiting the availability of oil or gas for new building hookups. While such a policy would be inflationary, it would be rationalized as a move to substitute, for example, nuclear energy for dwindling oil and gas. But to electrify space heat would achieve less reduction of oil and gas use than one might think. The reason for this is that the electrical output of a nuclear plant is poorly matched to space-heating demands. Because nuclear plants are expensive and their fuel is cheap, these plants are operated with a fairly steady level of electrical output. In contrast, space-heating demand is not constant, but weak early and late in the winter and very strong on the coldest days. Many utilities would have to hold in reserve less capital-intensive oil- and gas-fired power plants to meet the strong seasonal peak demand for electric heat.

The potential for saving oil by electrifying automobile transportation is even more limited. The problem is that storing electricity costs 1000 times as much as storing gasoline, and batteries are so heavy that electric cars typically have a range of only 25 miles between charges (see Chapter 9). Thus without major technological breakthroughs electric batteries will remain so heavy and costly that electric vehicles, if they come into wide use at all, will be used primarily for short trips and "second car" purposes. Electric cars, then, cannot be expected to have a major impact on oil consumption for many decades, if at all.

These examples illustrate the great difficulties in further electrifying the energy economy. In Appendix A we estimate an upper limit on electrification on the basis of a rather general argument. By taking into account the fuel forms required for various end uses, we estimate that in the long run the *minimum* requirements for fluid fuels in the overall economy—for transportation, for petrochemical feedstocks, and for some low-temperature-heat applications—would amount to about 40% of total energy use. Moreover, solid fuels and solar thermal energy would probably account for *at least* 10% of total energy consumption. Thus it appears that the energy economy could become at most about 50% electrified.

Consider the Business-as-Usual projection with maximum electrification. That is, assume a truly remarkable shift of consumers from fluid fuel to electricity so that by 2010 fuel consumption for electric-power generation rises to one-half of all energy use (see Figure 3.3). Under these conditions the use of fluid fuels would decline very slowly in the period to 2010, while electricity demand would increase about as fast as we have projected that GNP will grow, or 2.7% per annum. This means that electricity demand would continue to grow much more rapidly than total energy demand, but would grow only about half as fast as the 5.3–5.8% annual growth projected to the year 2000 by the Edison Electric Institute in 1976.[7]

While our projected electricity-demand growth rate is much lower than that of utility-industry and government bodies, it should be regarded as the *maximum potential growth rate* for Business-as-Usual conditions. With electricity demand growing this fast, the percentage of GNP spent on electricity would have to double to 6% of GNP by 2010 (with a doubling of electricity prices). The economy would tend to resist this dramatic a change. As Figure 3.4 shows, the percentage of GNP spent on electricity remained very close to 2% until 1970. This occurred despite the fact that electricity demand was growing twice as fast as GNP, because electricity prices were falling rapidly in this period (see Figure 2.4). Rapidly rising electricity prices in the 1970s drove this fraction up to the present level of 3%. As time passes, consumers will adjust to high prices by buying more efficient equipment. As this happens, we will probably see a new equilibrium level for the percentage of GNP spent on electricity that will be less than 6%, meaning that electricity demand will grow more slowly than GNP and electrification will not reach its maximum. In other words, the degree of electrification is becoming saturated in the U.S. economy. Fluid fuel will continue to be far more effective and economical to use in many applications than electricity. Thus sources of fluid fuel will continue to be developed, even at costs significantly higher than present replacement costs.

CONCLUSION

Largely because of demand saturation and rising energy prices, we can expect energy demand to grow much more slowly in the future than

Fig. 3.4

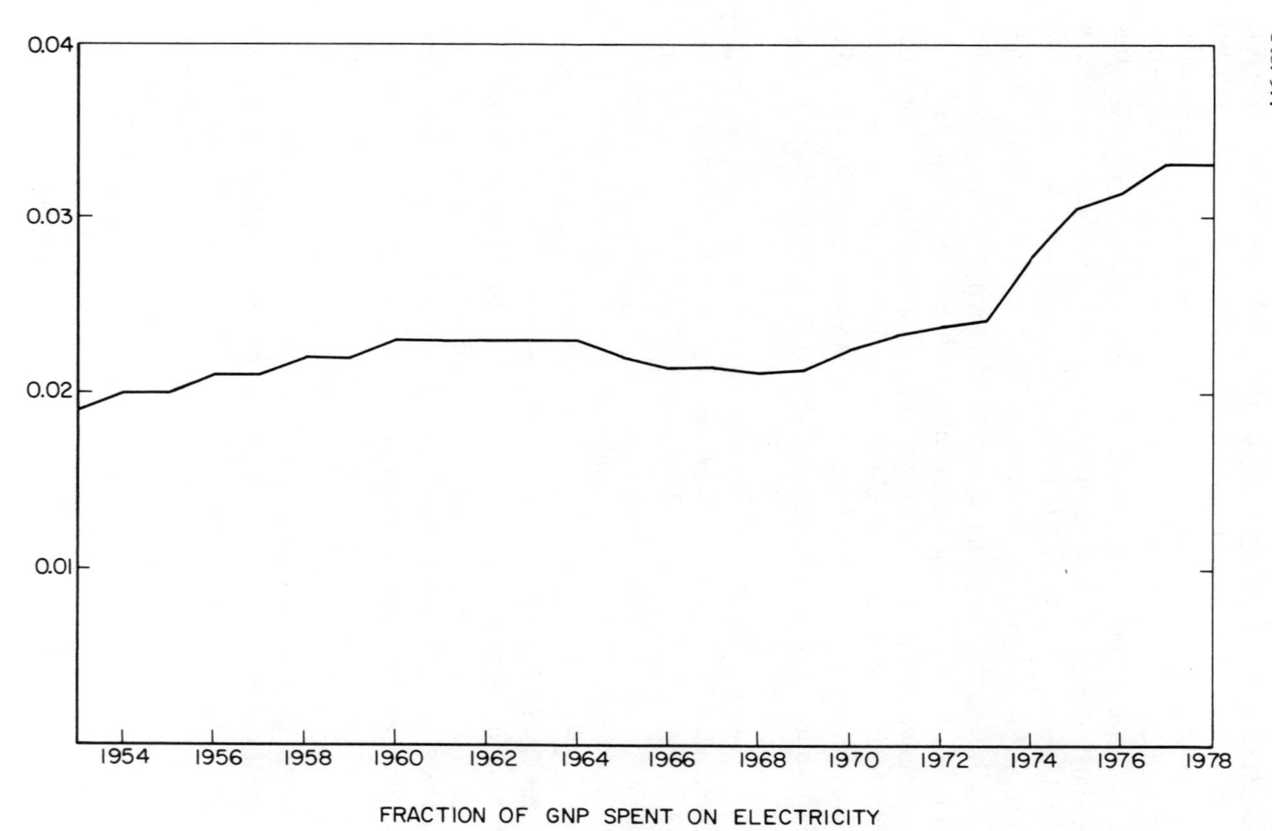

0.04
0.03
0.02
0.01
1954
1956
1958
1960
1962
1964
1966
1968
1970
1972
1974
1976
1978
FRACTION OF GNP SPENT ON ELECTRICITY

in the past and much more slowly than one would expect with the continuation of historical growth, as projected by others. While we would be very surprised if energy demand grew much faster than we have projected here, we cannot claim that our projection represents the *true* course for growth of energy demand under Business-as-Usual conditions—only a *plausible* course. Despite the considerable uncertainties underlying our projections, it is adequate for our purposes, because, in evaluating energy supply in the next three chapters, we will find that even at the levels of consumption we project, energy problems remain severe. The first problem is, Where will even this much energy come from?

CHAPTER 4

WHY ARE ENERGY PRICES RISING?

IT IS VIRTUALLY CERTAIN that energy consumption will grow much more slowly in the future than in the past. Yet even slow growth poses formidable challenges to the energy producers, because the easily exploited oil and gas fields are being exhausted, and because the present level of U.S. demand is so high that small fractional increments in energy production correspond to enormous absolute quantities of energy. For example, the additional energy needs alone by 2010 (compared to 1975) under Business-as-Usual conditions correspond to the carrying capacity of 10 Alaska pipelines or the fuel that would be consumed by over 700 large nuclear power plants.[1]

What we shall show in this chapter is that while there is no imminent danger of running out of energy resources, it does appear that low-cost energy resources are being exhausted, and that the shift to more costly sources is causing a rapid increase in energy prices.

CHEAP OIL AND GAS ARE RUNNING OUT

Consider first oil and gas. These fuels have come to dominate U.S. energy use (accounting for ¾ of the total) because they are truly remarkable energy forms: not only have they been cheap to extract, but also they are cheap to process, transport, store, and use. Domestic production of oil and gas peaked in the early 1970s (see Figure 4.1), leading many to conclude that the supplies of these fuels will soon be exhausted.

One of the most widely cited estimates of oil and gas resources, a 1974 estimate by the National Academy of Sciences, holds that of all oil and gas resources which will eventually be discovered and recovered, about ⅓ have already been consumed.[2] Unfortunately this estimate is very crude, because the knowledge of remaining resources is very poor.[3,4] For natural gas in particular there is growing evidence suggesting that official estimates of resources are much too low.[5] In the past gas has been found primarily as a by-product of exploratory efforts to find oil. When gas prices rise as a consequence of deregulation, gas will be sought in its own right. This reorientation may well lead to substantially

FIG. 4.1

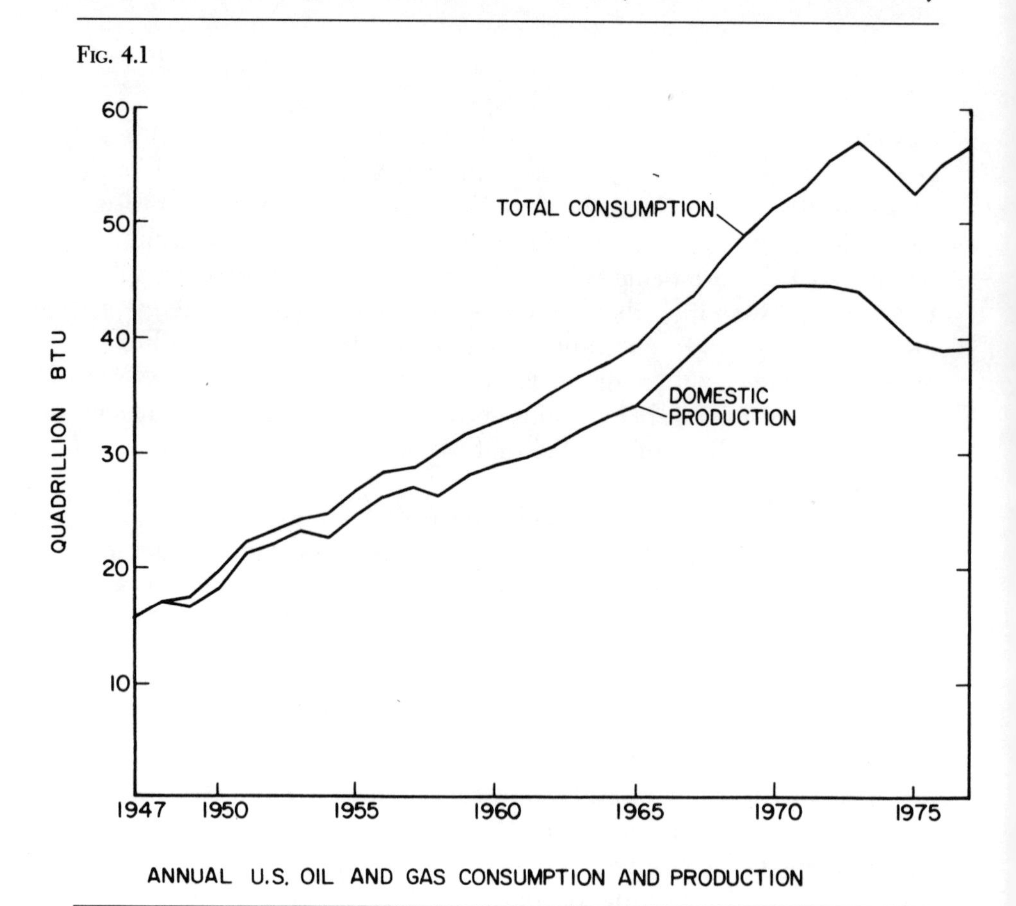

ANNUAL U.S. OIL AND GAS CONSUMPTION AND PRODUCTION

increased estimates of gas ultimately recoverable at higher prices. The notion of a "supply curve" for gas showing how much more gas would become available as the price increases, rather than the concept of a fixed quantity of extractable gas, needs to be developed.[5] United States energy policies must be flexible enough to take advantage of the possibility that substantial quantities of gas will become available as the price rises.

Despite the uncertainty in estimates of oil and gas resources, the situation is clear enough to draw a powerful conclusion about the length of time the United States can rely on *relatively low-cost* oil and gas. Suppose oil and gas use remains roughly constant in the future, as we have projected for Business-as-Usual conditions. Under these conditions the remaining oil and gas resources, as estimated by the National Academy of Sciences, would be adequate for 50 years.[2] (This depletion time would of course be considerably shorter if "historical growth" in oil and gas demand persisted.) Although 50 years may seem like a long time, this lifetime for relatively low-cost oil and gas resources is short—uncomfortably close to the minimum period that would be required to replace oil and gas with substitute fuels. That is why dislocations are already being felt.

THE CONVENTIONAL ALTERNATIVES: EXHAUSTIBLE BUT PLENTIFUL

The principal alternative fuels to which the nation and the industrial world are turning are coal, other fossil fuels, and uranium. Can these forms of energy take care of projected needs?

COAL

Total estimated coal resources are 8 times the estimated oil and gas resources.[6] Not only is coal abundant; also, strip-mined coal is now like Middle Eastern oil in that the cost of production is typically much less than the price. Not long ago one of us saw giant earth-moving equipment leveling hills in eastern Kentucky and loading coal for delivery to a nearby railroad, under a contract that paid $5 per ton of coal. Coal like this now often sells for $20 a ton, meaning that the owner of the coal and labor (through the collective-bargaining process) collects a $15 windfall. In that particular mine the seam was 10 feet thick and the $15 per ton would translate to a quarter of a million dollars per acre!

Because coal resources are enormous and because production costs are relatively low, however, the price of coal should be inhibited from rising in the future by competition within the industry, as well as (perhaps) by competition from other fuels, if the demand for coal does not rise too rapidly.

ALTERNATIVE FOSSIL FUELS

There are very large deposits of fossil fuels in the hard-to-extract forms of oil shale (especially in Colorado, Wyoming, and Utah) and tar sands (especially in Alberta, Canada). Discovered and measured resources of these two fuels in the United States and Canada approach total coal resources in energy content.

There are clear signs that a tar-sands industry is developing, as billions of dollars are being spent on tar-sands facilities in Canada.[7] However, efforts to extract a liquid hydrocarbon from shale have a long on-again, off-again history.[8] Even the new world price of oil has been slow in stimulating development. Indeed the price at which it has been predicted that production of oil from shale will become profitable has been consistently increasing as the price of oil has increased. It is now a joke in the energy community that the price at which oil from shale will be profitable will always be 33% above the price of oil.

For both these energy sources, capital costs are much higher than the present capital costs for oil; and both environmental[8] and water-shortage[9] problems may be severe. Thus progress toward supplying from these sources a substantial fraction of a fluid-fuel demand of 40 to 50 quads per year, as projected for the Business-as-Usual Scenario, is expected to be very slow. National mining capabilities and water availability in the mountain states may limit exploitation to a few quads per year at most.[10]

FISSION FUELS

The only nuclear fuel (i.e., the only chain-reacting nuclear isotope)* found in nature is uranium-235, which makes up 0.7% of natural uranium. Uranium-238, which makes up 99.3% of natural uranium, and thorium, while not directly usable as nuclear fuel, can be converted in nuclear reactors respectively to plutonium-239 and uranium-233, which are nuclear fuels like uranium-235. The energy value of natural uranium and of thorium depends sensitively on the nuclear technology employed to convert them.

* Each isotope of an element has a different number of neutrons in the atomic nucleus. The isotope number is the sum of the number of protons and neutrons. For example, uranium-235 has 92 protons and 143 neutrons in the nucleus.

Today's uranium-based light-water reactors (LWRs)* are not very efficient in converting uranium-238 to plutonium. They tap only about 1% of the nuclear energy stored in natural uranium. Because they create less chain-reacting nuclear fuel than they consume, they are called "burners." The second-generation reactor now under development is the plutonium-fueled liquid-metal-fast-breeder reactor (LMFBR).† It is called a breeder reactor because it produces more nuclear fuel than it consumes. For any such technology that can efficiently make use of thorium or uranium-238, resources are effectively inexhaustible; e.g., the uranium in a ton of average rock in the earth's crust contains the energy equivalent of 6 tons of coal. There is doubt, however, whether the breeder technology will be used in the foreseeable future, both because of risks (see Chapter 5) and because of high capital cost. Therefore the critical public policy question regarding fission-fuel resources is how long uranium supplies would last as fuel for burner reactors.

The extent of recoverable uranium resources depends sensitively on price; as the price increases, it becomes profitable to extract more extensive lower grade resources. Interpretation of estimated resources in terms of years of supply also depends sensitively on the estimated growth of nuclear power. As recently as 1975 the U.S. government was projecting that by the year 2000 nuclear power plants would be producing nearly 3 times as much electricity as all electric power plants in 1975 and that by the year 2025 they would be producing 11 times as much![11] With demand growing so rapidly low-cost uranium resources would be exhausted by 2010.[12] Such resource considerations have provided the major rationale for quickly bringing breeder reactor technology to commercialization.

The bottom has now fallen out of these projections.[13] Under Business-as-Usual conditions a heavy commitment to nuclear power would mean that by the middle years of the twenty-first century nuclear power would provide only about twice the total electricity production in 1975.[14] With this more realistic "high nuclear growth" projection, low-cost uranium resources[12] would be adequate for 70 years if used in today's reactors or for more than 100 years if uranium efficient advanced burner reactors were to be introduced after the year 2000—even without reprocessing spent reactor fuel so as to recover unused uranium

* Light-water reactors use ordinary water as both a coolant (to carry the heat generated in fission away from the nuclear fuel) and a "moderator." The neutrons emitted in fission are moving very fast and must be slowed down (moderated) by collisions in the moderator material for effective operation.

† Liquid-metal-fast-breeder reactors use liquid sodium metal as a coolant. Sodium, unlike water, is a poor moderator. However, these reactors are designed to utilize fast neutrons.

and plutonium.[14] Thus there is no longer any urgency to develop the breeder reactor.[15] Moreover, the breeder may *never* be needed with Business-as-Usual energy demands, because the inadequately assessed uranium resources at moderate to high costs may prove to be sufficient to keep advanced burner reactors competitive with breeders for hundreds of years, during which time it is virtually certain there will emerge new electricity generating technologies that are strong competitors with nuclear fission.

In light of these considerations of diminished nuclear electrical growth, the prospect of substantial high-cost uranium resources, and also the prospects for uranium efficient advanced burner reactors, there is no reason to expect that uranium resources will be a significant constraint on the development of nuclear power without the breeder reactor, for many decades or even centuries.

CAPITAL CONSTRAINTS

We have seen that although cheap oil and gas are approaching exhaustion, there is no imminent danger of running out of alternative exhaustible fuels. But energy will be much more costly.

In the early 70s, orders for nuclear plants were being placed at a dizzying pace, and there was significant activity in development and demonstration of methods of creating fluid fuels from the solid fossil-fuel resources—coal, oil shale, and tar sands. The quadrupling of world oil prices in 1973-74 should have brought a frenzy of activity to develop these alternatives to oil and gas. In the 6 years since that time this has not happened. Quite the contrary. Orders for new nuclear plants have become so rare that the manufacturers are warning that they may have to go out of the nuclear reactor business. Efforts to demonstrate gasification and liquefaction processes for coal and oil shale have not mushroomed; a number of ambitious plans have faltered.

The central reason for the present stalemate in investment in energy alternatives is that the capital costs of converting alternatives to oil and gas into convenient energy carriers have risen to levels which inhibit further construction. Even though there is a seeming slowdown in energy-related construction, actual capital expenditures for energy have been soaring. They accounted in 1977 for 43% of all expenditures for new plant and equipment, up from an average of 24% in the 1960s (see Figure 4.2).[16]

FIG. 4.2

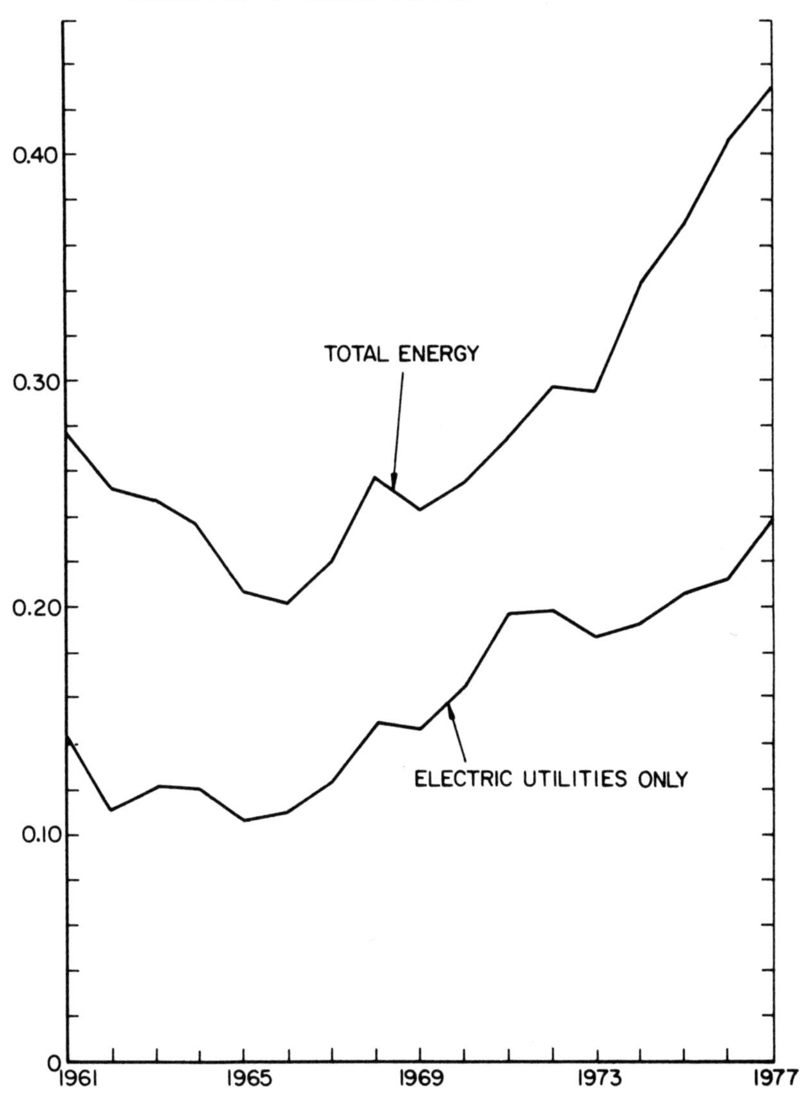

SOURCE: See note 16.

Do high capital costs put fundamental constraints on energy supply? Solar enthusiasts argue that sunlight is free. Fusion enthusiasts argue that fuel for fusion is inexhaustible. They believe that capital costs will not impose limits on their schemes. Unfortunately for their hopes, high capital cost can put just as effective a limit on energy as a scarcity of fuel resources. Capital formation is not just a matter of raising cash. It involves the organization of land, materials, and labor resources. Building huge new energy-supply facilities requires an abundance of technical and managerial skills and the general industrial support, or infrastructure, to provide specialized parts and construction equipment. This infrastructure itself involves a high level of capital.

In recent years energy-supply facilities have been plagued more and more by land-use and other environmental struggles which have caused long delays, by inadequate quality control in equipment manufacture and construction, by bottlenecks which have arisen because each big project has been in many ways unusual, and by high labor costs due largely to a shortage of qualified manpower and a decline in labor productivity. (Labor costs have risen sharply in large part because the required highly skilled labor force has been in a strong position to bargain on wages and working conditions.) These factors cause huge cost overruns, including interest and other charges which accumulate as construction programs are delayed. The cost overruns are much higher than typical inflation.

These high capital costs are not a surprising development. Consider some of the new endeavors the industry is attempting as old oil and gas fields run out: Instead of in places like east Texas, they're drilling in the open ocean from platforms served by helicopter. Instead of in places like Tennessee, they are building and maintaining pipelines above the Arctic Circle. Instead of oil tankers, they're building cryogenic (very-low-temperature) ships for liquefied natural gas. Instead of using a gas flame to make steam to generate electricity, they're creating heat with nuclear transformations.

Not only are these endeavors new; some of them are evolving in a manner which does not facilitate cost control: huge new facilities are being designed and created virtually from scratch, through the "pilot plant" and "demonstration" sequence instead of through the traditional successes and failures of a sequence of commercial ventures. The traditional approach is adapted to cost control; there is a "learning curve," as economists show. The "learning," or reduction in costs as experience is acquired, has not occurred with most of the very large new technologies. This may turn out to be a critical disadvantage for any technology which can be produced only on such a large scale that it cannot evolve gradually through a sequence of marketed products.

In addition to escalating costs, energy-supply investments are troubled by new uncertainties. There is uncertainty about world oil prices in the short term: will the OPEC cartel hold up? There is uncertainty about growth of demand for energy over the construction period. Finally, there is uncertainty about government energy policies—subsidies, energy pricing, environmental and other regulations.

These general ideas need to be examined in terms of the systems for providing the major energy carriers, electricity and fluid fuels.

CAPITAL PROBLEMS OF THE ELECTRIC UTILITIES

While the capital requirements of the energy industry are high relative to most other economic activities, capital requirements are especially high for the electric utility industry, which accounts for over half of all capital expenditures for the energy industry (see Figure 4.2). A measure of capital intensity is the capital-output ratio—the amount of capital investment required to produce a dollar of annual sales revenue. For electric utilities the capital-output ratio is $4.50* compared with about $1.00 for other energy sectors and about $0.60 for all manufacturing.[17]

Historically, however, capital intensity has not been a problem for electric utilities. Because of the rapid rate of technological improvements, the cost of a unit of capital for electric power generation (the cost per kilowatt of electrical generating capacity) declined rapidly until about 1970. The most significant technological improvement was in power-plant efficiency, shown in Figure 4.3. Between 1900 and the early 1960s a remarkable eightfold increase in the efficiency of converting fuel energy into electricity was realized. The efficiency improvements ceased in the early 1960s, and opportunities for further improvement are very limited.

As Figure 2.4 shows, the price of electricity continued to drop rapidly beyond the early 1960s. This occurred in large part because of the strong emphasis given thereafter to cost cutting through "scale economies," whereby savings per unit of electrical capacity are achieved by building larger plants. From 1930 till the early 1950s, the largest steam electric unit was about 200,000 kilowatts. Then the swing to larger units began, with the largest unit in operation passing 1 million

* One reason for this extreme situation is that utilities are a regulated monopoly. The price of electricity is set to allow a certain rate of return on capital. Thus utilities are rewarded in proportion to the amount of capital they have installed.

Fig. 4.3

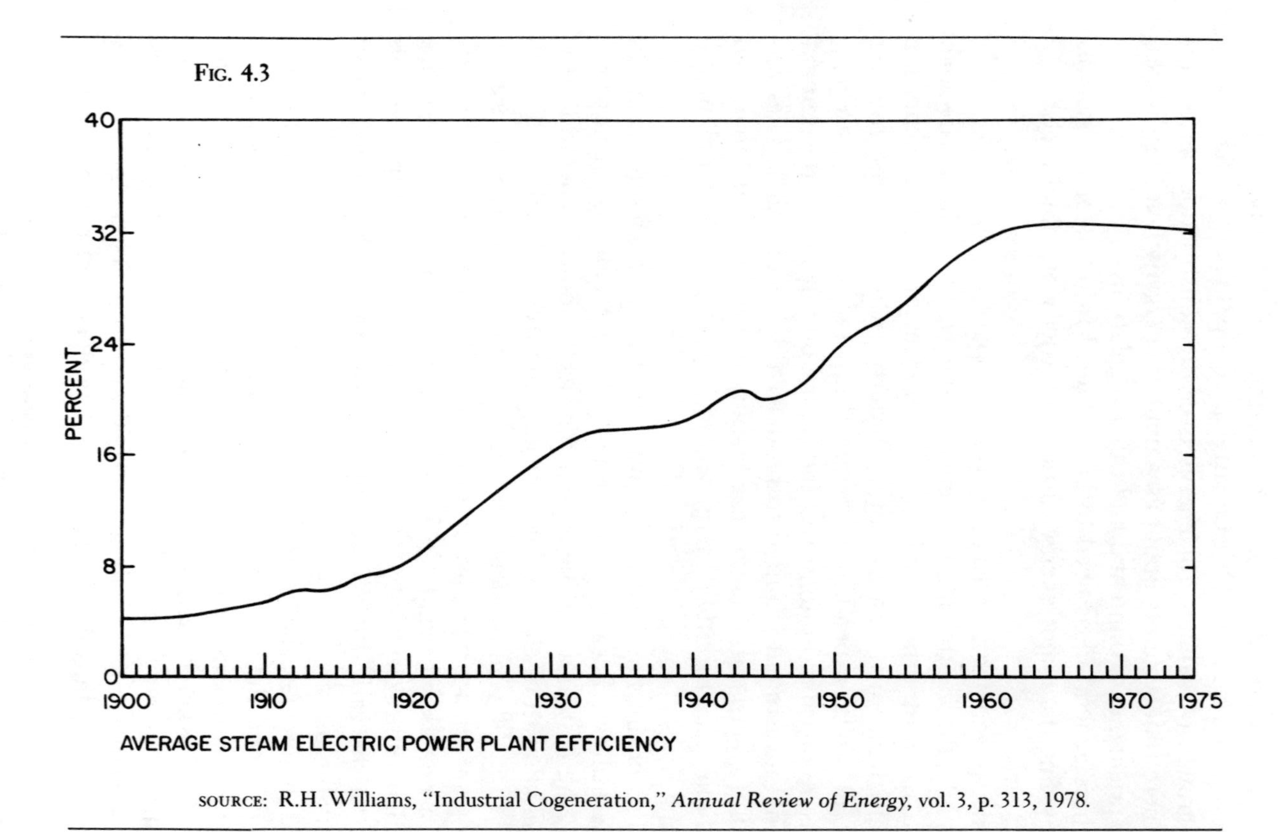

AVERAGE STEAM ELECTRIC POWER PLANT EFFICIENCY

SOURCE: R.H. Williams, "Industrial Cogeneration," *Annual Review of Energy*, vol. 3, p. 313, 1978.

FIG. 4.4

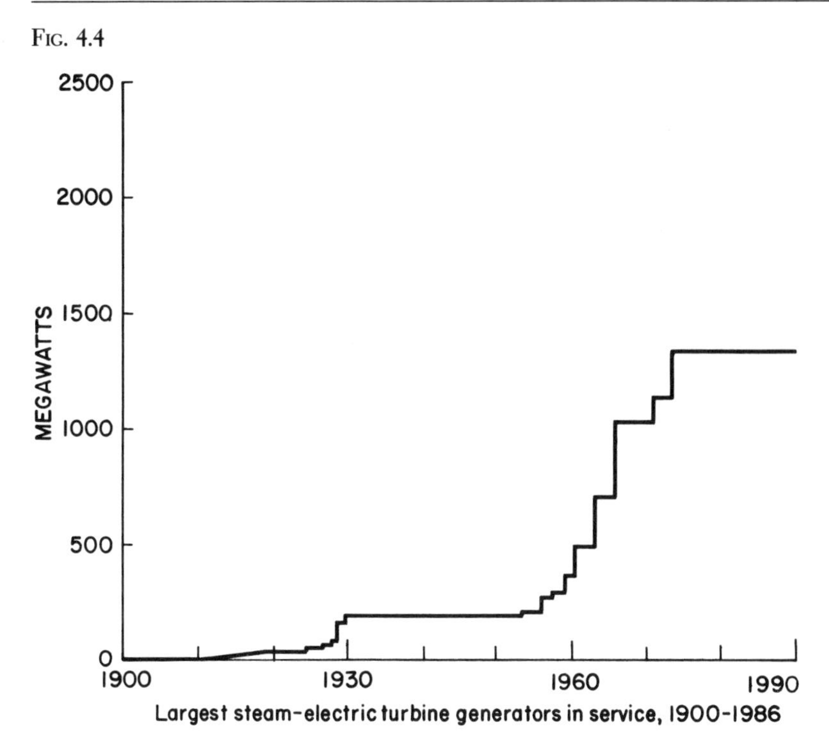

Largest steam-electric turbine generators in service, 1900-1986

SOURCE: R. H. Williams, "Industrial Cogeneration," op. cit.

kilowatts before 1970. This trend to larger plants is shown in Figure 4.4. Several years ago it was expected that continued cost cutting could be achieved with even larger units. However, power plants now being built may be larger than optimal; as suggested by Figure 4.4, larger plants are not now contemplated.

Figure 4.5 shows that in recent years there have been dramatic capital cost increases for both nuclear and coal-fired power plants (oil refineries are shown for comparison). In 1965 it was felt that large light-water reactors could be built for about $220 (1975 dollars) per kilowatt of capacity, written $220/kw(e).* The cost increased to about $720/kw(e) (1975 dollars) for plants coming on line in 1980 and is still rising. This cost increase by a factor of 3 or 4 beyond the increase due to general inflation has been ascribed by analysts to the kinds of problems we discussed early in this section on capital constraints.[18]

* The symbol kw(e), or kilowatt electrical, stands for a 1-kilowatt rate of flow of electrical energy, or, one could say, 1 kilowatt-hour per hour. A typical large power plant today has a capacity to deliver an electricity flow of a million kilowatts. If such a plant cost $220 million, it would have a unit capital cost of $220 per kilowatt electrical, or $220/kw(e).

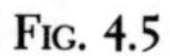

FIG. 4.5

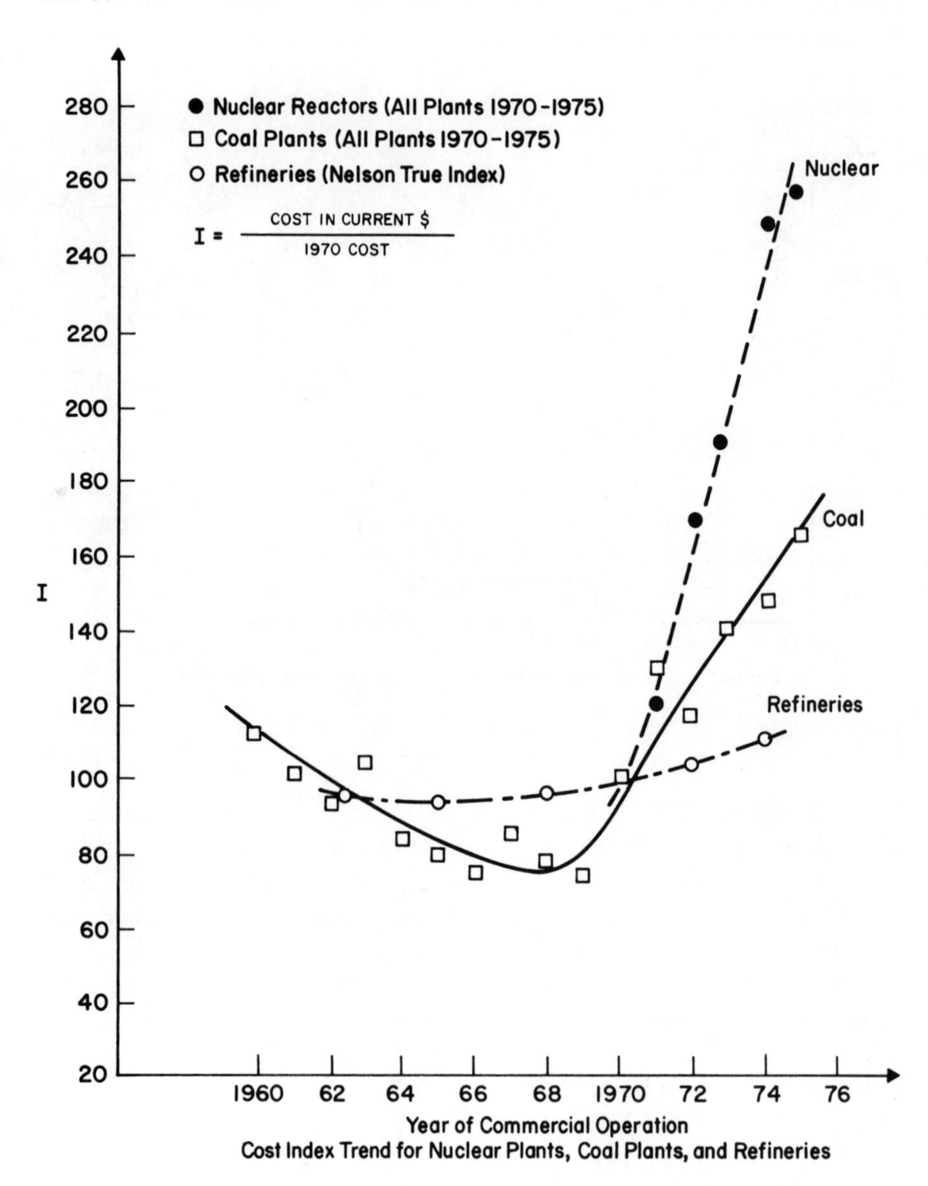

Cost Index Trend for Nuclear Plants, Coal Plants, and Refineries

SOURCE: Irvin C. Bupp et al., "Trends in Light Water Reactor Capital Costs in the United States: Causes and Consequences," Center for Policy Alternatives Report CPA 74-8, MIT, December 18, 1974.

Figure 4.2 shows the result of these rising costs: the electric utilities' share of total plant and equipment expenditures in the United States doubled over the decade 1967–1977 from 12 to 24%. This upward trend cannot continue. Given the many competing demands for capital, it is most unlikely that the electric utility-industry share of business capital could continue to rise. In the nature of economic relationships, consumers will help keep the lid on capital spending by electric utilities: they will consume less at the higher prices brought on by rising capital costs. In this manner capital costs are imposing strong limitations on the supply and use of electricity.

Of course if electricity demand grows much more slowly than projected recently by government and industry, this slower growth would help limit the upward trend in capital expenditures by electric utilities. But even with the slower growth projected in Chapter 3 for Business-as-Usual conditions, new electrical capacity corresponding to one large (million-kilowatt) plant would have to be built every 3 weeks, and each plant with supporting facilities would cost over $1 billion (1975 dollars)! Thus a huge effort would be required even for the moderate-growth conditions of this scenario. But the vast industrial infrastructure consisting of equipment manufacturers and engineering and construction firms which has evolved to provide electric-utility expansion could surely do the job. Or could it? Efforts to create support facilities for coal- or uranium-based systems (e.g., mines and transport networks for coal and waste processing and disposal facilities for spent nuclear fuel) are beset with many problems. We believe our Business-as-Usual projection with maximum electrification skirts the edge of what is feasible.

So much for electricity. The capital constraints on the other major energy carrier, fluid fuel, may be even more severe in the long term.

CAPITAL CONSTRAINTS ON SYNTHETIC FUELS

Synthetic fuels (or *synfuels*) is the general name for liquid and gaseous fuels created from primary sources other than oil and natural gas. The most important possibilities are synfuels made from coal. Liquid fuels from coal were widely used in wartime Germany and are widely used in South Africa. A low-quality, coal-derived gas called "manufactured gas" was widely used in the United States and other countries before natural gas displaced it.

Recently it was announced that the cost of a high-quality synthetic gas (methane) from coal to be made at the first American coal gasification plant would in 1983 cost 5 times as much in real terms as natural gas cost in 1979.[19] This example suggests that the capital problem is probably even more serious for synfuels than for electric utilities.

Because the projected costs of synfuels are so high, it is uncertain that they can ever compete with oil and gas extracted from the earth even where the extraction requires costly unconventional methods. In particular there may well be considerable quantities of natural gas at prices much higher than present prices but lower than those projected for synthetic gas from coal. (See the gas-supply curve in note 5.) This great uncertainty will discourage private-sector investments in synfuels. Another limiting factor is the sheer scale of physical effort required to substitute synfuels for oil and gas. One indication of the enormity of the problem is the rate at which coal would be mined if the United States were to provide all its fluid fuels in this way in the early decades of the twenty-first century: 8 times as much as all the coal mined today would be required to meet just this fluid fuel demand.[20] This would require a revolution of the mining industry—creation of skills, of a labor force, and of physical capital—quite beyond the concepts of the present industry.

And the availability of water, a problem of capital one step removed, would put severe constraints on synfuels production in most parts of the country.[9]

In spite of these constraints and uncertainties some synfuels may be needed eventually to meet fluid-fuel demands under Business-as-Usual conditions. The scale of their use will depend on whether it is possible to achieve major reductions in unit capital costs and water requirements.

POLICY IMPLICATIONS

While providing energy supplies would be easier under Business-as-Usual conditions than if historical growth were to continue, even slowed growth requires creation of enormous new sources of supply. And because the United States has already skimmed the cream of its energy resource base, endeavors to get new supplies involve huge capital investments and great technological and economic uncertainties. The high prices of the energy products from these facilities would clearly dampen demand—but by how much? This uncertainty about future demand compounds the financial risk involved with these projects. Such considerations lead us to conclude that the private sector is probably incapable of fueling a Business-as-Usual energy future with domestic supplies.

But a Business-as-Usual future could still evolve. This course could be realized from domestic sources *if* the government were to join industry as an active partner in the management of new energy technology, assuming much of the financial risk involved. The government-industry partnership in the development of nuclear-fission energy serves as a model in part. The partnership would have to be extended to include demonstration and commercial management of new energy technology. While a government-industry partnership is probably necessary to sustain this energy course with domestic supplies, this approach to the management of our energy resources would be fraught with great difficulties, not the least of which would be the problems of controlling the side effects of energy production and use, which we now discuss.

CHAPTER 5

CAN WE COPE WITH THE HAZARDS OF FOSSIL FUELS AND NUCLEAR POWER?

WHILE CAPITAL PROBLEMS for the energy industry are indeed severe, the most formidable problems facing energy planners in the coming decades may well be damage arising as a side effect of energy production and use. Many people believe that technical and economic change will make it possible to solve any problem. We disagree. While the application of "technical fixes" can be effective in reducing many risks to acceptable levels, some modern technologies pose risks that defy solutions. Thus for the long term it is useful to distinguish between *reducible risks*—those that can in principle be adequately controlled through technical fixes of moderate to substantial cost—and *irreducible risks*—those for which controls are not feasible even at substantial cost. In what follows we briefly survey some important representative reducible risks and the critical irreducible risks of fossil-fuel use and nuclear power for the United States.[1]

PUBLIC RISKS OF FOSSIL-FUEL USAGE

REDUCIBLE RISKS

Much of the concern of the environmental movement has been associated with the use of fossil fuels: small particle and sulfur oxide air pollution from stationary sources, automotive air pollution, oil spills at sea, land degradation from surface mining of coal, water pollution from mining operations, and the health and safety of underground coal miners. As serious as these problems are, there is no fundamental reason why they can't be substantially abated with technical fixes. The limiting factors for implementing these solutions are time, money, and concern.

There is an extensive literature of such fossil-fuel-related problems.[2,3] In the interest of brevity, we discuss three important reducible risks associated with coal which are representative of the problems encountered with all fossil fuels.

Sulfur Oxide Air Pollution: Sulfur oxides shorten lives,[4] damage property, and disrupt ecological systems.[5] One response to the problem is to burn relatively scarce low-sulfur coal. Another response is to burn high-sulfur coal, cleaning up the gases resulting from combustion as they go up the stack. Stack-gas scrubbers have been developed for that purpose. This type of control to meet today's standards will add less than 10% to the average price of electricity from a plant burning high-sulfur coal.[6] Considering how cheap energy has been, this cost does not appear to be excessive. Other approaches to the problem involve removal of sulfur (and other impurities) from the coal during or prior to combustion. Several such strategies, including fluidized-bed combustion (which removes pollutants during combustion) and solvent refining of coal (which removes pollutants before combustion), are now being explored.

Land Degradation: Land degradation from surface mining of coal is another intense concern. In Appalachia degradation has extended far beyond unreclaimed mine sites, through pollution of streams. The reclamation record to date is poor. The problem is often one of enforcement rather than technology or economics; in many areas it is not difficult to restore land to useful functions within a period like a decade at a cost of only a few cents per ton of coal. In other areas, such as arid regions of the West, reclamation is difficult if not impossible. However, banning surface mining in such areas would not jeopardize the using of coal. For example, 60% of the Western surface-minable coal reserves lie in areas where reclamation has been judged feasible at low cost.[7]

Occupational Health and Safety: Underground coal mining is one of the most hazardous occupations. Coal miners have suffered more than twice as much disability as workers in other high-injury industries such as metal mining and milling, lumber and wood products, and primary metals. At the rate of fatal accidents occurring in the 1960s, a man who works a lifetime in an underground coal mine stands a one-in-ten chance of being killed on the job.[8] Moreover, black lung is an incurable, disabling occupational hazard of coal mining which is believed to be a contributory cause in some 3000 to 4000 deaths in the United States each year.[9]

These are outrageous occupational hazards. Experience suggests that underground coal mining can be made much safer. For example, since the passage of the Federal Coal Mine Health and Safety Act of 1969 there has been about a 50% reduction in the rate of underground coal-mine fatalities, and enforcement of the coal dust and silica standards in this act is expected to drastically reduce the number of future cases of black lung. Automated mining equipment could lead to further improvement.[10] The problem here is like that with land degradation: political factors and the narrow economic interests of producers are more limiting than technology in efforts to make coal mining a much safer occupation.

AN IRREDUCIBLE RISK: CARBON DIOXIDE AND CLIMATE CHANGE

In contrast to the more commonly discussed hazards, it appears that at least one problem associated with fossil-fuel combustion, the potential for global climatic change arising from the buildup of carbon dioxide in the atmosphere, may be inherently insoluble. The global risks resulting from this buildup may be manifest in just a few decades.[11]

Perhaps the most striking man-induced change in the *global* environment is the established fact that the atmospheric level of carbon dioxide has increased between 10 and 20% over the past century.[12] There is good evidence that most of the increase going on now is due to the burning of fossil fuels, with some of the past increase having arisen also from clearing of forests. As Figure 5.1 shows, the CO_2 buildup is accelerating rapidly. And even if there should be no further increase in the rate of per capita use of fossil fuels, the level of CO_2 in the atmosphere would rise to 50% above the preindustrial level in 50 years.

What does atmospheric CO_2 do to the earth's climate? Let us digress for a moment to discuss the earth's general energy balance. A rough energy equilibrium is maintained at the earth's surface between the energy flowing in as sunlight* and the energy flowing away from the earth as *earthshine*. Earthshine is infrared light, the kind of light emitted by warm objects. (Although infrared light is not visible, it can be

* There are also other relatively small flows of energy: heat from radioactive decays within the earth, tides caused by the moon, and man's conversion of fossil and nuclear fuels.

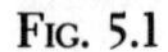

Fig. 5.1

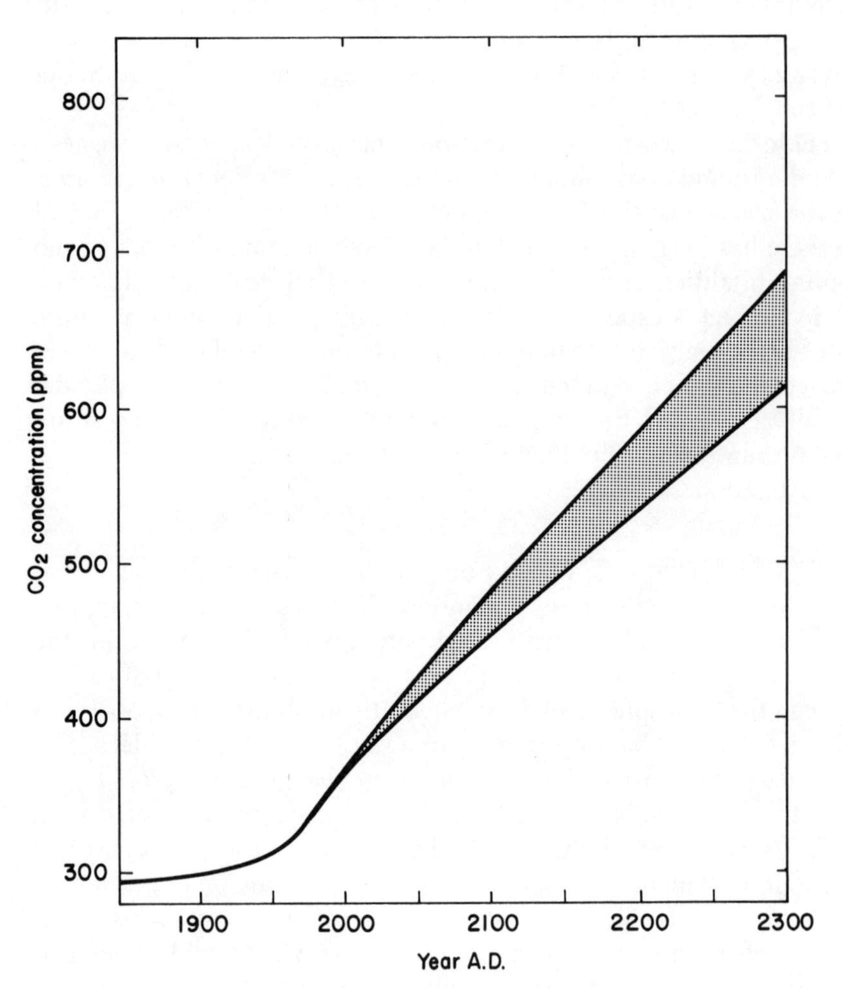

Historical increase in the carbon dioxide in the atmosphere and projection of the future concentration assuming the rate of production of CO_2 remains fixed at the level of 1975. The projection is uncertain; the shaded area shows the concentration range considered "most realistic" by the physicists making the projection.

SOURCE: Siegenthaler and Oeschger, ref. 12.

photographed, and when it is intense, it can be felt. The warmth radiated by the deep red coals of a fire is the infrared light they emit.) In Figure 5.2 the flows of sunlight and earthshine are illustrated.

The key notion about earthshine is that it is the energy waste product of the earth. The disposal site is outer space. If the earth didn't rid itself of this waste energy, the surface would become hot. Carbon dioxide in the atmosphere is a regulator of the emission of earthshine. Largely because of the presence of carbon dioxide and water vapor in the atmosphere, most of the infrared radiation emitted by the earth's surface does not escape directly into space, but first is absorbed by the atmosphere and then is reradiated upward and downward (see Figure 5.2). These gases thus act as a thermal blanket for the earth. Increased CO_2 in the atmosphere "thickens the blanket," increasing the surface temperature. Detailed calculations show that the effect of a 50% increase in the atmospheric CO_2 concentration would be to increase the global average temperature by over 1°C (2°F).[13] At high latitudes the surface heating is calculated to be twice as great as the global average.[14]

Unfortunately climatologists don't know for sure that this is what would happen. There may be feedback mechanisms at work which either mitigate or enhance the heating effect of CO_2 predicted by today's theories. Or there may be compensating natural climatic changes taking place simultaneously.[15] Global climate modeling is so difficult and the atmospheric level of CO_2 is rising so fast that we might learn the consequences by experience before it becomes possible to make credible predictions. Despite all the uncertainty there is a growing conviction in the climatological community that the CO_2 buildup must be taken very seriously.

Just how serious would a 1°C global warming be? The Little Ice Age, which persisted from about 1600 to 1860, was a period when the global mean surface temperature may have been just 1°C colder than at present. W.W. Kellogg has pointed out that a 1°C global average temperature change should be regarded as significant because it would probably be accompanied by marked change in rainfall patterns. Today's climatic models, however, cannot predict what would happen to particular regions. Kellogg suggests that one way to get a feeling for what a warmer earth might be like is to study a time when the earth was warmer than it is today. In most regions for which data are available, there was more precipitation in the warmer Altithermal Period, 4000 to 8000 years ago, than today. It is noteworthy, however, that the Midwest of the United States, our "breadbasket," is one of the areas that was drier and thus worse off.[16]

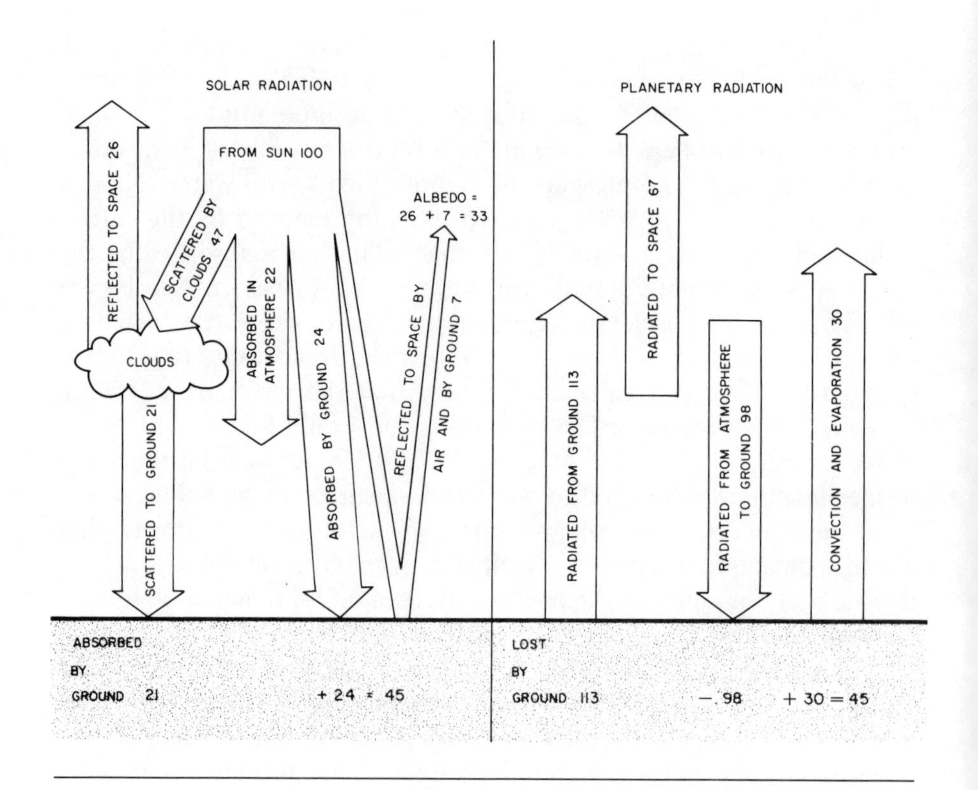

Fig. 5.2

THE HEAT BUDGET OF THE EARTH AND THE ATMOSPHERE

About 33% of the solar radiation (sunlight) striking the earth is reflected directly back into space by the ground and by clouds; 67% is absorbed. That which is absorbed is subsequently reemitted into space as long wavelength planetary radiation (earthshine).

SOURCE: R.M. Goody and J.C.G. Walker, *Atmospheres*, Foundations of Earth Science Series, Prentice-Hall, Inc., Englewood Cliffs, New Jersey, 1972.

Despite the uncertainties surrounding the understanding of the CO_2 problem, it is clear that man has the potential to substantially alter climate in a matter of decades through the continued combustion of fossil fuels.* Soil and agricultural capital could not readily adapt to major shifts in rainfall on such a short time scale. The dislocations in agriculture might well have a catastrophic effect on man.

In the face of this threat to climate, global society has two principal alternatives: (1) It can accept the uncertain but probably very serious consequences to climate of continued growth in fossil-fuel combustion. (2) It can mitigate the threat by limiting the use of fossil fuels. (It would be desirable to pursue extensive climatological studies in order to establish a basis for a rational global limit on the burning of fossil fuels.)

PUBLIC RISKS OF NUCLEAR POWER

The risks posed by nuclear power are qualitatively different from those of other energy systems. In their day-to-day operations, nuclear plants produce little environmental degradation in the usual sense.

One distinctive type of nuclear risk arises from unintended releases of hazardous nuclear materials. Exposure to the intensely radioactive products of nuclear fission and elements heavier than uranium (such as plutonium) is a threat to life. Nuclear fuels are also the material nuclear bombs are made of; such material could be diverted to military applications or get into the hands of criminals or political extremists who might use it to make weapons.

The hazards of nuclear power have been discussed at considerable length by others.[18,3] As in the case of fossil fuels, we shall briefly discuss here a few important risks in the reducible and irreducible categories.

As with fossil-fuel reducible risks, our judgment that a nuclear risk is reducible reflects merely what we feel is technologically and institutionally possible. Whether or not appropriate corrective measures are taken depends on many factors. In the case of nuclear reducible risks the consequences of not implementing appropriate fixes could be particularly disruptive and long lasting. Furthermore, human society at

* Technological controls of this problem are conceivable but highly unlikely. Cesare Marchetti has suggested recovering CO_2 from combustion processes and pumping it to the bottom of the ocean, which does have the capacity to absorb enormous quantities of CO_2.[17] But this would require an incredible CO_2 transport effort. Roughly 3 tons of CO_2 would have to be hauled to the bottom of the ocean for each ton of coal burned.

various times and places will be prone to gross incompetence or extended disorganization and violence, so it must be accepted that any and all controls will from time to time be ineffective. For this reason the concept of reducible risk in the case of nuclear power should not be interpreted too optimistically.

REDUCIBLE RISKS

Nuclear-Reactor Accidents: Significant releases of radioactivity to the environment could occur in accidents at various parts of the nuclear "fuel cycle" (see Figure 5.3). We focus here on the possible consequences of a serious reactor accident.

A large release of radioactivity could occur if the reactor fuel melted and the released radioactivity breached the reactor containment structure. And even accidents that involved failures short of a "meltdown" could pose serious public risks. Consider the accident at Three Mile Island near Harrisburg, Pennsylvania. If the containment building at Three Mile Island had been breached, then the radioactive iodine released from the partly damaged nuclear fuel into the containment building could have been released into the atmosphere, causing hundreds or thousands of cases of thyroid cancer among those living downwind.[19]

Theoretical risk analyses [20] are simply not adequate at this time to adequately assess the likelihood of large-scale system failures in systems as complex as today's reactors.[21] Despite this uncertainty, there is no fundamental reason why reactors can't be designed so that if they failed, the hazards to the population would be relatively low. It is likely that emergency planning procedures, modifications of reactor containment structures so as to greatly reduce atmospheric releases of radioactivity in the event of meltdown accidents, and schemes for isolating reactors and improving control over them could be developed, and would, perhaps, not be very costly. Some measures along these lines were suggested in a 1975 study on reactor safety by the American Physical Society.[22] Many in the nuclear industry believe, however, that reactors have been "overdesigned" with respect to safety. Because of this attitude a tough regulatory stance would be required to bring about needed safety improvements.

Radioactive-Waste Disposal: High-level radioactive wastes generated in nuclear-power production must be isolated from the biosphere.[23] These high-level wastes need to be sequestered at least several hundred

FIG. 5.3

Current once-through and proposed plutonium breeder fuel cycles

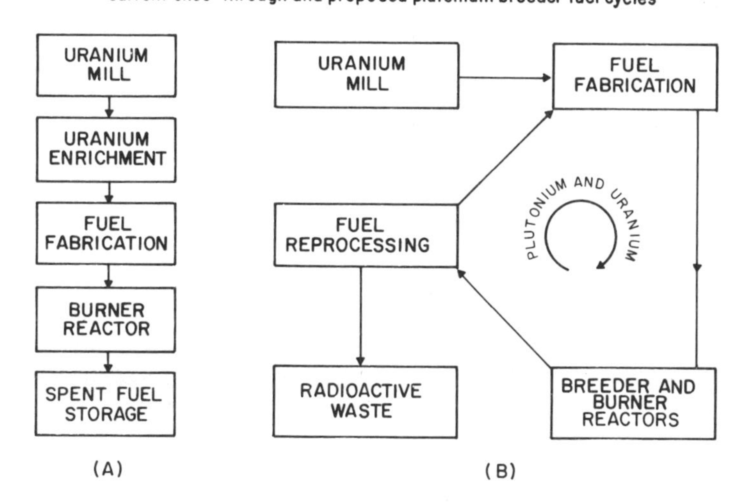

These figures show the contrast between the current "once-through" fuel cycle of commercial power reactors in the United States (see A) and the proposed plutonium breeder fuel cycle (see B).

In the once-through cycle the uranium is mined; a large fraction of the uranium-238 is removed, thus enriching the uranium by raising the percentage of uranium-235 from its natural level of 0.7 percent to approximately 3 percent; the "enriched" uranium is used to make reactor fuel; the fuel is used to sustain a chain reaction in the reactor until most of the uranium-235 has been consumed; and, then the "spent" fuel is stored. This spent fuel contains some plutonium which has been produced by neutron bombardment of the uranium-238 in the fuel.

In the proposed plutonium-breeder economy, the primary fuel is derived from the element uranium-238, which makes up 99.3 percent of natural uranium. Breeder reactors would produce more chain-reacting plutonium out of this uranium-238 than they consume. Since uranium-235 is not the primary fuel, no enrichment plant is required. Natural uranium is mixed with plutonium in the reactor fuel. After being irradiated for some time the fuel is removed from the reactor core, processed chemically, and the uranium and plutonium are recycled into fresh fuel.

SOURCE: H.A. Feiveson, T.B. Taylor, Frank von Hippel, and R.H. Williams, "The Plutonium Economy: Why We Should Wait and Why We Can Wait," *The Bulletin of the Atomic Scientists*, December 1976.

Table 5.1 Some Milestones in the History of Planning for the Disposal of High-level Radioactive Wastes

DISPOSAL OF SOLID RADIOACTIVE WASTES IN BEDDED SALT DEPOSITS [36]

[November 1970] At the request of the Atomic Energy Commission, The Committee on Radioactive Waste Management of the National Academy of Sciences–National Research Council has examined the technical feasibility of burial of solid radioactive wastes in bedded salt deposits [and has concluded that]

1. The use of bedded salt for disposal of radioactive wastes is satisfactory. In addition, it is the safest choice now available. . . .
2. The site near Lyons, Kansas, selected by the AEC is satisfactory. . . .

NUCLEAR WASTE: KANSANS RILED BY AEC PLANS FOR ATOM DUMP [37]

[April 1971] Plans by the AEC to set up its first graveyard for radioactive wastes in the middle of Kansas have . . . raised a good deal of controversy in the state, fueled by reports from the Kansas State Geological Survey, which doesn't want the AEC to purchase the 1000 acre site and the 1700 acres of underground rights until further studies prove that all risks have been eliminated. . . .

AEC SAYS SITUATION WELL IN HAND: . . . COMMERCIAL WASTE MANAGEMENT STATUS [38]

[August 1973] The AEC's multi-pronged program for storage and disposal of commercially-generated radioactive waste . . . include[s]a near-term program adequate for safekeeping of high-level wastes "for as long as the human race wants to," and intermediate and longer-range programs that would improve the economics by lessening surveillance and maintenance burdens.

The near-term program calls for retrievable surface storage, and AEC is planning construction of such a storage facility to be ready to accept the first commercial wastes. Development is continuing meanwhile on a possible successor concept: placement in a deep salt bed. . . . Despite the fiasco at Lyons, Kan., which was partly technical and partly political, AEC still feels bedded salt has the best potential. . . . An area in southeastern New Mexico appears, from surface exploration and information supplied by the U.S. Geological Survey, to be satisfactory. AEC plans call for drilling into the salt bed in New Mexico for further exploration. . . .

ERDA SHELVES A NUCLEAR WASTE STORAGE PLAN [39]

[April 1975] ERDA has abruptly shelved its controversial plan—inherited from the extinct AEC—to build a $55 million surface storage facility for the nation's nuclear wastes. ERDA apparently will proceed with its long-range plan of developing a permanent burial site for nuclear waste in southern New Mexico. . . . Critics said the AEC seemed to have its priorities upside down in that it was emphasizing a temporary fix to the problem of work on a long-range solution. . . .

RADIOACTIVE WASTE SITE SEARCH GETS INTO DEEP WATER [40]

[October 1975] The search for a permanent disposal site for radioactive wastes . . . hit another snag recently. Sandia Laboratories, of Albuquerque, New Mexico, which is managing the search for an underground repository in the remote areas of southeastern New Mexico, reports that the latest test hole has discovered unexpected geologic conditions that may render the immediate area under investigation unsuitable.

Table 5.1 (Continued)

NUCLEAR REPORT CASTS DOUBT ON . . . DISPOSAL [41]
[July 1978] . . . Even deep salt beds, a leading contender as a permanent storing place for atomic wastes, got a poor evaluation in the report prepared by the President's Office of Science and Technology Policy. . . . Deutch, head of the panel, told a House Interior subcommittee that a solution to the problem of waste disposal was still years away. . . .

CARTER TO CONSIDER DELAYING EARLY DEMO OF COMMERCIAL NUCLEAR DISPOSAL [42]
[March 1979] President Carter will be urged to forego DOE's plans for early demonstration of spent nuclear fuel disposal and to launch instead a wide-ranging search for alternative geologic sites, according to White House officials working on a draft presidential review memorandum on the nuclear waste issue. These sources said a Carter decision on the option to search for alternatives could push the target date for the nation's first permanent waste repository beyond 1993.

years, and perhaps tens of thousands of years. An effective long-term waste-storage scheme has yet to be developed. Eventually the wastes will have to be disposed of in such a way that for a very long period and without continuous safeguarding there is a negligible probability that they will be released to the environment. Various seriously flawed proposals have been advanced. The one proposal that appeared for many years to be promising, burying the wastes in salt beds, is still subject to serious question.[24]

Nevertheless, because relatively small storage volumes are required and because it should be possible to package the wastes to safeguard the environment, we see no obvious technological obstacle to permanently disposing of high-level radioactive wastes.[25] In this sense we classify the waste-disposal problem as a reducible risk. But while there are hopeful prospects, the chances of being unsuccessful are by no means negligible. As shown in Table 5.1, the history of radioactive-waste-disposal efforts in the U.S. gives little indication that a solution to this problem will be found soon. Given the high stakes involved, society may be making a mistake in moving ahead with nuclear power without a proven solution to the waste-disposal problem.

Nuclear Theft: Uranium-235, uranium-233, and plutonium are all reactor fuels and materials from which nuclear explosives can be made. Nuclear weapons are relatively easy to make once such materials are in hand.[26,27] With a few kilograms of weapons material a very small clandestine group could build a crude bomb with the power of at least 100 tons of chemical high explosive. In a typical suburban area such a bomb might kill 2000 people. The same explosion in a parking lot beneath a very large skyscraper might kill 50,000 people and destroy the entire building.

The primary fuel of today's light-water reactors is uranium containing such a low concentration of uranium-235 that nuclear weapons *cannot* be made from it.* But these reactors discharge plutonium in their spent fuel, enough plutonium each year to make some 20 nuclear explosives per reactor. The problem of diversion of plutonium from the nuclear-power system by criminals or terrorists is not serious with today's "once-through" fuel cycle, where the spent reactor fuel is not reprocessed to recover and recycle the "unburned" fuel (see Figure 5.3), because the plutonium is protected from diversion by the intense radioactivity of the spent fuel. Nuclear theft would be a serious problem, however, if the plutonium were extracted and recycled.[26]

The nuclear industry wants to recycle plutonium in today's reactors because so doing would reduce uranium requirements by 30% and would help the industry gear up for a "plutonium economy" based on breeder reactors. The latter is the more important motivation because there is little economic incentive to recycle plutonium in light-water reactors.[25] Nevertheless, the Europeans and Japanese are developing the capability to do so.

Two major suggestions have been made for "fixing up" plutonium-based nuclear-reactor systems to prevent nuclear theft:[28] establishing extremely tight security measures wherever plutonium is handled or transported, and "spiking" all fresh plutonium with radioactive contaminants. Both approaches appear to us to be quite inadequate.

Elaborate security systems are not likely to be good enough. Achieving a truly high degree of security is frustrated by the fact that, for some individuals or groups, the incentives for nuclear violence may be very high, in terms of either political or financial gain. Extensive experience with items such as heroin and precious jewels clearly shows that where incentives are sufficiently high, security will be breached. Similarly, the idea of "spiking" plutonium is fundamentally flawed be-

* Light-water-reactor fuel contains about 3% uranium-235 and 97% uranium-238. Uranium-235 is a weapons-usable material, but uranium-238 is not. For the mixture to be weapons usable, the uranium-235 concentration must be in excess of 20%.

cause it creates an artificial hazard which would be a nuisance to the industry. Handling and processing would be much easier if the spiking were eliminated or substantially reduced. Because the incentive to do away with spiking would always be present it is likely that at many places and times it would be done away with.

But while it is impractical to "fix up" the plutonium-based systems so as to reduce the risk of nuclear theft to a level we find acceptable, one can chart a course for nuclear power which would be no more vulnerable to nuclear theft than today's once-through fuel cycles. The idea is to develop nuclear systems which continue to employ no weapons-usable material in fresh reactor fuel. This can be done in a manner consistent with any constraints related to uranium resources (see Chapter 4) by shifting over time to advanced burner reactors which, though less efficient in their use of uranium than breeder reactors, would be more uranium efficient than today's burner reactors. It has been shown[29] that such a strategy, in which these reactors are operated on once-through fuel cycles that do not involve the recycling of plutonium, would be economically viable for the U.S. for at least 100 years.[30]

In summary it is our view that a high degree of security against nuclear theft could be achieved while maintaining nuclear power as a major energy option, but this would require directing the course of nuclear power away from the plutonium economy. Since a nuclear theft occurring anywhere in the world is a global problem, this redirection would require coordinated changes in commitments that have already been made in several national programs.

AN IRREDUCIBLE RISK: NUCLEAR-WEAPONS PROLIFERATION BY NATION-STATES

While a nation can acquire nuclear weapons without having civilian nuclear-power technology, having nuclear power provides an easy route to nuclear weapons. This approach to nuclear weapons requires little time and resources beyond the nuclear-power system, so it is inherently ambiguous and does not force a nation to signal or even decide its intentions in advance.[31]

The possibility of acquiring nuclear weapons is of interest to many nation-states.[32] As the example of India demonstrates, the spread of civilian nuclear power puts nuclear arms within the reach of many countries. If present trends persist, about three dozen non-nuclear-weapons states will have large nuclear-power or research reactors by the mid-1980s (see Table 5.2). These developments make control of the prolif-

Table 5.2 Nuclear-Power and Nuclear-Proliferation Capabilities

COUNTRY	NPT STATUS	OPERATIONAL NUCLEAR-POWER CAPACITY[a]	FORECAST NUCLEAR-POWER CAPACITY MID-1980s[a]	ANNUAL BOMB EQUIVALENT[b]
Nuclear-Weapons States				
United States	Party	37,600	208,400	4,168
USSR	Party	4,600	14,400	288
United Kingdom	Party	5,300	11,800	236
China	Nonparty	?	?	?
France	Nonparty	2,800	21,300	426
Insecure States				
Israel	Nonparty	0	?	?
South Africa	Nonparty	0	?	?
South Korea	Party	0	1,800	36
Taiwan	Party	0	4,900	98
Yugoslavia	Party	0	600	12
Status-seeking States				
Brazil	Nonparty	0	3,200	64
India	Nonparty	600	1,700	34
Iran	Party	0	4,200	84
Spain	Nonparty	1,100	8,300	166
Rivals to States in Preceding Categories				
Argentina	Nonparty	300	900	18
Egypt	Signatory	0	?	?
North Korea	Nonparty	0	?	?
Pakistan	Nonparty	100	100	2
Politically Constrained Major States				
Czechoslovakia	Party	100	1,900	38
East Germany	Party	900	2,700	54
Italy	Party	500	5,200	104
Japan	Party	5,100	15,500	310
Poland	Party	0	400	8
West Germany	Party	3,300	23,300	466

eration of nuclear weapons extremely difficult. It is indeed late in the history of this technology to be discussing adequate control.*

* The present system for controlling proliferation began in 1957 with the creation of the International Atomic Energy Agency and continued with the Non-Proliferation Treaty of 1970. The stated purpose of the system is "timely detection of diversion of . . . nuclear materials from peaceful nuclear activity to the manufacture of nuclear weapons . . . and deterrence of such diversion by the risk of early detection." Even if early detection should be achieved, effective penalties could not, realistically, be invoked. For an example see ref. 33.

Table 5.2 (Continued)

COUNTRY	NPT STATUS	OPERATIONAL NUCLEAR-POWER CAPACITY[a]	FORECAST NUCLEAR-POWER CAPACITY MID-1980s[a]	ANNUAL BOMB EQUIVALENT[b]
Other Developed Countries				
Australia	Party	0	?	?
Austria	Party	0	700	14
Belgium	Party	1,600	5,400	108
Bulgaria	Party	900	1,800	36
Canada	Party	2,500	11,800	236
Finland	Party	0	2,200	44
Hungary	Party	0	1,800	36
Luxembourg	Party	0	1,300	26
Netherlands	Party	500	500	10
Romania	Party	0	400	8
Sweden	Party	3,200	8,400	168
Switzerland	Signatory	1,000	5,800	116
Other Developing Countries				
Chile	Nonparty	0	0	0
Greece	Party	0	0	0
Indonesia	Signatory	0	0	0
Mexico	Party	0	1,300	26
Philippines	Party	0	1,400	28
Thailand	Party	0	?	?
Turkey	Signatory	0	0	0

NOTES:

[a] Capacity in Megawatts derived from "World List of Nuclear Power Plants, December 31, 1975," *Nuclear News*, February 1976. Many countries' programs have undergone change since then, but the overall picture is reflected here.

[b] "Annual Bomb Equivalent" is a rough approximation which assumes each 1,000 MWe of forecast capacity is operated in such a way as to produce 200 kg of plutonium annually as a byproduct, and 10 kg of plutonium are required for one bomb.

SOURCE: *Nuclear Power Issues and Choices*, ref. 18.

Many thoughtful people feel that proliferation of nuclear weapons among nation-states is a less important concern than the huge stocks of weapons held by the great powers. We share the concern about existing weapons and agree that negotiations to disarm and systematically reduce this reign of terror are essential. The creation of weapons capability (almost at moment's notice) among many states, some of them highly unstable, creates a different danger which is also very serious. Both of these problems should be controlled as well as possible. But unfortunately it appears that there is no technical fix to *prevent* proliferation by nation-states through civilian nuclear power.

The proliferation risks would be extreme in a plutonium economy. Preventing diversion of plutonium fuel would require tight security maintained by an international police force that attended all plutonium processing, storage, and shipments. Many nations would not relinquish the degree of national sovereignty implicit in such an arrangement. "Spiking" would not be effective either, because any nation-state bent on acquiring nuclear weapons could construct crude reprocessing facilities and recover plutonium from irradiated fuel in a matter of weeks. Further, the weapons potential of a single diversion "incident" would be very high. The diversion of the plutonium contained in the annual discharge of one plutonium breeder reactor would be enough for a hundred nuclear explosives.

Fuel cycles that involve no weapons-usable material in fresh reactor fuel are, as mentioned above, more proliferation resistant.[29] The most promising situation is where only once-through fuel cycles are involved. Nations should be more willing to relinquish spent fuel to international control if there is no reprocessing than if there is, because without reprocessing spent fuel has no economic value. However, even a once-through proliferation-resistant fuel cycle is vulnerable to nations *determined* to acquire nuclear weapons. Any nation that wanted a few nuclear weapons in a hurry could build a crude clandestine reprocessing plant to recover plutonium from spent fuel. In the future the amount of plutonium circulating in global commerce with this fuel cycle could rise to 250,000 kilograms per year.[34] While this plutonium would be present *only in spent fuel* and in absolute terms would be far less than the more than 10 million kilograms per year of fissile plutonium that would circulate in *fresh fuel* in a plutonium-breeder-based nuclear economy of the same size, the magnitude of the plutonium at risk would still be huge. A diversion of 1% of 1 year's flow would be enough for hundreds of nuclear explosives. International commerce in plutonium would be maintained under international safeguards, but we cannot imagine internationally based human institutions capable of maintaining tight control over essentially 100% of these flows, year in and year out.* Eventually there would be diversions.

* The institutional obstacles to effective controls against proliferation are fundamentally different from those that today inhibit implementation of effective measures to improve reactor safety or dispose of radioactive wastes. In these latter cases the barriers involve mainly institutional inertia. In the case of proliferation the problem is institutional determination to violate safeguards.

Nuclear-fission technology cannot be made proliferation proof. An inherent feature of worldwide development of nuclear power is that it would make nuclear weapons available to nation-states almost at moment's notice. The potential for proliferation would be so great that it would ultimately lead to the use of nuclear weapons. The history of violence among men permits no other conclusion. The fear and confusion arising from such use, even in isolated instances, could well trigger a large-scale nuclear exchange.

The United States cannot indifferently accept the possibility that bomb capability will be placed in the hands of many dozens of nations as a result of a relatively minor commercial activity—an international commercial activity initiated largely by the U.S. Atoms for Peace program and based largely on General Electric and Westinghouse technology. It is gratifying that the Carter administration has taken the proliferation problem seriously and has taken initiatives to delay commitments to use of plutonium fuels.[35] Our analysis shows, however, that the Carter initiatives, even if accepted by other nations, are not enough: what is needed is to move away from nuclear-fission power altogether. So doing would not entirely eliminate the proliferation risks of nuclear power. Significant quantities of nuclear materials have already been produced, and many reactors are producing more. It is, indeed, very late to take action. Nevertheless we do have the choice of whether or not to live with the greatly increased quantities of nuclear materials and greatly increased traffic in them implicit in present plans to expand nuclear-generating capacity throughout the world. The problem of controlling weapons materials will be far more severe with a pervasive, growing nuclear industry than with a relatively small and moribund industry.

POLICY IMPLICATIONS

Not only is it becoming ever more difficult to provide more energy, but also planners must deal with the environmental, health and safety, and security implications of energy systems. These are relatively new issues in our society. Only recently have levels of natural resource exploitation reached the point where adverse impacts have been severe over wide areas. Only recently has the level of affluence so far exceeded subsistence levels that citizens feel secure enough to become actively concerned about the secondary impacts of technology.

If we continue our present energy course, its adverse impacts will rapidly become more severe. Human society will have to accommodate in a matter of decades to the irreducible global risks we have described here. And to deal effectively with the myriad reducible risks will require ever more pervasive government controls over technology and the industrial institutions that develop and manage it.

Public concern about adverse impacts such as these has led to the creation in just a few years of a new generation of regulatory agencies in the U.S. with the nominal responsibility to bring offending technology under public control—the Environmental Protection Agency, the Nuclear Regulatory Commission, the Occupational Safety and Health Administration, and many others. The challenge to these new agencies is to limit to acceptable levels the adverse impacts of risky technologies *without* seriously inhibiting industrial production. Will the proposed mechanisms of control, the skills brought to bear, and the power allocated to these efforts be adequate to protect broad public interests? Because these efforts are truly in an embryonic stage, it is too early to know. But for reasons we set forth in Chapters 14 and 15, we are skeptical that regulatory controls can be effective without seriously constraining production.

The prospect of only limited regulatory success in dealing with serious reducible risks associated with the present energy-supply strategy provides a strong motivation for seeking an alternative energy course. Moreover, to avoid the irreducible risks we have described here, the nation and world would *have* to adopt a new energy course. Of the various alternative energy-supply technologies we have not yet considered (solar, nuclear fusion, geothermal), only solar energy appears well enough developed and of wide enough availability to become really important in the foreseeable future. What is its potential?

CHAPTER 6

WHAT ABOUT SOLAR ENERGY?

THERE IS A LAW of nature which says that things tend to get more mixed up as times goes by. If cream is poured into a cup of coffee, the cream and coffee eventually become uniformly mixed, and it is difficult to separate them. This tendency toward greater disorder is often illustrated by the rhyme

> Humpty Dumpty sat on a wall
> Humpty Dumpty had a great fall
> All the king's horses and all the king's men
> Couldn't put Humpty Dumpty together again.

The law that disorder must increase is called *the second law of thermodynamics;* and the quantitative measure of disorder is called *entropy.*[1] Highly ordered systems have low entropy, while disordered systems have high entropy, as illustrated in Figure 6.1.

Everything that happens, whether it be a thought flitting through your mind or a thunderstorm, is associated with an overall increase in entropy, a tendency toward greater disorder or lack of pattern. How is it, then, that we see such marvelous patterns about us—the structure of the atmosphere and oceans, the varied forms of life, man's creations? The answer is sunlight. The earth is bathed in sunlight, a very low-entropy and hence potent form of energy.

Fig. 6.1

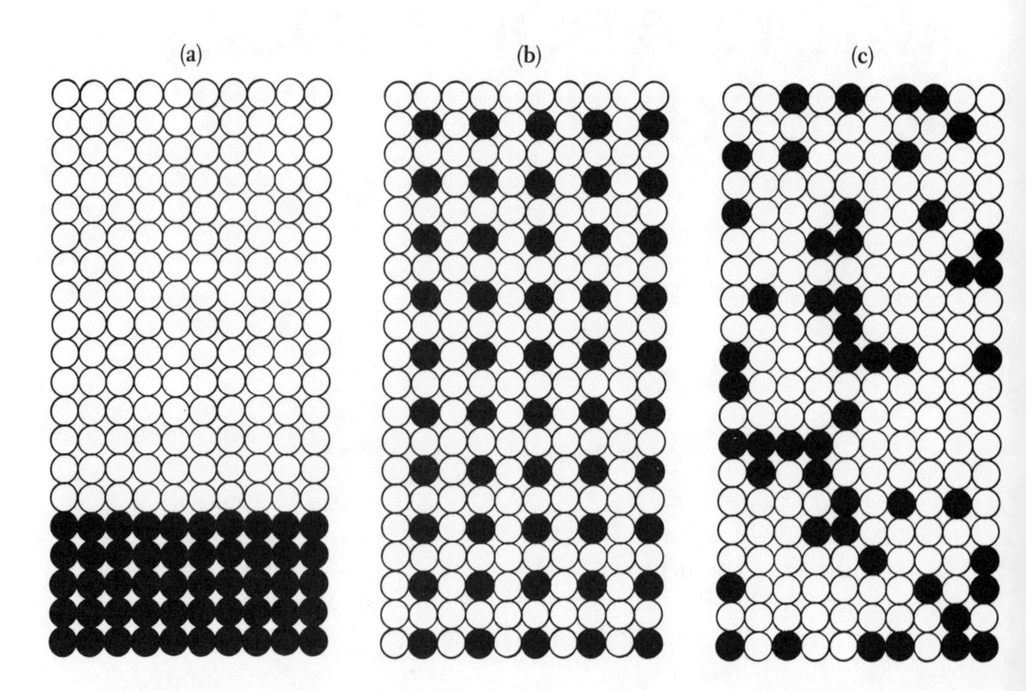

Illustrative examples of low and high entropy systems. The organized, patterned arrangements of black and white marbles at left and center have low entropy. At right is a random arrangement; such arrangements have high entropy. The second law of thermodynamics states that entropy must increase, i.e. a system like (a) or (b) tends in time to become randomlike (c). The law does not require, however, that all parts of an interacting system move together to higher entropy. If the entropy of one part of an interacting system increases, that of another part may decrease. Thus the huge entropy increase which occurs when sunlight is converted to warmth in various ways on the surface of the earth enables many-ordering or entropy-decreasing processes to occur, such as the growth of living things.

Low-entropy sunlight creates order as it flows into the earth's atmosphere and onto its surface. Sunlight shining on a simple molecule can change its structure. Indeed sunlight created life from simple molecules and sustains it. Sunlight created the atmosphere and its structure and maintains it. Sunlight created the fossil fuels.*

Not only is solar energy a very potent form of energy; it is abundant as well. The flow of solar energy through the earth's natural systems is about 10,000 times greater than the flow of fossil-fuel energy into man's machines. Even the rivulets of solar energy which flow through the earth's systems as the winds and as photosynthetic production are respectively 100 and 10 times greater than man's rate of energy use. (In this book, the term *solar energy* encompasses all solar-related forms such as wind and growing plants as well as direct conversion of sunlight.)

These considerations of the high *quality* and enormous *quantity* of the solar-energy resource suggest that it may be reasonable and desirable for man to try to use his ingenuity to rechannel a small amount of the flow of solar energy so as to reduce or remove his dependence on fossil and nuclear fuels. A shift to solar energy would also be desirable because many (though certainly not all) of the more promising strategies for tapping the solar-energy resource would have much less damaging impacts on the environment, public health, and security than the irreducible risks of fossil and nuclear fuels described in Chapter 5.

It would take us beyond the scope of this book to attempt even a survey of the possibilities for harnessing solar energy, because there is such a rich diversity of options. For an overview of the possibilities the reader can consult a number of monographs that are now available.[2] Despite the great diversity of technological possibilities and the rapidity with which new inventions are being introduced, some conclusions can be drawn. Most applications of solar energy involve its conversion to heat, chemical fuels, or electricity. In this chapter we briefly discuss the prospects and some fundamental problems associated with exploiting solar energy in each of these three forms at levels necessary to meet the future demands projected in Chapter 3.

* The order created by sunlight does not violate the second law. In the sun the order in sunlight is dwarfed by the amount of disorder created by the nuclear-fusion reactions inside the sun which produce sunlight. Similarly the earth as a whole automatically satisfies the rule that the amount of disorder must increase because the entropy of the earthshine flowing away from the earth greatly exceeds the entropy of the sunlight flowing to the earth. See Figure 5.2.

SOLAR HEAT

Sunlight can be absorbed in collectors mounted on rooftops or in open spaces to provide domestic hot water, space heat,[3] or industrial process heat.[4]

DOMESTIC HOT WATER

The commercial use of most solar-energy devices is plagued by high capital cost. Because solar energy is diffuse, large collector areas are required. Because it is intermittent, substantial storage capacity (or a back-up energy supply) is needed for when the sun doesn't shine. Both collectors and storage can be very costly. These problems are not insuperable for solar water heating, however.

Solar domestic water heating is already an established industry. Millions of rather rudimentary solar hot-water heaters are in use in Japan; solar hot-water heaters with insulated storage tanks are widely used in Israel and Australia; and tens of thousands were in use in Florida when electricity was expensive and before cheap natural gas became available.[5] Now that fuel and electricity are becoming more expensive again, solar hot-water heaters, including swimming-pool heaters, are being installed in considerable numbers throughout the U.S. During a 12-month period in 1977–78 over 6 million square feet of pool-heating collectors and over 2 million square feet of water-heating collectors (for roughly 50,000 dwelling units) were installed, with the lion's share in Florida and California.[6]

Solar water heating is catching on because the nature of domestic demand for hot water is such as to mitigate the capital-cost problem. Because hot water is needed throughout the year, capital is used quite efficiently in a solar water heater. Moreover, half of domestic hot-water demand can be met with an efficient unit less than 40 square feet in area. This is sufficiently small that installation is simple and cheap and water-heater production can benefit from the economies of mass production. Finally, the solution of the storage problem is facilitated by the fact that there already is some short-term storage with conventional water heaters.

These favorable conditions mean that solar water heaters can compete with electric water heating in especially sunny parts of the country today and that with relatively modest innovations solar water heating should be able to compete both with new sources of electricity and with new synthetic-fuel sources, even in moderately sunny areas such as New York City. (Our references to the City here and below are to its climate and not to its suitability for energy systems from other points of view; for instance, when comparisons are made with other energy systems, the cost estimates for such systems represent national average values.) However, solar water heating will not soon become competitive with new conventional oil or gas sources.[7]

SPACE HEATING

Solar space heating via rooftop collectors is not as attractive economically as solar water heating.* The main reason is that demand is highly concentrated in the winter months, while the best sun is available in summer. (We estimate for New York City that 85% of the space-heating demand occurs between November 1 and March 31, while about 70% of the available sunshine comes during the rest of the year.)[11] Thus space-heating capital equipment will be used much less efficiently than water-heating equipment. Another reason for the less favorable economics is that solar collectors for space heating require a large fraction of the typical available roof area. Thus installation is not so simple, and the cost benefits of mass-production techniques are not nearly so great as with water heaters.

For example, if collector costs should fall to the point where solar water heating became competitive with synfuel in a New York City climate, then either collector costs would have to fall by an additional ⅓ or synfuel costs would have to be ⅓ higher than estimated to make solar space heating competitive.[7,12,13]

However, solar space heating has special advantages when compared with electricity, because electric space heating poses a serious peaking problem for the electric utilities which drives up the cost of providing space heat with electricity. For space-heating applications the peak demand for electricity near New York City is about 6 times the average demand.[14] This peak demand would occur on the coldest day, when there would be a coincident peak demand for electricity from all space-heating customers of the utility. The utility would have to meet this peak demand with reserve generation, transmission, and distribution capacity. If costs were equitably allocated, these space-heating customers would have to pay their fair share of the costs of holding this extra capacity in reserve for these peaking periods.† Under these condi-

* "Passive" solar heating, which involves designing a house so that windows can serve as solar collectors, is discussed in Chapter 8 as a fuel-conservation measure; the "active" solar heating considered here is regarded as an energy-supply option.

† This statement applies to a system having a sufficient number of space-heating customers that the collective space-heating demand gives rise to a sharp peak demand on the utility system.

tions the cost of the solar space (plus water) heating system would be competitive at about the same collector cost as for a solar water-heating system.

These judgments are based on a solar-energy system providing about ⅔ of the total space- and water-heating demand. It would not be worthwhile to design a solar rooftop-collector system to meet a much larger fraction of the total heating demand because costs per kilowatt rise sharply as one increases the collector capacity so as to meet a larger and larger fraction of demand.* This means that a conventional energy source must be used for backup. Should that backup source be electricity or oil?

Consider first electricity. The cost per kilowatt for an electrical system that would be needed only when the supply of solar energy was inadequate would be even greater than the cost per kilowatt for a system providing 100% electric heat, because the electrical backup system would tend to be needed only on the coldest days. The peaking problem would be even more severe than if electricity provided all the heat (as discussed in Chapter 8).[16,17] Thus a solar-electric supply mix is inherently unpromising.

If instead the consumer could obtain oil as the backup fuel, then he would be motivated to choose a 100% oil system over "going solar" in the first place, because the solar option would be more costly. Of course this problem could be resolved either by taxing away the cost advantage of oil or by subsidizing solar-heating investments. The government in fact has done some of the latter by allowing for large tax credits for solar-heating investments.[18] The rationale is that such subsidies are justified by the social need to lessen our dependence on oil. But is this the most efficient use of taxpayers' money to achieve this social goal?

We think not. The solar energy system we have considered here that would be competitive with electric heat would require the homeowner to invest more than $7000 for a system having a 300-square-foot rooftop collector to reduce his fuel use by ⅔. In Chapter 8 we show how a homeowner could reduce his fuel use this much or more, with a much smaller investment, simply by making his house much more fuel efficient.

We don't mean to imply that solar space heating is inherently impractical—only that fuel-efficiency improvements are much more important. Once extensive fuel-efficiency improvements have been made,

* This sharp increase in unit capital costs arises because more collector capacity must lie idle during much of the year or because large storage capacity would be required in order to use collectors more efficiently.

solar energy may still be used to reduce space-heating fuel requirements even further. But just how much solar heating would be used is very uncertain because future costs for solar heating are uncertain. Many new designs are being invented and tried. Some of these new approaches could be successful and thus alter our conclusion, based on present technology, that solar space heat is typically very costly. Already, owner-built solar systems are being constructed with materials costs well below the total costs of typical contractor-installed systems.[19] Theodore Taylor has proposed systems based on simple low-cost plastic collectors.[20] Solar ponds[21] and other low-cost designs are also being explored.[3] We describe in detail a particularly promising but largely unfamiliar technological arrangement in Chapter 11.

INDUSTRIAL PROCESS HEAT

Industrial process heat poses a very different challenge from that posed by space heating. Because industrial heat is used throughout the year, the economics of providing process heat could in some applications be as good as or even more favorable than the economics of domestic water heating.[22] However, the energy intensity of industrial process-heat applications poses a serious problem. In the case of domestic space and water heating we have seen that the area of a typical solar collector would be several hundred square feet—an area which could well be available on the roofs of many households. A typical plant in an energy-intensive industry with round-the-clock operations, on the other hand, would require a solar collector field of some 1 or 2 square miles.[23] Since much of American industry is located in or near metropolitan areas where open spaces are scarce, it appears that the potential for providing direct solar heat for large-scale industrial operations is quite limited.

The area constraint would not be so formidable for relatively small industrial plants. Consider a plant that operates only in the daytime and consumes 100 barrels of oil per day for its boilers (resulting in an annual oil bill of about $0.75 million). Such a plant would require less than 50 acres for a solar collector field.[24] While this is still big enough that land-use constraints would be important in many areas, collector areas of this size could be accommodated at some locations.

CONCLUSIONS REGARDING HEATING APPLICATIONS

Table 6.1 shows that nearly half of U.S. energy use is for heating purposes in buildings and industry. Thus a major task for solar-energy technology would be to provide these heating needs.

Table 6.1 Distribution of U.S. Energy Consumption by End Use in 1973

END USE		PERCENTAGE OF TOTAL ENERGY USE
Water heating		4
Space conditioning		19
Industrial process heat		24
low temperature (process steam)	14	
high temperature (direct heat)	10	
Cooking and clothes drying		1.5
Transportation		25
urban auto use and small trucks	11	
rail	1	
rural auto use and large trucks	6.5	
air	2.5	
other	4	
Miscellaneous electric		19.5
electric drive	9	
refrigeration	3	
lighting	5	
electrolytic processes	1	
other	1.5	
Feedstocks and other nonfuel uses		7
TOTAL		100

NOTE: In this table electricity consumption is measured in terms of the fossil fuel energy (or its equivalent) required to generate the electricity.

SOURCE: M. H. Ross and R. H. Williams, "The Potential for Fuel Conservation" (Albany: Institute for Public Policy Alternatives, State University of New York, July 1975).

As we have suggested, domestic water heating by solar energy looks promising for most areas of the country. (Solar water heating would be limited in cloudy areas like the Pacific Northwest and certain Great Lakes areas, in highly urbanized areas, and on shaded lots.) Solar space heating is much less attractive economically, but electric space heating (the conventional alternative to oil or gas space heating) appears to be equally unattractive in most areas. As we shall show in later chapters, solar energy can be made more attractive for space heating, but only in combination with fuel-efficiency improvements and perhaps then only in configurations that are largely unfamiliar today.

Solar heating for industrial processes would appear to be impractical for large energy-intensive firms in or near metropolitan areas, even in the sunniest parts of the country.[25] The best opportunities for direct use of solar heat in industry appear to be in relatively small, rurally located industrial firms, for which most of the energy demand occurs in the daytime.

Use of direct solar heat is not the only way for solar energy to serve industry, however. Chemical fuels derived from solar energy could be burned to provide industrial heat instead. We now examine the prospects for deriving chemical fuels from the sun.

SOLAR-DERIVED CHEMICAL FUELS

One approach to chemical-fuel manufacture would be to use solar energy to "crack" water into its constituent elements, hydrogen and oxygen, and use the hydrogen for fuel.* This could be done today via electrolysis, using solar-generated electricity. However, even with improved electrolysis technology electrolytic hydrogen would be too costly—perhaps 50% more costly even than synthetic methane from coal, which is an especially costly fossil-fuel based replacement for natural gas.[27]

The major alternatives to electrolysis for solar hydrogen production are use of chemical cycles driven with solar heat (in which all of the chemicals involved other than water would be regenerated),[29] and photolysis, in which the sun's rays are used directly with appropriate catalysts to split the water molecule.[30] So far neither of these approaches has advanced beyond the research stage.

A more promising strategy for producing chemical fuels from the sun, at least for the foreseeable future, involves converting biomass, i.e., green plant matter, into useful fuel forms. Fuels derived from biomass differ from hydrogen (or hydrogen-based fuels) in that they contain carbon. Thus their combustion adds carbon dioxide to the atmosphere. However, combustion of fuels derived from renewable biomass, unlike combustion of fossil fuels, leads to no *net* increase in atmospheric CO_2, because the CO_2 released in combustion is just balanced by the uptake of CO_2 from the atmosphere by growing plants.

While photosynthesis is very inefficient (typically only a fraction of 1% of sunlight reaching ground level is converted into the chemical energy of plant matter), it offers unique advantages as a solar energy converter: both collection and storage are very cheap. For space- and water-heating systems, costs are expected to be about $20 per square foot, which corresponds to $800,000 per acre of collectors. Photosynthesis is typically only 1% as efficient as a rooftop collector, so that land would have to cost less than $8000 an acre to compete with a rooftop

* Hydrogen might subsequently be converted into a more convenient fuel such as ammonia. See ref. 26.

collector in producing useful energy. But $8000 an acre is a land price that is easy to beat in most areas. Moreover, the recovered solar energy—the chemical energy stored in plants—is high-quality (low-entropy) chemical energy rather than low-quality (high-entropy) heat. The fact that plant matter provides naturally a concentrated energy-storage medium is extremely important, because it means that the contained solar energy can be consumed remotely in both time and space from where it is produced.

Plant matter can be used as fuel in a variety of forms. The most straightforward approach is simply to burn wood or even nonwoody plant matter directly. Wood of course was the dominant fuel in the United States a century ago. Fuelwood still accounts for about 1% of U.S. energy use, and the fact that fuelwood consumption, after a long period of decline, has been growing since 1972 at 5 times the overall rate of growth in energy consumption is testimony to the fact that biomass energy can compete with fossil fuels in the new era of high-cost energy.[31]

For many applications, however, a fancier fuel is required. Various relatively straightforward technologies are available for making any one of a number of fluid (gaseous or liquid) or solid fuels from biomass: industrial gas of low energy content, hydrogen, methane, methanol, ethanol, gasoline, charcoal, to name some of the more interesting possibilities.

How much chemical fuel is needed? In Chapter 3 and Appendix A we argue, using very general considerations, that at least about 40% of the total energy use must be in the form of fluid chemical fuel in a future U.S. energy economy based on conventional energy sources. In an economy where solar energy is a major energy source, an even greater percentage of total energy use must be in the form of fluid chemical fuel, because of the extra "back up" fuel requirements for when the sun doesn't shine.[32] For the Business-as-Usual energy-demand scenario set forth in Chapter 3, this means at least 50 quads of fluid chemical fuel would be required in the year 2010. Unfortunately, as we shall now show, putting together a chemical-fuel supply of 50 quads from biomass sources is not practical.

The most promising biomass-energy source is organic wastes—urban refuse (mainly paper), pulp waste, wood waste in forests, crop residues, and feedlot manure. However, as we show in Table 6.2, organic wastes can potentially provide no more than about 6 quads of fluid fuel.

To supplement organic-waste resources, one could grow plants on "energy plantations." Competition for scarce land will limit the extent

Table 6.2 The Maximum Potential Supply of High-quality Fluid Fuels Derived from Biomass in 2010 (Quadrillion Btu per Year)

	BIOMASS RESOURCE[a]	FLUID FUEL[b]
Organic Wastes		
Crop Residues	6.5	2.6
Manure	0.8	0.3
Waste Wood in Forests	4.5	1.1
PulpWaste	1.5	0.8
Urban Refuse	3.3	1.7
		6.5
Biomass Plantations		
Nonwoody	11.3	5.6
Woody	4.5	2.2
		7.8

SOURCE: Ref. 35

NOTES:

[a] Crop residues and manure from confined animals are scaled by population growth. Pulp wastes and urban wastes are scaled by GNP growth.

[b] A nominal efficiency of 50% is assumed for creating high-quality fluid fuels from all biomass resources. It is further assumed that only 80% of crop residues and manure is collectable and 50% of waste wood in forests is collectable.

of energy plantations, however. Energy crops simply have too low a value to be able to compete on lands where food crops could be grown instead. What the farmer is paid for his corn, for example, is 7 times the price of coal today.[34] Similarly, lumber and pulp production would represent a more cost effective use of most forest lands than energy production.

This does not mean that the energy-plantation idea does not make sense—only that energy crops must be grown on marginal lands. Are there any marginal lands that can be photosynthetically productive? Much vacant land is available in the Southwest, but such land is entirely inappropriate for energy plantations simply because water supplies are inadequate. One promising category is lands that are characterized by periodic waterlogging. Such lands are quite productive since they are wet, but they are unsuited for agricultural purposes. Poole and Williams estimate that such lands may be about 25 to 35% as extensive as U.S. croplands.[35] If such lands were exploited to raise energy crops, then the useful chemical-fuel supply could be roughly doubled, as shown in Table 6.2. This must be regarded as more or less an upper limit on the amount of land that could be exploited for energy plantations in the U.S. Even though such land is inappropriate for agriculture, much is needed for forestry, for wilderness, and for wildlife preservation.[36] Even if heroic efforts were made to collect and process organic wastes and these efforts were combined with very extensive use of wetlands, the chemical-fuel potential from both organic wastes and energy plantations would provide less than 30% of the need for fluid chemical fuel in the year 2010 under Business-as-Usual demand conditions.

Another possibility for biomass production is ocean plantations, or mariculture, which is free of the principal constraints of terrestrial systems, limited land and water.[37] While biomass production via mariculture is an embryonic technology, so that meaningful cost data are not available, it is clear that this approach would be much more capital intensive than terrestrial production, largely because of the grid structure needed to attach the plants.

There may be fundamental difficulties with mariculture as well. For example, nutrients such as nitrogen and phosphorus would have to be supplied to mariculture crops. One way to provide nutrients is to apply a fertilizer. Poole and Williams show that if this were done, world demand for phosphorus could be increased by nearly 20% for the production of just 5 quads of useful energy.[35] Since phosphorus is an essential nutrient for which world supplies are limited, it would be imprudent to allow such a large increase in phosphorus demand in return for such a small increase in energy supplies. Alternatively, artificial upwellings could be created to bring nutrients into the surface waters from the ocean depths. But if this were done, serious environmental problems could result.[35]

These judgments about mariculture are of course tentative and speculative; nevertheless they show that not enough is known to count on mariculture as a significant source of biomass energy.

In summary, the limited availability of solar-derived chemical fuel means that without major technological breakthroughs most of this essential energy form will have to be obtained from nonsolar sources if energy is required at Business-as-Usual levels.

SOLAR ELECTRICITY

The renewed interest in fuelwood in the United States is proof that at least one form of biomass energy is competitive; similarly the fact that hydroelectricity today accounts for 34% of world electricity production is proof that solar electricity, at least in one form, is competitive.[38] (Hydroelectricity is a form of solar energy because the sun drives the earth's hydrological cycle, in which water is raised from the oceans to high elevations, thereby making possible the generation of electricity from falling water.)

Among other possibilities for solar-electricity production are the options based on heat engines. Just as the heat of combustion from fossil fuels can be used to drive a steam or gas turbine, or a gasoline or other engine, so can solar heat be used as a heat source to drive a heat engine. Such engines could be driven either by solar heat recovered in a collector[39] or by solar heat stored in the warm surface waters of the tropical oceans (in what are called ocean-thermal-gradient electricity generators).[40] Other possibilities for solar electricity have no counterpart among conventional energy sources: the conversion of solar energy in the form of wind or wave energy into electricity, using windmills[41] or wave machines,[42] and the direct conversion of sunlight into electricity using solar cells.[43]

This brief listing suggests that a rich diversity of options exists for producing solar electricity. Of course one always explores many alternative approaches in the research and development phase of any new technology, but usually only a very small number of different systems survive to become commercial. Indeed, the U.S. nuclear-reactor development program in the 1950s explored many reactor types as potential power reactors, but only the light-water reactor survived in the U.S. as a commercial venture, and throughout the world today only the liquid-metal fast-breeder reactor is being pursued for the second generation of reactor types.

But solar energy is different. There will not be a corresponding "narrowing of the options" as the technology matures. The reason for this is that a given solar technology, unlike a nuclear or oil-based "universal" energy technology, will not work equally well everywhere. Different solar technologies will be favored in different regions so as to best exploit the local solar-energy resource—for example, technologies that involve direct sunlight in the sunny Southwest; windmills on the Great Plains, offshore in the Pacific Northwest and New England, and on certain lake shores and mountain ridges; and ocean-gradient generators off the coast of the Gulf of Mexico.

We have identified physical constraints of a fairly fundamental nature limiting industrial direct-solar-heat and biomass-energy technologies, but there do not appear to be physical constraints limiting solar-electricity production. This is because even though any particular source of solar electricity may have limited potential at a given location, one has considerable flexibility in putting together an electricity supply from multiple sources in different locations. This flexibility arises because electricity need not be produced near where it is consumed. Like fuel but unlike heat energy, electricity is a high-quality (zero-entropy) energy form that can be economically transported long distances.

This is not to say that particular solar electric options are not faced with formidable physical constraints. Von Hippel and Williams[44] point out that the glamorous proposal to generate electricity using solar cells at an orbiting space station[45] could lead to a serious disruption of the ozone layer of the upper atmosphere during the construction phase. (There are also other serious objections to this proposal.)[46]

Less serious than the problems posed by the solar satellite but by no means trivial, are the land-use and aesthetic impacts of wind power. Large numbers of windmills would be needed to make a significant contribution. For example, about 2500 windmills with a blade diameter of 200 feet would be needed to provide as much power, on the average, as a conventional 1000-megawatt coal- or nuclear-fired power station.[47] In the extreme situation where windmills would provide all electricity in the U.S. in 2010 under Business-as-Usual demand conditions with maximum electrification, some 2 million large windmills would be needed. If lined up side by side, these windmills would extend 90,000 miles. Of course it is extremely unlikely that so much wind power would be developed. Although wind power may soon be competitive in windy areas, its development will be geographically limited, first to areas with high winds and within such areas to locations where there is minimal competition with alternative land uses and aesthetic values.

While physical constraints to the development of solar electricity in general do not appear to be formidable, serious financial constraints will very likely limit the rate of development. The financial constraints arise in part because of the stiff economic competition faced by solar-energy systems. If the capital costs of electricity escalate no more than they have already, then to replace our present electrical system with a new conventional power system would cost about $2800 per kilowatt of average delivered electric power.[50] To compete with this, the cost of a complete solar electric system—generation, storage or standby facilities, and, if necessary, transmission and distribution—would have to be some $6 per square foot.[51] This is a very tough cost goal to meet. Systems involving simple rooftop collectors to provide domestic water and space heat today cost several times as much.[7,12]

But even if solar electricity could compete with electricity from conventional sources, that would not be enough to enable it to capture a large share of the energy market. Electricity, and solar electricity in particular, will seize the new markets implied by maximum electrification in our Business-as-Usual scenario only if capital costs can be brought down from present levels. The reasons for this were pointed out in Chapter 3. Electricity production has always been very capital intensive. Because of soaring capital costs, electricity capital investments have been climbing dramatically in recent years and in 1977 accounted for 24% of *all* new plant and equipment expenditures in the U.S. (see Figure 4.2). Meeting the relatively modest growth rate of the demand for electricity set forth for the Business-as-Usual scenario in Chapter 3 (2.7% per year) would still require 20% of all new plant and equipment expenditures in the period 1975–2010 even if there were no further capital-cost escalations.[52] Such "capital hogging" by the electric utility industry cannot continue if the U.S. industrial economy is to remain viable. *Financial* constraints on electricity production overall confirm that it would be impractical to achieve primary reliance on solar energy under conditions of Business-as-Usual growth.

CONCLUSIONS

In summary, we see that solar energy is not *the* answer. It would be desirable to move decisively toward a solar-energy economy in the next several decades to avoid the serious irreducible environmental risks associated with fossil and nuclear fuels. However, formidable physical and financial constraints make this all but impossible if the U.S. continues on its present course of growth of energy demand.

We are not, however, pessimistic about solar energy. The physical constraints on solar energy we have identified in this chapter are constraints which arise in large part from the prodigious levels of energy demand that would characterize a Business-as-Usual energy future. We shall show in the following chapters that there are substantial opportunities, through fuel-efficiency improvements, to sustain a viable economy with much less energy than is consumed today. If these opportunities were exploited, then solar energy could meet a substantial share of total energy requirements without running up against the physical constraints described here.

Moreover, there is such a diversity of solar-energy sources and applications and there is such a creative ferment in solar-energy technology today that it is inappropriate at this time to draw broad negative conclusions about the *economics* of solar energy. We *can* say that the kinds of solar-energy systems which will prove to be successful may be different from the technologies with which we are now familiar. In particular, solar energy is unlikely to be harnessed with "universal" technologies (technologies which provide for a wide variety of energy services at all locations).

In Chapter 11 we show in a detailed example how the pursuit of energy-efficiency improvements and innovative design strategies could make feasible the wide adoption of solar energy.

PART II

ESCAPE FROM ENERGY GROWTH

CHAPTER 7

SAVED ENERGY AS THE MAJOR ENERGY RESOURCE

RECAPITULATION

Our analysis of the energy problem has been based on a Business-as-Usual projection of energy needs. Even though these needs are substantially less than those projected by most other prognosticators, the problems of supplying these increasing quantities of fuels and electricity over the next several decades are severe:

- Fuel resources are in principle very extensive, but cheap oil and gas resources are not. The direct cost of conventional energy supplies will become very high. Threefold real increases in oil and gas prices and more than a doubling of real electricity prices to final consumers are in the offing, relative to prices in the early '70s. Less than half of these price increases have as yet been felt by customers.
- These cost increases are associated with severe difficulties of capital formation. Enormous capital resources were needed even to bring forth conventional energy supplies. The fossil fuel and nuclear energy alternatives to oil and gas pose far greater financial challenges. Not only are these energy sources inherently more costly, but also their exploitation is proving to be far more expensive than anticipated. The severe cost overruns and performance failures which are occurring now can be expected to continue. Because of the large financial risks involved, it is unlikely that the private sector would be willing to bring forth domestic energy supplies adequate to fuel a Business-as-Usual course. To meeting growing energy

demands would require a new government-industry partnership to manage new energy technology, with government assuming much of the financial risk involved.

- The level of damage caused as a side effect of this energy-supply activity would be high. The most serious hazards are certainly beyond effective control: climatic change due to atmospheric CO_2 buildup and nuclear-weapons proliferation.
- Even for risks which are controllable in principle, effective controls are becoming more difficult. There has been a veritable explosion in regulatory activity. The regulatory apparatus is becoming so pervasive and unwieldy that it could in itself prevent the creation of successful energy systems. Moreover, the growing role of government as regulator is on a collision course with the growing role of government as entrepreneur in the management of energy technology.
- The more benign substitute supplies of energy, i.e., the various forms of solar energy, are limited or exceedingly costly. They could provide only a fraction of the energy required in a Business-as-Usual future.

Our prognosis for energy with Business-as-Usual policies is, in a word, pessimistic. The tail is wagging the dog; energy technology is forcing American economic and political development off course.

In Part II of this book we address the question, Can energy systems be designed so as to avoid major supply constraints and the need for stifling regulations? Briefly, can energy technology be brought to heel? We approach our answer by abandoning the Business-as-Usual future and reexamining energy needs. These needs are, after all, technically determined in part, so that they can be examined from a technical viewpoint.

TAKING A CLOSER LOOK AT ENERGY DEMAND

How much energy is really needed to run the economy and to support a comfortable standard of living? Until about 1973 this question was rarely asked in discussions of energy problems, and little detailed information was available as to how energy is used in homes, factories, and transport systems. But since the problems of providing energy have become so acute, increasing attention has been given to how energy is used.[1]

Energy growth can be curbed by conserving fuel in many ways. A common concern about fuel conservation is that it would hurt the economy. The utility ad shown in Figure 7.1 represents an attempt to play on this fear and associate fuel conservation in people's minds with irresponsible protests against the establishment. Indeed this conventional wisdom would be correct if the nation, while planning a course of continued growth in energy use, were suddenly confronted with cutbacks in energy supply. With a physical plant designed for a high level of energy inputs, there would undoubtedly be plant shutdowns and widespread unemployment in response to unexpected shortages in energy supply. But this is not what we mean here by fuel conservation. Instead we have in mind the situation in which government, corporations, and individuals throughout the economy acquire over time technologies which consume less fuel for specific activities than the technologies now used for these activities. With new investments and new energy policies a transition to reduced energy growth and perhaps an absolute reduction in energy use could be achieved with minimal dislocations and indeed with major economic benefits.

FIG. 7.1

To show the possibilities for reducing energy consumption, it is useful to express the fuel consumption associated with any activity or process as the product of two factors:

$$\left(\text{fuel consumption}\right) = \left(\begin{matrix}\text{demand for the} \\ \text{product or activity}\end{matrix}\right) \times \left(\text{energy intensity}\right) \quad (7.1)$$

where the energy intensity is the energy required to perform the task of providing each unit of the product or activity. For example, the fuel used to drive automobiles is the product of the number of miles driven and the energy intensity. The energy intensity in this case is the average fraction of a gallon consumed per mile, the reciprocal of the familiar miles per gallon. Energy can be conserved both by curbing demand for energy-intensive activities and by reducing energy intensities. The demand for products depends on considerations like personal goals and income. In this book we shall focus our attention on conservation opportunities associated with reducing the energy intensities involved in providing goods and services, that is, on reducing energy use through improved design of products and processes.

TWO KINDS OF ENERGY EFFICIENCY

We need to introduce a technological efficiency measure for energy-consuming processes which can be used to point up the possibilities for efficiency improvements. Unfortunately, the efficiency concept commonly used is an inadequate indicator of the long-term potential for fuel savings. A couple of examples will illustrate this point:

Household furnaces are typically described as being about 60% efficient, which means that 60% of the heat released in fuel combustion can be delivered as useful heat to the rooms. This measure suggests that a 100% efficient furnace would be the best you could do; this is incorrect, however, because devices exist for actually delivering more heat. A heat pump, which is an air conditioner turned around, extracts heat from the local environment (thereby cooling the out-of-doors) and delivers this energy plus the energy needed to run the heat pump as heat at a useful temperature. Thus the heat pump delivers as heat more than 100% of the energy needed to run it.

Air conditioners are rated by a coefficient of performance (COP), which is the heat extracted from a cooled space divided by the electrical energy consumed. A typical air conditioner might have a COP of 2. Unfortunately this measure provides no hint as to how this performance compares to the maximum possible, which is a COP much greater than 2.

The efficiency used in these examples is called first-law efficiency and is defined as

$$\frac{\text{energy transferred to the purpose of the system}}{\text{energy input to the system}} \quad (7.2)$$

It is called first-law efficiency because it is based on the first law of thermodynamics, which holds that energy is neither created nor destroyed. This efficiency concept enables one to keep track of energy flows and is thus useful in comparing devices of a particular type. However, it is wholly inadequate as an indicator of the general potential for fuel savings. A much more meaningful efficiency is one which measures actual fuel consumption in relation to the theoretical minimum amount needed to perform a task. For example, in heating a house, the task of the heating system might be to provide warm air to maintain rooms on a certain temperature schedule for a season, given particular heat losses from the house. The task of the engine and transmission of a car might be to maintain a 55-mile-per-hour speed for 1 mile, given the car's air drag and tire losses. The task determines the theoretical minimum fuel consumption without reference to the actual equipment used (that is, without reference to use of a furnace for heating or use of an internal-combustion engine for a car). Thus we can define an efficiency, called *second-law efficiency*, as[2]

$$\frac{\text{theoretical minimum fuel consumption for a particular task}}{\text{actual fuel consumption for a particular task}} \quad (7.3)$$

It is called second-law efficiency because the minimum amount of fuel required to perform a task is determined by the second law of thermodynamics.

WHY THE SECOND-LAW APPROACH IS IMPORTANT

The second-law concept enables one to take into account the quality of energy in estimating fuel requirements for particular tasks. If appropriate technology is used, far less fuel is required to provide a unit of low-quality energy than to provide a unit of high-quality energy. This is important because a large fraction of total fuel consumed today is used ineffectively; it is simply burned to provide low-temperature heat.

The quality of a unit of energy is high (i.e., it has low entropy) if a large fraction of that energy can be converted into useful work (see the brief discussion of entropy in Chapter 6). Highly organized energy (such as electrical energy, chemical or fuel energy, or the energy of motion of a falling weight or of falling water) is high-quality energy because all or nearly all of it can be converted into useful work.

In contrast, thermal energy, or heat—i.e., the energy of the random motion of molecules—is disorganized energy which can be converted only partially into useful work. While thermal energy at flame temperature has high quality, thermal energy at the ambient or out-of-doors temperature has very low quality. In principle, up to 70% of the energy in combustion gases can be converted to work and, in practice today, up to 40% is converted to work.[3] But you can't get any work from thermal energy at the ambient temperature. This is not surprising since that energy is free; you can get some by opening a window. Indoor air at 70°F on a winter's day with 40°F ambient temperature has somewhat higher quality: the second law says that up to 6% of the heat in this air can *in principle* be converted to work.

Just as one can get a lot less useful work out of a source of low-temperature heat than one can get out of a high-temperature source, the provision of a given quantity of heat at low temperature via a heat pump requires much less work (and hence less fuel) than the provision of the same quantity of high-temperature heat.* The second law says that the theoretical *minimum* amount of *fuel* required by a (thermodynamically ideal) heat pump is much less than the amount of *heat* provided. This is because part of the heat provided by a heat pump is heat extracted from the ambient environment and "pumped" up to a useful temperature (see Figure 7.2).

We can now apply this analysis to a concrete example to illustrate the difference between the first and second laws. Consider space heating, in which heat is delivered to a building at 86°F by a gas furnace with a 60% first-law efficiency. For this application the second-law efficiency is 5% when the ambient temperature is 40°F. Thus while the first-law efficiency for a gas furnace (60%) gives the misleading impression that only a modest improvement is possible, the second-law efficiency (5%) correctly indicates a 20-fold maximum potential gain in theory (see Figure 7.2). It is correct to draw the conclusion from this analysis that the existing stock of equipment which provides low-temperature heat via fuel combustion is quite inefficient compared with the stock of equipment which could eventually replace it.

* There is an alternative to the heat pump which also permits efficient provision of low-temperature heat: the combined production of electricity and heat, or cogeneration. Examples of this approach are discussed in Chapters 10 and 11.

FIG. 7.2

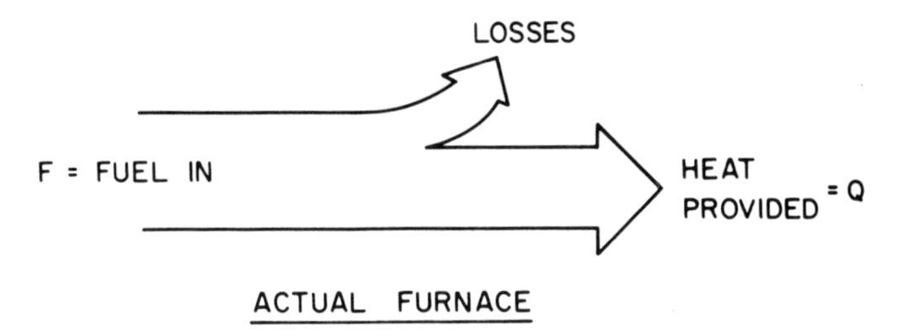

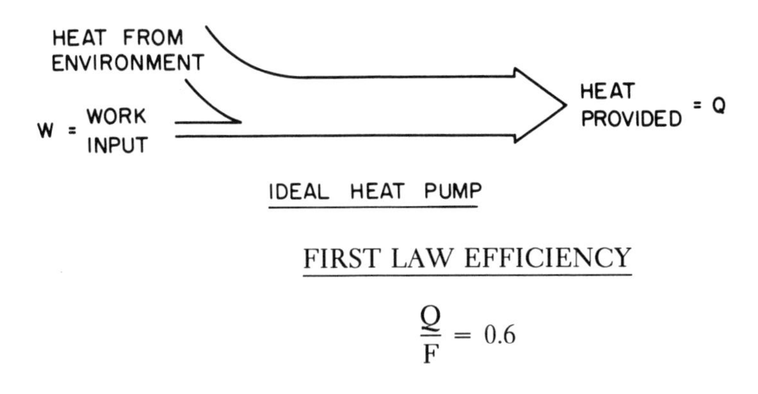

FIRST LAW EFFICIENCY

$$\frac{Q}{F} = 0.6$$

SECOND LAW EFFICIENCY

$$\frac{W}{F} = \frac{W}{Q} \times 0.6 = 0.05$$

Schematic of energy flows for efficiency evaluation of a furnace system. The first-law efficiency is the ratio of furnace output to input. The second-law efficiency compares the work input for a thermodynamically ideal heat pump to the fuel input for the actual furnace.

THE TECHNICAL POTENTIAL FOR SAVED ENERGY

In Table 7.1 we list typical second-law efficiencies in significant energy-consuming areas of the U.S. economy. These values suggest that throughout the economy energy is used very inefficiently.

Table 7.1 Second-law Efficiencies for Typical Energy-consuming Activities

SECTOR	SECOND-LAW EFFICIENCY (PERCENT)
1. Residential/commercial:	
Space heating:	
Furnace	5
Electric resistive	2½
Air conditioning	4½
Water heating:	
Gas	3
Electric	1½
Refrigeration	4
2. Transportation: Automobile	12
3. Industry:	
Electric-power generation	35
Process-steam production	28
Steel production	23
Aluminum production	13

Just how far can we expect to go toward achieving the theoretical maximum of 100% second-law efficiency? In practice 100% efficiency is never achieved. This maximum is limited by both available technology and economics. At some point the fuel savings associated with a further efficiency gain are not worth the additional capital cost. Our judgment, which is based on the study of a variety of devices and processes, is that over the long term goals of 20 to 50% are reasonable for typical practical systems.[4] The values at the high end of this range would be characteristic of highly engineered devices designed for specialized tasks (mainly in industry), and the values at the low end would be representative of what could be achieved with more flexible, less sophisticated devices suitable for wide applications in our homes, in buildings, and in transportation. Thus there is considerable room for efficiency gains through innovation, starting from today's technology.

While the second-law efficiency measure suggests potentially enormous opportunities for savings, it does not tell the whole story, because the efficiency given is for a specific task, which can often be modified without adversely affecting the utility of the product provided. For example, Table 7.1 indicates that the second-law efficiency for aluminum production is 13%. But this is the efficiency for producing aluminum from virgin ores, where the theoretical minimum energy requirement is

25 million Btu per ton of aluminum, compared to 190 million Btu used today. If the task is redefined to allow for recycling, the potential for fuel savings is even greater, since aluminum production from scrap requires less than 10 million Btu per ton. Similarly, for space heating, the efficiencies listed in Table 7.1 are for heating systems; these efficiency measures do not reflect the potential for improvements in the characteristics of the building—characteristics that include the degree of insulation, whether or not there are storm windows, etc. Adding insulation and improving furnace efficiency are complementary approaches to reducing fuel consumption.[5]

The structure of the argument as to how fuel might be conserved is summarized in Figure 7.3. One factor of energy consumption is the final demand—a matter, mainly, of income, lifestyle, and population. The technological factor in energy consumption can be separated into two parts: a part that depends on task definition, which, in many cases, is subject to major technical improvement, and an efficiency. The concept of second-law efficiency is useful for policy analysis because it is a factor whose scope for improvement is known: the maximum efficiency rating is 1, and practical maxima are reasonably subject to estimation.

This analysis of second-law efficiency shows that *in theory* enormous reductions in energy consumption are possible without lifestyle changes. This truly remarkable result is a reflection of the fact that energy has been so cheap that there has been little incentive to seek to improve the efficiency of the technology with which we use energy. Thus at this point in history there is still great scope for technological innovation.

In practice the potential for conservation at any particular time is limited by available technology, by economic considerations, and by

Fig. 7.3

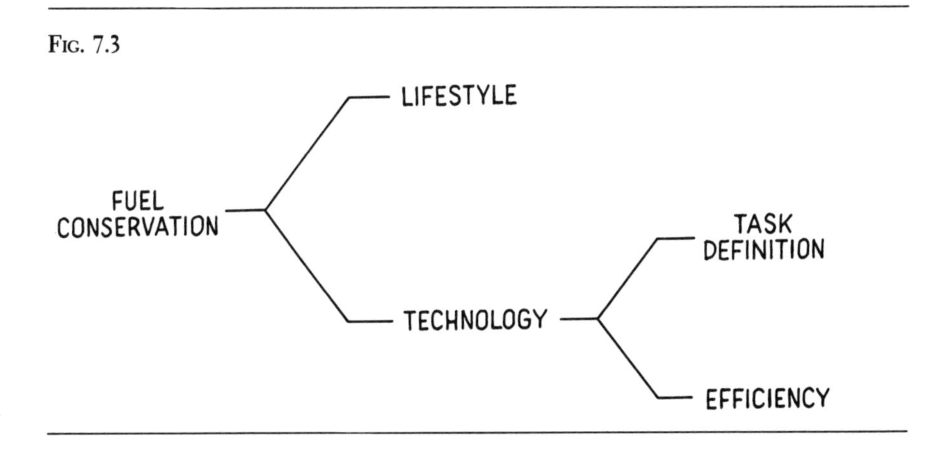

institutional obstacles to implementing economically justified innovations. Despite these constraints, the practical potential for fuel conservation is still very large, as we shall show in the following chapters. If policies were carried out to reduce the constraints—to encourage technical innovations, to move toward pricing of energy based on replacement costs, and to reduce the economic regulation and subsidization of energy supply which inhibit conservation—then one would find that the potential is so large that saved energy can be regarded as *the* major energy resource, a prospect that completely alters the energy-supply picture.

CHAPTER 8

FUEL CONSERVATION IN THE HOME

INTELLIGENT INVESTMENTS AIMED AT achieving a high level of fuel savings in houses would provide a return to the householder that would be tax free and better than the return from Xerox or Polaroid stocks in the decade after they "took off." But while there are great opportunities for fuel savings in the residential sector, which accounts for ¼ of total U.S. energy use, there are also many institutional barriers which inhibit the realization of the rich returns to investments in fuel conservation in housing. Because these opportunities and obstacles, characteristic of all fuel conservation, are so well illustrated in the housing sector, our exposition in this chapter is especially detailed. We shall give particular attention to space heating, which accounts for about ½ of residential energy use (see Figure 8.1).

SPACE HEATING IN EXISTING HOUSING

While greater fractional fuel savings are possible with improvements made in the construction of new houses than with conservation "retrofits" (i.e., renovation with conservation measures) in existing houses, potential fuel savings in existing housing are important because the housing stock turns over so slowly; even in the year 2000 today's houses will account for about half of all housing.[1]

FIG. 8.1

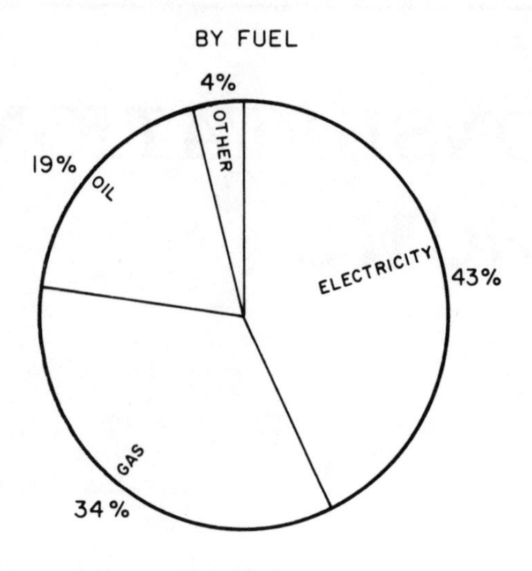

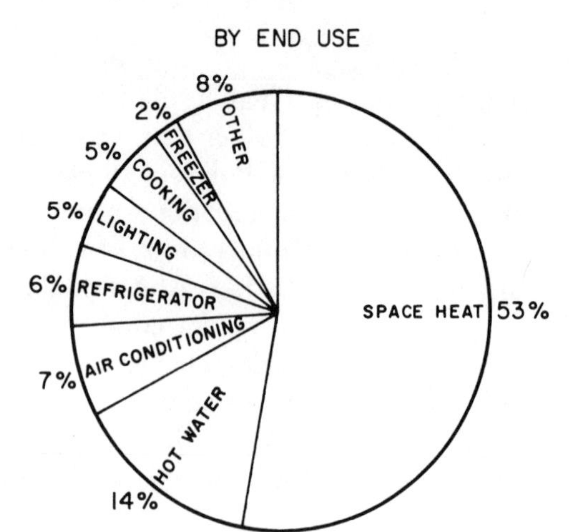

DISTRIBUTION OF RESIDENTIAL ENERGY USE (1975)

NOTE: Electricity is measured as the fuel consumed at the power plant. Thus generation, transmission, and distribution losses are included.

SOURCE: Hirst and Carney, ref. 1.

It is widely believed that only 1⁄10 to 1⁄3 of fuel consumption for heating can usually be saved by retrofitting existing houses.[2] As we shall now show, far greater energy savings are technically feasible and would be economically justified.

IMPROVING AN UNINSULATED HOUSE

Consider first a very favorable illustration of investment in conservation: a hypothetical *completely uninsulated* house in Oakland, California.[3] In Table 8.1 we list a set of rather conventional retrofit measures, their cost, and the estimated fuel savings.[4] If he were to carry out all these measures, the homeowner would have to invest $1560 in return for an 80% reduction in his fuel bill. It would not, however, be worthwhile for him to carry out all these measures. The more he saves, the more costly it becomes to save more. At some point the additional fuel savings are not worth the cost.

The consumer would be motivated to continue investing in additional conservation measures as long as so doing would lead to a further reduction in his total energy bill—the loan payment plus the fuel bill. To estimate his total annual energy bill, he needs to know both the conditions of the loan and the price of the fuel (which we assume is gas) in addition to the information provided in Table 8.1.

Suppose that the homeowner can get a 10-year-term home-improvement loan at 12% interest. For such a loan his annual payment expressed in constant dollars would be $14.20 for each $100 borrowed. (The interested reader can refer to Appendix B for the details of how to calculate this annual loan payment.)

Table 8.1 Conservation Measures Considered for an Uninsulated House in Oakland, California

CONSERVATION MEASURE	INVESTMENT REQUIRED (1976 $)	ANNUAL FUEL SAVINGS (10^6 BTU)*	CUMULATIVE PERCENTAGE OF FUEL SAVED
Night Setback Thermostat	100	46	26
6″ of Ceiling Insulation	360	43	51
3½″ of Wall Insulation	610	38	72
Storm Windows	490	13	80

* Before making the conservation investments the total annual fuel requirement was 176 million Btu.

SOURCE: Ref. 3.

Estimating the appropriate price of gas is harder than calculating the loan payment, because the price of gas is changing. For this discussion it is important to distinguish three gas prices: the current price, the price based on replacement cost (see Chapter 2, "Prices" subsection of "A New Era for Industrial Society"), and the average price over the life of the improvement, which we shall call the average future price. It is the average future price that the consumer should use in his calculations. The average future price of gas is much higher than the current price because the cost of new gas supplies, i.e., the replacement cost of gas, is much higher than the current price. We estimate that the replacement cost of gas in Oakland, California, in 1976 was about 70% higher than the 1976 price.[5] We expect that because of gas-price-deregulation legislation the price will rise to the replacement cost in about 10 years. As a result the average future price turns out to be about 30% higher than the 1976 price.[6]

The economics of conservation from the perspective of the owner of the Oakland house are illustrated in Figure 8.2. The upper half of this figure shows that the combined loan payment and fuel bill (expressed in constant dollars) would decrease upon making the first two conservation investments listed in Table 8.1, would be essentially unaffected by wall insulation, and would increase with an investment in storm windows. Thus while it's to the homeowner's advantage to carry out the first two measures, he is not motivated to carry out the second two.

It's not necessary to calculate the total energy bill to find out what measures are economical. It is easier to calculate the "cost of saved energy" and compare this cost with the price of gas, as shown in the lower half of Figure 8.2. The consumer's total cost would be minimized if he invested in fuel conservation up to the point where the cost of saved energy equaled the average future price of gas.

The cost-of-saved-energy concept can be illustrated with the first conservation measure shown in Table 8.1, the night setback thermostat. The device costs $100 installed, and borrowing that sum as part of a home-improvement loan leads to a net annual loan payment of $14.20 for 10 years. According to Table 8.1 the device would reduce fuel requirements by 46 million Btu in an average season. In this case the cost of saved energy is

$$\frac{\$14.20}{46} = \$0.30 \text{ per million Btu}$$

which is only 15% as large as the average future gas price of $2.24 per million Btu.[6] This is cheap energy indeed.

As shown in Figure 8.2, the cost of saved energy is less than the average future price of gas for the first two investments, it's about the same as the average future price for the third, and it is greater for the fourth. Thus the cost-of-saved-energy concept, like the total energy bill, shows that the consumer would be motivated to make only the first two investments.

These conclusions are valid for a loan with a term of 10 years. Whereas a 10-year period is a good estimate of the effective life* of the home improvements considered here, a typical homeowner today might not be able to secure a 10-year-term loan. If the best he could get were a 4-year loan, then the cost of saved energy calculated for this 4-year payback period, even for ceiling insulation, would be greater than the average price of gas. In this circumstance many homeowners would not make the investment even though it would be to their long-term advantage to do so.

This example illustrates an institutional barrier to conservation investments: the unavailability of long-term loans inhibits the consumer from making his investment decisions according to the criterion that the "lifecycle cost" (the average cost of fuel plus the cost of the investment appropriately averaged over the expected life or "lifecycle" of the investment, as shown in Figure 8.2) be minimized. One of the most important challenges to energy policy is to make such financing available.

The cost-of-saved-energy concept is useful in examining the economics of conservation from the national as well as the individual consumer's perspective. From the national point of view any conservation investment would be worthwhile if the cost of saved energy were less than the cost of fuel from new sources, that is, the replacement cost of fuel. For the Oakland house considered here it would be in the nation's best interest for the owner to insulate his walls as well as his ceiling, because our illustrative estimate of the replacement cost of gas is ⅓ higher than the cost of gas saved with wall insulation. It would still not be in the national interest for him to install storm windows, however, because so doing would cost about 75% more than our estimate of the replacement cost of gas (see Figure 8.2).

This example illustrates a serious pricing obstacle to conservation

* While the expected life of the improvement would probably be substantially longer than 10 years, the average homeowner moves every 7 years and probably could not recover the full value of his investment in selling his house, so 10 years is perhaps a fair estimate of the "effective financial life" of the improvements.

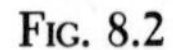

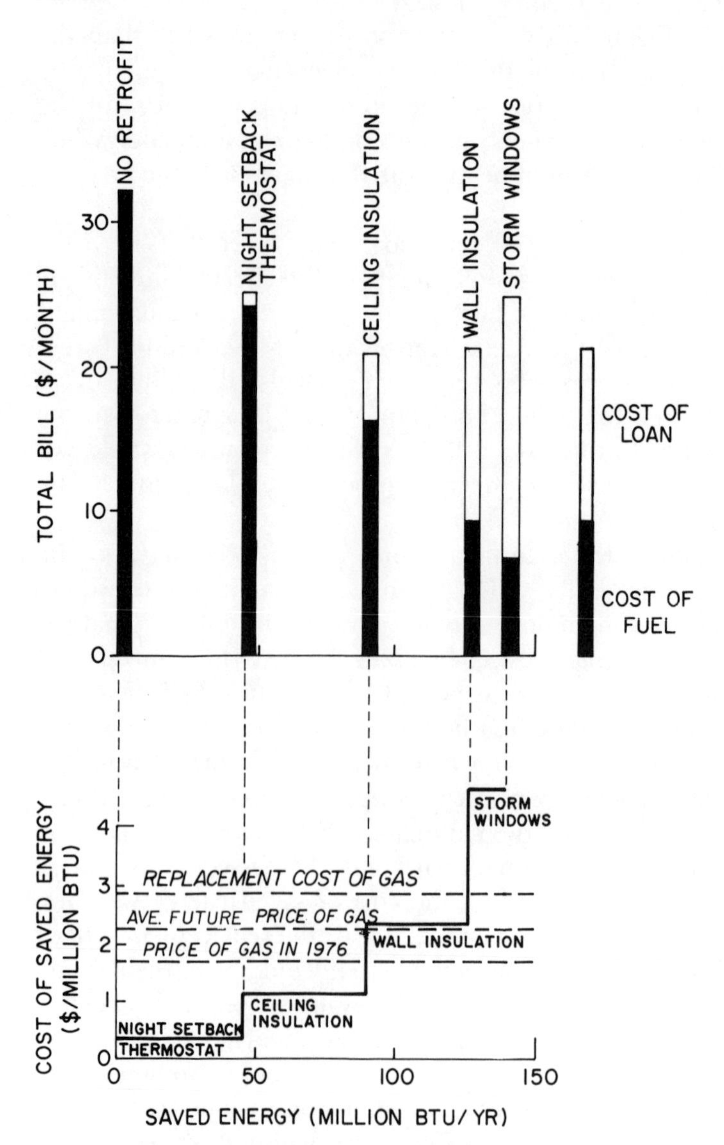

THE ECONOMICS OF SAVED ENERGY FOR AN OAKLAND HOUSE

These figures illustrate the economics of conservation for the Oakland, California house described in the text. The top figure shows the total monthly energy bill (loan payment plus fuel bill based on the average future price of gas) for the initially uninsulated house on the left and for the situations where four fuel saving measures are adopted in sequence. This figure shows that the consumer's total bill would be reduced if the first two measures were adopted, that the consumer would not save more money

by insulating his walls, and that his bill would be slightly higher than the minimum if he adopted all four measures. This figure shows therefore that from the consumer's perspective it would be economic to adopt the first two measures but not the last two. The cost of saved energy for each of these measures shown in the lower figure also indicates which of these measures are economic. Since for the first two measures the cost of saved energy is less than the average future price of gas, these measures are economical; but since the cost of saved energy is approximately equal to this price in the case of ceiling insulation and is much greater than this price in the case of storm windows, the consumer would not be motivated to carry out these latter two investments. Note, however, that adding wall insulation is still less costly than producing new gas, the cost of which is indicated by the replacement cost. Thus it would be in the nation's interest to give the homeowner an incentive to insulate his walls.

---

investments. Consumers are insulated from the full impact of costly new fuel sources because the cost of fuel from such sources is "rolled in" with the cost of fuel from less expensive older sources, so that the price of fuel rises only slowly toward the replacement cost. Thus the consumer today is not motivated to make some investments that are economically justified from the point of view of the nation. A major challenge to public policy therefore is to create incentives that bring the consumer's interest in line with the public interest of minimizing the cost of energy to the nation.

IMPROVING AN ALREADY INSULATED HOUSE

The above analysis is incomplete because it does not indicate what the fuel-saving opportunities are in houses that already have some insulation, as most houses do.[7] Fortunately, large percentage reductions in fuel use are possible even in houses that are already insulated, but more imaginative conservation measures than those just considered here have to be employed.

Recent work by a Princeton University group retrofitting a townhouse in the community of Twin Rivers, New Jersey, illustrates what is possible.[8] Typical townhouses in this community are already fairly "tight." The average fuel consumption for space heating per household is about half of the national average when corrections for floor area and climate are made. Typical Twin Rivers townhouses already have double glazing (e.g., storm windows or "thermopane") and ceiling and wall insulation. In Table 8.2 additional retrofit measures adopted in a test house in this community are shown. Only one of these measures, the added attic insulation, is conventional. The specially designed insulated movable interior shutters and window covers are important heat savers

Table 8.2 Retrofit Measures for Twin Rivers Townhouse

	ESTIMATED FRACTION OF TOTAL ENERGY SAVINGS
· Insulated movable indoor shutters on south-facing windows for nights	0.10
· Translucent covers on north-facing windows for the winter season	0.21
· Perimeter insulation in basement, extended to floor of basement	0.13
· Attic insulation increased to nine inches	0.06
· Extensive caulking to reduce air infiltration and special measures to reduce air flow to attic	0.50

because even double glazed windows have no more insulating value than fiberglass batt insulation which is ⅔ of an inch thick. (Window heat losses can also be significantly reduced with specially designed drapes, blinds, or shades.)[9] The insulation of basement walls on the inside is usually valuable because basements are usually heated. The plugging of air flows into attics has recently been shown to be a very important conservation measure.[10] Various features of construction create pathways for warm air to rise into the attic from the living space (e.g., stairwells, unfinished ceilings, and recessed lighting) or from the basement (e.g., air passages around flues and plumbing pipes and wiring, and in balloon walls). In the Twin Rivers house, researchers plugged such major air flows into the attic and also plugged pathways for fresh air to infiltrate into the living spaces. (At the low air-infiltration rates achieved in this house a number of health issues might arise, as discussed in Chapter 12.) One important type of retrofit was not made at Twin Rivers: heating-system improvements.

It was estimated that the net effect of the retrofit measures made would be to reduce fuel use for space heating by ⅔—that is, the retrofitted house would consume only ⅓ as much fuel as the typical Twin Rivers townhouse. This theoretical estimate of fuel savings was corroborated by actual measurements of gas use for this test house versus "control" houses.[11]

The cost of these retrofits was estimated to be about $1250. If this investment were financed by a 10-year loan at 12% interest, the cost of the saved fuel would be about the same as the replacement cost of gas in New Jersey.[12,13] Thus it would be in the nation's interest to make this investment now. Without an incentive, however, the homeowner could not justify making this full investment now, since the cost of the saved fuel would be higher than the estimated average future price of gas in New Jersey.[13]

Both the Oakland and Twin Rivers examples are suggestive of the enormous potential for saving heating fuel by retrofitting existing houses. Ross, by studying a variety of cases, concludes that if energy prices rise by the year 2000 to the replacement costs, then by that time an average fuel savings of about ⅔ would be economically justified for pre-1976 buildings.[14] But because of the present difference between the replacement cost for gas and its actual price, the average consumer would not make the full investment today that is justified by replacement-cost considerations, unless he was provided an incentive. The average consumer would nevertheless be justified in making retrofit investments over the next decade that would reduce his fuel use by about ⅓, without any incentive.

If saved energy is valued at the full replacement cost for fuel, then an average investment of about $2000 per housing unit or a national average investment rate of over $4 billion per year for the period 1975–2000 would be justified. New public policy initiatives would be needed to ensure that this large amount of capital would be available for this type of investment in long-term loans at reasonable interest rates. We shall return to this matter subsequently.

SPACE HEATING FOR NEW HOUSING

There are great opportunities for making low fuel consumption a design feature in new housing. Let us consider a new house for a family of 4 (with 1500 square feet of living area) located in central New Jersey, where the climate is about the same as the U.S. average.

Heat losses can be reduced both by reducing the amount of "building skin," the surface area needed to enclose a given living space, and by reducing the thermal leakiness of this skin. The amount of building skin can be reduced by shaping a building more like a ball than a matchbox or pack of chewing gum, because a sphere has the least area for a given enclosed volume. For our model house, therefore, we choose a colonial instead of a ranch-style design. The ranch-style design with the same features as our model house would have 10% greater heat losses, but would give rise to a 22% greater demand on the heating system than the colonial! This surprising result arises because only 45% of the heat loss from the thermally tight colonial is balanced by heat from the heating system. The other 55% is balanced by sunshine coming in the windows and by the heat generated by appliances and people.

We shall assume better construction practices for our model house than were required for a house to qualify for an FHA-approved mortgage in the mid-1970's.[15] The energy-conserving features of our model house are compared with those of an FHA-approved house in Table 8.3. While the FHA house is already much tighter than a typical house in the United States, our model house is much better still—mainly because (1) it leaks only half as much air as the FHA house, (2) it has 2 × 6 instead of 2 × 4 studs, so that the walls will accommodate 5 inches instead of only 3 inches of insulation, and (3) like the Twin Rivers townhouse described above, it has basement-wall insulation. Attention to design details is also important, e.g., at corners and around windows.

In a thermally leaky house most of the required heat must come from the heating system. But as a house is tightened thermally, internal heat sources and sunlight in windows become very significant. Both appliances and people contribute to the internal heat. Because our model house is equipped with much more efficient appliances, the heat from appliances is only about ⅓ as great as in a conventional house. One way to help compensate for this loss is to install winter heat-recovery systems. We chose two—one involving recovery of the exhaust air from the clothes dryer in winter (requiring use of a good filter to recover

Table 8.3 House Characteristics

	FHA STANDARD	MODEL HOUSE
Ceiling	7 inches insulation	same
Walls	3 inches insulation	5 inches insulation
Doors	ordinary	ordinary and storm
Basement	basement-wall insulation 1 inch thick 18 inches high	basement-wall insulation 4 inches thick 48 inches high
Window placement	45 ft² per side	100 ft² on south 40 ft² on west 40 ft² on east none on north
Window structure	single glazing poorly fitted storm	single glazing tight-fitted storm indoor shutters at night
Air infiltration	1.0 exchange/hour	0.5 exchanges/hour

the clothes lint from the dryer exhaust) and the other a waste-water holdup tank for the dishwasher. With these waste-heat-recovery systems the heat generated by appliances in our model house is increased, although appliances still contribute only about half as much heat as they do in the conventional house.[15]

Demand on the heating system is also reduced by utilizing windows effectively as solar collectors. Windows on the north side of the house are pure energy losers, because there are no direct solar gains to offset the large conduction losses. Double-glazed windows on the east and west sides of the house capture about as much solar heat as they lose through conduction. But south-facing windows can capture about twice as much heat as they lose. Moreover, by employing indoor shutters or other window coverings at night, window losses can be reduced to ⅓ of the solar gain on the south side. Thus south-facing windows in conjunction with indoor shutters at night are effective solar collectors. In our model, 100 square feet out of a total of 180 square feet of window area is located on the south side* and exposed to the winter sun, no windows are located on the north side, and all windows have night shutters.[15] To prevent overheating from too much sun at midday and to provide solar-energy storage for nighttime, some floors and/or interior walls in this house are made of heavy masonry or concrete. Moreover, there have to be overhangs, awnings, or other adjustable shading for the solar windows, so as to let in the winter sun, which is low in the sky, and keep out the summer sun.

The net effect of adopting all the innovations described here is illustrated in Figure 8.3.[15] Heat losses are reduced about 45% below what they would be for a similar house designed to meet only FHA construction standards, but the heat required from the heating system is reduced more than 60%.

The design we have described here is one that could be widely adopted today. It is far from the limit of what is feasible. A house built recently in Massachusetts with innovative construction techniques and an amount of south-facing window area comparable to that in our model house consumed only ⅙ as much fuel for space heating as a conventional "well insulated" house, even though this house was only slightly

* Much greater solar gains could be achieved by putting more glass on the south side. However, such extensive use of "passive" solar collectors may not appeal to most people: the house would be especially sunny and there would be considerable fluctuations in the indoor temperature. Moreover, south-facing glass may not be feasible in many locations. The house considered here would still be very energy conserving even with uniform placement of the windows around the house: for the all-electric version of the house we describe below the total household electricity requirements would go up only 10%.

FIG. 8.3

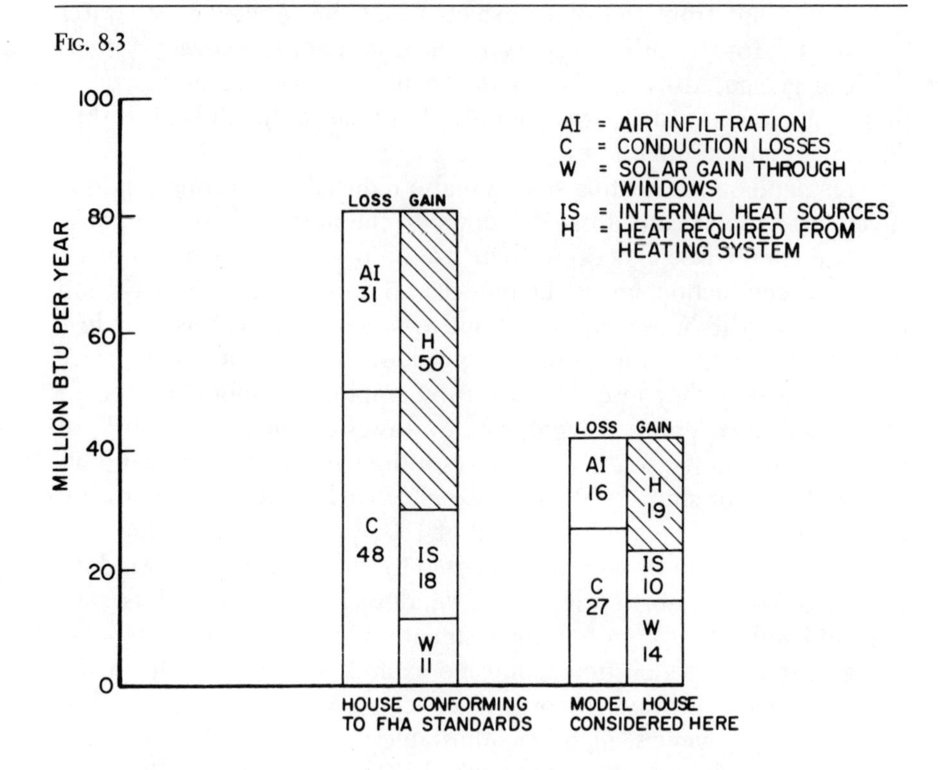

HEAT LOSSES AND GAINS FOR ALTERNATIVE HOUSE DESIGNS

more expensive than a conventional house.[16] Even the Twin Rivers retrofit described above requires less heat from the heating system than our model house.[17] To suggest what is feasible over the longer term, the American Physical Society designed a target house so tight that the heat gains from internal sources and windows would actually exceed the heat losses on all but the coldest winter days.[18] The heating system would be needed only when the outdoor temperature fell below 39°F. The annual heating requirement could be met with a small oil furnace consuming 25 gallons of heating oil per year.[15]

Good construction is not the end of the story of what can be achieved with conservation in new housing. There are also attractive possibilities for more efficient heating systems. Relatively simple modifications of gas furnaces such as proper sizing (furnaces typically have about twice the needed capacity), replacement of the pilot light with electronic ignition, reduction of the thermostat setting at which the air-circulating fan comes on, and use of a sealed combustion unit would in combination reduce fuel consumption 25 to 30%.[19] Advanced gas-fired heating systems would reduce consumption even more. For example, the pulse-combustion furnace and the gas-fired heat pump would lead to 35 and 50% reductions respectively in fuel consumption relative to typical consumption with a conventional furnace.[20]

It is also possible to achieve large fuel savings with electric heating systems. Whereas electric resistance heating requires twice as much fuel as a conventional gas furnace, electric heat-pump systems can be about as efficient. In what follows we describe how one could go much further and achieve remarkable energy savings in an all-electric house using advanced heat-pump technology. The analysis of a gas heat-pump system, which could provide comparable overall fuel savings,[21] would be similar.

EFFICIENT ELECTRIC SPACE CONDITIONING

As discussed earlier, the heat pump is a device that extracts heat from a lower-temperature "heat source" and delivers it to a higher-temperature "heat sink." Heat pumps used for space heating today heat the house by cooling the out-of-doors (i.e., by extracting heat from low-temperature outdoor air and "pumping it" up to a useful indoor air temperature). An ordinary air conditioner is a heat pump that cools a house by heating the out-of-doors. Similarly, the heat pump in a refrigerator extracts low-temperature heat from the refrigerator and dumps higher-temperature heat into the room.

Because heat does not naturally flow "uphill" (from low to high temperatures), a heat pump must be driven. The most common heat pump today is driven by electricity. The measure of efficiency conventionally used to describe heat-pump performance is the "first law" coefficient of performance (COP) discussed in Chapter 7:

$$\text{COP} = \frac{\text{heat transferred}}{\text{electricity input}}$$

The heat pump has the following two interesting features: First, the heat transferred is usually greater than the electricity input; that is, the COP is greater than 1. For example, if one of today's heat pumps is used to heat a house, the heat delivered over the heating season typically amounts to about twice as much as the heat equivalent of the consumed electricity. Thus heating with a heat pump is twice as efficient as electric resistive heating. The other interesting feature of heat pumps is that they can provide both heating and cooling.

THE ANNUAL CYCLE ENERGY SYSTEM

An interesting advanced heat-pump system called the Annual Cycle Energy System (ACES) enables one to make *simultaneous* use of both the heating and cooling provided by a heat pump, by storing the cooling produced whenever heating is provided. Thus summer air conditioning becomes a by-product of winter heating.[22] In the ACES the heat source for the heat pump is a large tank of water—10,000 gallons, or about half the size of a typical backyard swimming pool. This tank is buried in the basement or backyard and maintained year round at 32°F, the freezing point of water. With this water tank as a heat source, a very efficient heat pump can be used—one that delivers, with today's technology, 2.7 units of heat energy for every unit of electrical energy consumed.[23] As heat is extracted from the tank, ice is formed and stored for air conditioning in the summer. By springtime, the tank is nearly filled with ice. By fall, the ice is melted and freezing begins anew.

If the energy savings from the model house proposed above and from the use of the ACES were combined, electricity requirements for space heating and cooling would be reduced more than 85% relative to electricity use for these purposes in a house built to FHA standards and using resistive space heating and conventional central air conditioning.[24] The use of the ACES accounts for about ⅓ of the total savings.

Space heating and cooling are not the only energy-consuming activities that can be integrated into the ACES. The water heater and refrigerator can be integrated into it as well, thereby reducing electricity requirements by 60% for these appliances.[24]

Although the ACES is not yet commercially available, so that its true cost cannot be accurately given at this time, we estimate that the ACES would cost about $3,000 more than a resistive-heat–central-air-conditioning system and about $2500 more than a conventional heat-pump system.[25] Relative to these conventional alternatives the cost of saved energy would be respectively 3.8¢/kwh and 4.7¢/kwh (the difference between these two costs arising because the more one saves, the more costly it becomes to save more).[26] These costs are only slightly more than half the replacement cost of electric service, the cost of service from new central-station power plants to houses with conventional heating systems.[27-31]

We estimate that over ⅔ of the extra capital required for the ACES unit is for the ACES storage tank.[25] The economics of ACES might therefore be improved if the ACES storage were carried out at a centralized location so as to capture the scale economies of centralized thermal storage—i.e., a unit of storage capacity is less costly in a large than a small facility. If community-scale thermal-energy-storage technology is developed, however, the ACES described here is just one of many interesting possibilities for exploiting the scale economies of seasonal energy storage. In Chapter 11 we describe a solar-energy system involving community-level thermal storage that appears to be roughly competitive with conventional energy sources on a replacement-cost basis.

A PEAKING PROBLEM

If all or even most lights, appliances, and electrical equipment owned by any utility's customers were turned on simultaneously, the result would be "brownouts," where the utility's electricity supply falls short of demand. Fortunately for utilities, most electrical devices in an area are not usually used simultaneously, so that brownouts are very infrequent. However, use of electricity for space heating and cooling does pose a potential peaking problem for utilities because the need for space conditioning is especially great on the coldest and hottest days and this need occurs at the same time for all customers.

Electric resistance heating would create an especially severe peaking problem if this type of heating became common-place because the electricity needed for heating on the coldest days would be several times the average level of demand during the heating season.[15] The cost of holding in reserve the capacity to generate and deliver enough electric power to meet these infrequent demands is high. It would require an investment of more than $6000 to provide the model energy-conserving house we discussed above with resistive heat and central air conditioning; we estimate that more than one third of this total is associated with utility peaking facilities.[27] Moreover, this peaking problem reduces one of the main advantages offered by electric heat—the opportunity to provide space heat with fuel sources other than scarce oil and gas. Oil and gas are the fuels usually used by utilities to provide peaking power because it is uneconomic to operate much more costly coal or nuclear plants on a part-time basis (see Appendix A).

The peaking problem would be even worse with conventional heat pumps, because even though the average level of demand for electricity would be cut in half, the peak-demand level would be about the same—heat pumps are no more efficient than resistive heaters on the very coldest days. Therefore the ratio of peak to average demand would be higher, and more stand-by capacity would be needed.

ACES alleviates the peak-demand problem because the ACES requires much less power at times of peak demand than conventional technologies. Its seasonal storage acts like a flywheel to level out the peaks. Thus the electric utilities should be very much in favor of ACES development and deployment.

ACES is one response to high peaking costs. Asbury and his associates have analyzed schemes involving thermal storage at the residence (such as electric heating of bricks during off peak hours with heat recovery during the peaking period) and a scheme involving a backup fossil-fuel furnace as alternative approaches to reducing the peaking cost associated with electric heat.[32] It is our tentative judgment that such storage or backup systems would be more economical than ACES in climates where prolonged cold spells don't occur. In continental climates where long periods of extreme cold occur, ACES may be preferable if electric heating is to be widely utilized. Further study of these issues is needed.

This discussion of the peaking problem posed by electric space heating highlights another institutional issue relating to energy-conserving technology. At present most consumers who require a lot of electricity at the time of the utility peak do not pay the extra costs for providing peak power, because prices of residential electricity provided by most utilities do not vary with time to reflect these costs. If electricity prices appropriately reflected the cost of service, high-capital-cost systems like the ACES, which mitigate the peaking problem, would be more attractive to the consumer.

EFFICIENT ELECTRIC APPLIANCES

As shown in Table 7.1, second-law efficiencies are very low not only for space conditioning but for household appliances as well. The most significant consumers of electricity among the household appliances are those which generate significant amounts of heat and are often on. Many appliances—gadgets like electric toothbrushes and devices that do not generate much heat such as telephones—do not require much power at any time. Other appliances are relatively unimportant because although they are high powered, they aren't used very much—the power lawn mower is an example. To show some of the possibilities for improving efficiencies, we now briefly discuss water heating, refrigeration, and lighting, which account for about 30% of electricity use in the conventional all-electric house.

WATER HEATER

Energy is wasted in the conventional electric hot-water system four ways: (1) the hot-water temperature is typically much higher than it needs to be, (2) there are uneconomical heat losses through the water-heater walls and distribution pipes, (3) resistive heating (now used in nearly all electric water heaters) is especially wasteful, and (4) more hot water is used than is really needed.

The following series of measures makes it possible to save 80% of the electricity used for water heating. Three simple steps typically reduce the electricity required for a water heater by a total of almost 15%: lowering the thermostat setting, placing extra insulation around the tank, and insulating the pipes.[33] Integration of the water heater with other major appliances yields a larger reduction. Particularly large savings would be possible if the ACES were available. Incorporating the water heater into the ACES replaces resistive heating with a heat pump and results in savings of an additional 55% relative to present usage.[33] A further 10% savings relative to present usage can be achieved by measures to reduce water usage: use of low-flow showerheads and a warm-water-wash–cold-water-rinse laundry cycle.[34]

Of the measures considered, only the last one involves a modest "lifestyle change." It is noteworthy that this demand reduction accounts for only ⅛ of the total 80% savings. If the demand-reduction measure had been instead the only measure considered, it would have represented a savings more than 3 times as large. It is a general result that after technical fixes are carried out, further absolute gains possible with lifestyle changes are much diminished relative to what they would be if lifestyle changes alone were carried out.

REFRIGERATOR

The refrigerator is a classic example of a device for which energy consumption of new models increased dramatically over time while electricity prices tumbled (see Figure 8.4). Little attention was given to energy efficiency as new features were added. Not only did refrigerator designers develop "thin wall" styles with less insulation, but also they added half a dozen *heaters* within the refrigerator for such tasks as butter warming and inhibition of condensation! Consequently, for very little extra investment and insignificant change in performance, large efficiency improvements are possible today. Consider a 16-cubic-foot refrigerator-freezer with automatic defrosting. Seven relatively simple technical changes enable cutting refrigerator electricity use by more than 50%,[35] to the level for new refrigerators 20 years ago (Figure 8.4), for an increased initial refrigerator price of about $75. This increased first cost translates into a cost for saved energy of about 1¢/kwh.[36]

FIG. 8.4

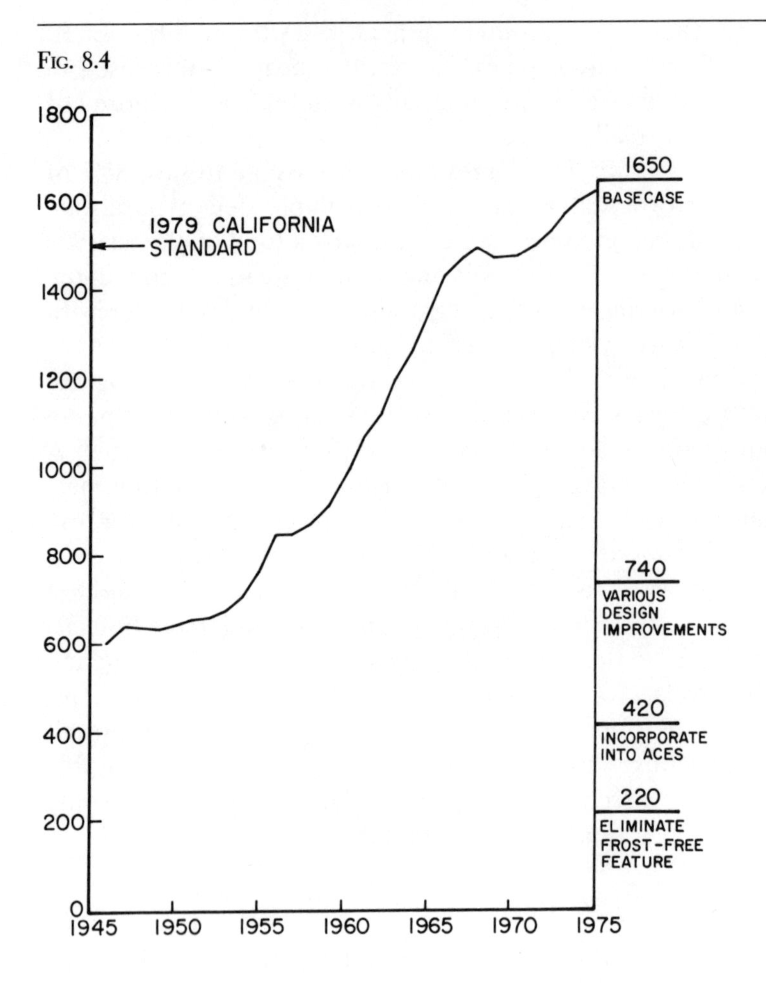

ELECTRICITY REQUIREMENTS FOR REFRIGERATOR
(KwH PER YEAR)

For the period 1946–1975 the sales weighted average consumption for new refrigerators is shown. The numbers on the right represent consumption figures for a 16 cubic foot frost free refrigerator-freezer (with a top freezer) before and after various design changes are made. See text.

If, in addition to the adoption of these measures, the refrigerator-freezer were incorporated into the ACES, the overall energy savings would be 75%.[37] Since with the adoption of all these measures the level of electricity use is so low, most consumers would probably not be motivated to go further and give up the frost-free feature (Figure 8.4).

LIGHTING

Conventional fluorescent lights consume only about a third as much electricity as incandescent bulbs but will probably not be used very much in residences because of their size and shape and their "harsh" white light. New fluorescent bulbs are being developed, however, which resemble the conventional incandescent bulb, fit standard sockets, and produce a warm white light similar to that of the familiar incandescent bulb. But the new bulbs will probably cost 15 times as much as incandescent bulbs. This technology provides a rather extreme example of the dilemma of whether or not to invest now to save energy later. For most consumers the initial cost of this bulb (perhaps $7.50) would be a shock. Would the consumer go for it? He should, because with one of these bulbs he could save $2.20 a year, and the bulb is expected to last 10 years in typical use. The cost of saved electricity with this bulb would typically be only 0.7¢/kwh![38]

TOTAL ENERGY REQUIREMENTS FOR THE MODEL ALL-ELECTRIC HOUSE

The energy requirements for the average U.S. household in 1975, for an all-electric house built to FHA standards with conventional appliances, and for a model all-electric energy conserving house with the features described here (a thermally tight house equipped with an ACES and other saving appliances[24]) are compared in Figure 8.5. This figure shows that shifting to the all-electric house built to FHA standards would result in a large increase in energy consumption, whereas shifting to the model energy-conserving all-electric house would result in a ⅔ reduction in energy use relative to the average household. Indeed, the model all-electric house would consume even less electricity than the average household in 1975. Thus a shift to energy-conserving all-electric houses not only would lead to a large reduction in total residential energy use but could result in no growth in electricity demand by the residential sector.

FIG. 8.5

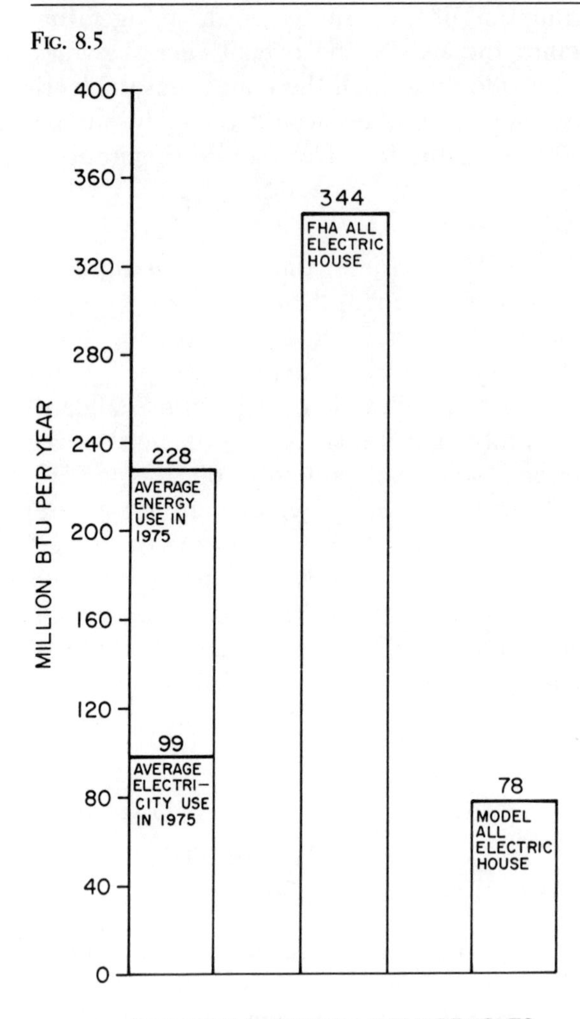

This figure indicates that the model all electric house would consume 80% less energy than a conventional all electric house with the same amenities and 65% less energy than the average household in 1975. Our model house is not strictly comparable to the average household in 1975, however. The model house is larger (1500 vs. 1100 ft² for an average size dwelling unit), it is single family unit (30% of housing units were multifamily units in 1975), it has a greater occupancy (4 vs. 3 people), and it has more appliances than the average household in 1975. The savings for a more typical model dwelling unit would be greater than 65%.

NOTE: Electricity use is expressed here in terms of the fossil fuel energy or its equivalent required to generate the electricity.

OVERVIEW OF THE TECHNOLOGICAL AND ECONOMIC OPPORTUNITIES

We have touched on opportunities for household fuel conservation for a variety of cases: for existing housing and for new construction, for both space conditioning and appliances, and for both gas-heated and all-electric housing. If skillful design and additional investment in buildings and appliances are combined, then remarkable reductions in both fuel use and costs can be achieved. In an overview of housing-conservation opportunities in California, Arthur H. Rosenfeld comes to the same conclusion.[39]

Some results of our analysis for new houses are summarized in Figure 8.6, which shows the annual fuel requirements and the annual energy cost for a new house built to FHA standards and for the new model house described in this chapter, with alternative types of space-conditioning systems.* (Here the cost includes energy supply evaluated at its replacement cost, the annualized cost of energy-related improvements, and energy-related operation and maintenance costs.)

Figure 8.6 shows that with either gas or electric heat, savings of more than 50% in total household energy consumption are achieved with the energy-conserving house relative to a new house built to FHA standards, while annual costs are reduced about 40% (over $300 per year for the gas-heated house and double this amount for the all-electric cases). This figure also shows that a gas-heated energy-conserving house would be considerably less costly than any of the energy conserving all-electric options, all of which would be about equally costly.

In discussing savings to the nation, it is often useful to express the results another way—in terms of capital investments required. Consider the energy efficiency improvements described in this chapter that are applicable to a gas heated house. To reduce annual energy consumption by about 120 million Btu requires a net investment of about $1500 by the homeowner.[43] But when this investment is made, the investment required in energy supply is reduced from $6900 to $3300,[43] *resulting in a net capital savings to the nation of $2100 per household.*

These comparisons have been between a model energy-conserving house and a house built to FHA standards. If instead comparisons were

*The particular features we have selected for each house do not include all the fuel-saving ideas described above, for if all were adopted, costs would not be minimized. It is beyond the scope of this book to carry out a systematic cost-minimization analysis for each house as we did for space heating with the initially uninsulated Oakland house. We have instead selected a level of improvements which we believe is reasonable (see notes 40 and 41).

Fig. 8.6

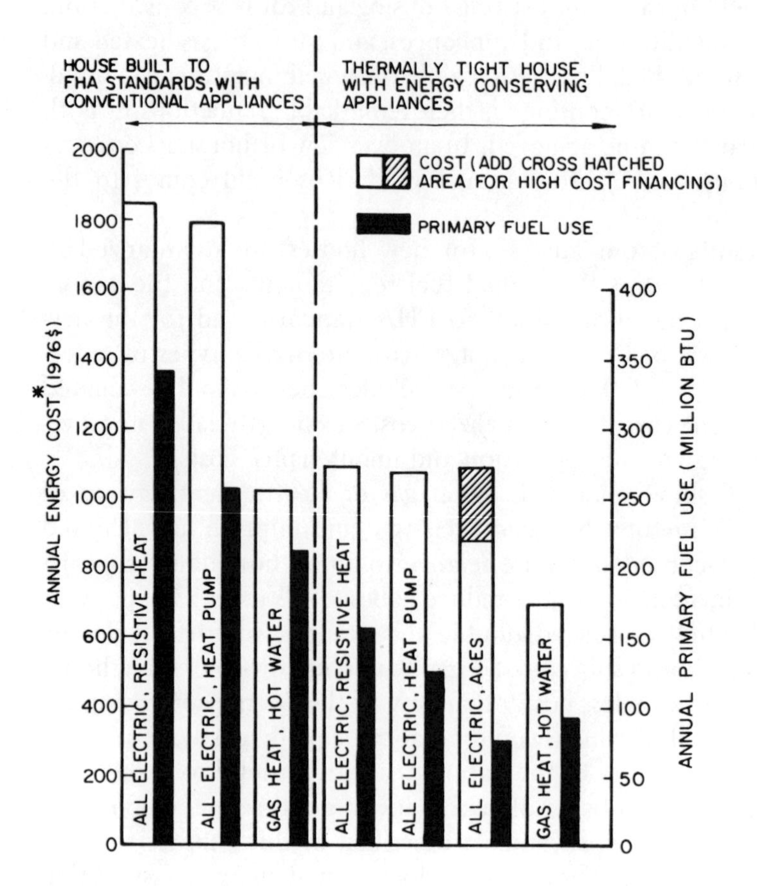

*Includes fuel cost, operation and maintenance cost, and loan payment on capital investment

SOURCE: See notes 40 and 41.

made between the model house and a more typical house, the fuel and dollar savings would be greater. Our model house requires only ⅓ as much fuel as the average household in the U.S., even though the model house is somewhat larger than average and has a larger set of appliances.

PUBLIC POLICY ISSUES

It would be nice if effective investments in energy-conserving technology were made expeditiously so as to minimize lifecycle costs with saved energy valued at the replacement cost of energy. In the normal course of events this will not occur. There are many barriers to such investments in all different kinds of housing, including:

- information problems
- technical inertia in the building industry
- the obstacles to basing economic decisions on the replacement cost of energy
- the first-cost problem

In what follows, these problems will be surveyed and solutions for overcoming them proposed.

INFORMATION PROBLEMS

Perhaps the most fundamental problem inhibiting conservation is that most consumers do not know how to invest, and there are presently no really satisfactory means to answer their questions. There is such a variety of buildings, building conditions, climate conditions, and owners' tastes that standardized solutions to problems of buildings' energy waste are likely to be far from optimal.

The House Doctor: The "house doctor" concept proposed by Robert Socolow[44] would be a means to break the information barrier regarding conservation in existing housing. The house doctor would be a professionally trained inspector, equipped with some simple diagnostic equipment, who would make "house calls" to diagnose the needs of houses from a thermal standpoint and advise householders on the basis of these inspections as to the most cost-effective conservation investments.

Because he would base his recommendations on an actual physical diagnosis of the house, the advice of a competent house doctor would tend to be far more reliable than advice based on theory alone—as might be provided by a typical insulation salesman or an auditor using only a checklist or a computer program which simulates housing energy performance as a basis for recommendations. The trouble is that theoretical models are often wholly inadequate to provide the optimal advice for making conservation investments. Such models assume that a building is like an idealized box, but a real building cannot always be treated that way. The pathways for heat flow are often complex. The magnitude of the problem is illustrated by the escape of warm air into the attics of houses—heat flow that bypasses the ceiling insulation. Recent measurements of attic heat losses for a large number of houses showed that these losses often far exceed the theoretically predicted heat losses.[45] Various construction features create pathways for heat to leak from the basement and living space into the attic, bypassing the ceiling insulation. Stairwells, recessed lighting fixtures and unsealed surfaces behind dropped ceilings are examples of possible pathways from living space to attic, while spaces around flues and wiring and balloon walls may provide pathways from basement to attic. Anomalies such as this can be discovered only by expert inspection.

For this approach to be practical on a wide scale, the house doctor must be able to provide useful and reliable information on the basis of a relatively short inspection—a couple of hours—that would cost $50 to $100. To keep the time and costs this low and still be able to provide dependable information would probably require that much more thorough inspections involving detailed measurements be made first on a few houses of all important housing types in a given area.

The house doctor's responsibilities for a given house must not end with a single diagnosis. Rather he should follow through with a "postoperation checkup" to see if "the patient is cured." Inspections to verify the quality of work and analysis of fuel purchases before and after the retrofit are needed to assure the success of the conservation effort.

The critical element in a program to create house doctors would be training. Experience must be provided in the use of modern diagnostic equipment. A pair of diagnostic tools which together can quickly identify air flows that cause heat loss are the blower door, a fan set in an exterior door frame which pressurizes or depressurizes the house, and the infrared viewer, which is sensitive to surface temperature.[46] One application of this system is to detect losses to the attic by pressurizing the house on a cold day and looking for hot spots in the attic floor with the viewer. Experience must also be provided in retrofitting a variety of types of buildings to learn and demonstrate how to achieve much higher levels of saving than have been feasible with conventional techniques.

In addition to auditors, the house doctor training program should be made available to vocational instructors, construction workers and others. Government should expedite this training.

House doctors serving as auditors could be either private practitioners or public or quasi-public officials supported by a small tax on energy.[47]

An Energy-Performance Index: When a prospective house buyer is shopping for a house, he is rarely able to obtain good information about energy performance. He will be able to find out what kind of fuel is used, and if he is seeking a "used" house, he may be fortunate enough to see the previous winter's fuel bills. But fuel prices are changing rapidly, and fuel bills reflect individual habits[48] to a considerable degree. It would therefore be highly desirable to be able to characterize a building's energy performance by a simple, objective performance index that would enable the prospective house buyer to compare houses according to their fuel economy.

The energy-performance index par excellence is the miles per gallon of gasoline determined systematically for cars by the EPA (Environmental Protection Agency). This index has, to a surprising degree, the acceptance of buyer, manufacturer, and federal regulator. Its great power has been particularly evident with manufacturers—e.g., Volkswagen is setting a stiff pace for the industry by using the index—and with regulators. It is important to recognize that the index does not accurately portray the fuel use of each car and driver. Individual driving habits differ widely; cars of a given model differ; and even the measurements themselves involve simulations and approximations which deviate significantly from experience in the field. In spite of such discrepancies the index is reliable enough to be quite effective. It puts different cars in about the same order with respect to fuel economy as any driver would. If a simple, inexpensive, and reliable performance index (or indices) could be invented and systematically determined for buildings, it would probably capture people's imaginations and have a powerful impact on conservation investment. If the energy performance of buildings must, instead, continue to be stated in terms of description of the insulation and storm windows—a big step away from a performance index—then conservation investments may not be pursued aggressively.

A universal energy-performance index akin to miles per gallon might be kilowatt-hours (or million Btu) of heating fuel energy per typical heating season at the actual location.[49] That is, the dwelling could be rated according to the fuel that would be used in a typical winter by typical users. But could a meaningful index, perhaps defined this way, be easily measured, or be reliably computed from secondary information? Because of variations in building design, construction, use, and climate, the answer is uncertain. Research aimed at defining a meaningful performance index and a simple reliable means of actually measuring it should be given high priority.[50] Such research should be much less costly than research and development on problems of comparable importance relating to energy supply.

If a reliable performance index could be developed, then regulations should be developed to require that performance indices *be measured* for new buildings—and for existing buildings at the time of resale as well. While it might be impractical to require that all new construction meet a tough standard based on such an index,[51] having building-performance information available to the prospective house buyer would lead to better energy performance in housing through competition in the marketplace.

Feedback Mechanisms: Closely related to performance indices are mechanisms that provide feedback to householders regarding building energy performance. The present situation—for example, monthly bills based on estimated use rather than actual meter readings, or heating-oil bills after tank refills—gives the resident feedback which is long delayed and hard to interpret and thus of limited value in helping him experiment to determine the efficacy of alternative actions. A good feedback system would help establish the householder as master over his house energy machine.

One proposal along these lines involves a conveniently placed display showing fuel use per typical heating season as implied by the fuel use in recent days.[49] The availability of equipment to display such information on a routine basis in houses might assist the development and acceptance of a performance index for policy purposes. Experiments have shown significant impacts from this kind of feedback.[52]

OBSTACLES TO CONSERVATION IN THE BUILDING INDUSTRY

Good conservation practice in buildings requires innovation in design and construction practice. Yet innovation is difficult in the building industry, which is fragmented and horizontally stratified; new technology must fit "the existing distribution, sales and service system or, alternatively, be capable of establishing a parallel, equally effective system."[53] The industry is largely craft based, with workers in each union applying separate skills to the construction process; this tends to make the industry tradition oriented. These features suggest that the industry needs to be given a powerful external stimulus to foster innovation. A fruitful role for government would be to create new markets for advanced building technology through government procurement. In addition the industry should be encouraged to innovate by policies which motivate the consumer to seek energy-efficient housing—policies which educate the consumer about investment opportunities, policies which sensitize consumers to the cost of bringing forth new energy supplies, and policies which make financing more readily available for investments in conservation.

THE FAILURE TO BASE ECONOMIC DECISIONS ON REPLACEMENT COSTS OF ENERGY

As we have emphasized, the costs for new energy supplies are much higher than today's energy prices, and under Business-as-Usual conditions many years may pass before energy prices rise to the level of today's replacement costs. A public policy is needed which would encourage making decisions about investments in energy efficiency as if the price for energy reflected the full cost to society of producing new energy supplies. This could be accomplished by economic incentives, e.g., grants or income-tax credits, or it could be accomplished by enabling the price of energy to rise to the full cost level. But the incentive approach, so popular with American leaders at this time, is fundamentally flawed because it guides the development of technology down the wrong paths.

Adding more ceiling insulation to a layer of insulation already there, for example, is much less effective in many houses than plugging anomalous air flows through the ceiling, yet the plugging of air flows is not recognized in many government programs which reward ceiling insulation.

Specific incentives also inhibit multipurpose investments, which tend to be especially effective because the costs can be shared among different activities. Consider incentives to promote solar heating. A rooftop solar collector is a good candidate for a government incentive because it has an unambiguous purpose. A properly designed south-facing window, on the other hand, would usually be a much more cost-effective solar collector, but an investment in such a window is not a good candidate for a government incentive, because there is no unambiguous way to divide costs between the "window" function and the "solar heating" function of the installation and because it is not the government's intention to encourage people to have more windows! Thus eligibility for solar tax incentives is usually restricted to collectors.

Difficulties such as these with specific incentives are not due to inattention to detail. They are inherent. Incentives require simple fixed specifications. Improved technology of energy use requires changing and multipurpose approaches in highly diverse situations.

We find that proper development of technology requires that energy price be brought in line with the full costs of new energy supply. This means not only that energy price controls should be phased out but also that subsidies to energy suppliers should be ended, and an energy tax should be levied to cover social costs not accounted for in ordinary market transactions. (The appropriate level of taxation should be adequate to reflect the social costs associated with both the adverse side effects of energy production and use and the fact that the Earth's non-renewable energy resources belong to future generations as well as the present generation.) A policy which relies on price in this manner offers the advantage that improved energy efficiency would be encouraged through the workings of the marketplace with a minimum of bureaucratic involvement, thereby allowing maximum flexibility for innovation.

In Chapter 15 we propose an energy tax as a source of general revenues, *offsetting some existing tax*—that is, a tax shift. We recognize that a proposal for a new tax, even one that would substitute for an existing tax, will not be popular. One of the main objections to an energy tax is that it is regressive. In many low-income households consumers are already responding to higher energy prices by lowering thermostats in winter; still higher prices could cause great suffering in such households. A policy for preventing such hardships in low-income households in the face of rising energy prices is outlined later in this chapter.

THE FIRST-COST PROBLEM

Even if consumers were well informed about energy-conservation-investment opportunities, if the building industry were well equipped to introduce fuel-saving innovations, and if energy prices reflected the true costs of bringing forth new energy supplies, the efforts of millions of the nation's building owners to improve their buildings would fall short of what is cost justified, because building owners would be overly sensitive to first costs and too insensitive to future savings. In other words, they would fail to base investment decisions on lifecycle costs.

A Role for Utilities: One reason for failing to base decisions on lifecycle costs is that for many building owners it would be difficult if not impossible to secure long-term loans on favorable terms that would enable clear and immediate benefits like those shown in Figure 8.2. One way to overcome this problem would be to encourage energy-supply companies to invest in energy conservation. This strategy would facilitate lifecycle costing by exploiting a major advantage of the energy industry, its skill in raising large sums of money for long terms at favorable interest rates (recall that in Chapter 4 we pointed out that the energy-supply industry accounts for over 40% of new plant and equipment expenditures in the United States). The involvement of these companies in building conservation efforts would establish a direct competition between investments in conservation technologies and investments in new energy supplies. Given the striking economic advantages of conservation investments, this strategy should lead to a shift in investment priorities away from energy supply toward energy conservation.

Electric and gas utilities are prime candidates for energy-industry involvement with building conservation efforts, because they offer an existing administrative structure for channeling capital for conservation investments to nearly all buildings in their districts. An intriguing way to provide an effective role for utilities has been established in Oregon, where electric utilities are allowed to finance conservation investments, (for residential electric-heating customers,) and introduce them into the rate base.*[54] An electric-heating customer may request an audit of his dwelling. Subsequently the utility will offer to finance and contract for that part of the work recommended by the auditor which the homeowner actually wants done. Since the cost of the conservation investment goes into the rate base, the homeowner does not pay for the retrofit outright; but the customer does repay the utility the cost of the conservation investment (without interest), if and when the house is sold.

The Oregon plan has a number of attractive features. For the utility, "rate basing" of investments to improve the thermal performance of buildings would provide a new avenue of growth distinct from construction of new power facilities. In effect the utility would become a seller of comfort conditioning rather than of electricity for heating and cooling as such. Utility earnings would grow while participating customers would pay much less. Because the interest payments on conservation investments would be spread out over all customers, the energy bill for participating customers would typically drop much more than is sug-

*The value of a utility's capital investment on which a state's utility commission bases the earnings to which the utility is entitled.

gested by Figure 8.2. Moreover, it is a startling fact that this arrangement would not penalize nonparticipating customers.[54] The utility's expenditures for conservation would be approximately balanced by the savings to the utility in deferring new supply investments. In other words, even though the utility would provide "free" retrofits to some customers, the price of electricity would rise only about as fast as it would have without the conservation program, i.e., if capacity to provide electricity had been expanded instead. This remarkable result arises because the cost of saving energy is expected to be much less than the cost of expanding energy supply.

Rate basing of housing retrofits should be even more attractive to gas utilities because their rate bases—the value of their distribution systems—are hardly growing today. By becoming, in effect, sellers of comfort conditioning instead of sellers of gas, the gas utilities would have renewed opportunities for growth.

Not only would the conservation investments themselves add to the rate base; but also, the saved gas could be used for other customers. Saving half of the gas used in the U.S. for space heating would free up gas supplies equivalent to 40% of the total amount of oil used today by American industry. Thus gas saved in buildings could be used to substantially reduce industry's dependence on oil.

Of course the Oregon plan (or its equivalent for a gas utility) is not perfect. The extent of rate-basing is restricted in this plan by a "no-losers" test, in which rate increases resulting from conservation investments would be no greater than if there were an equivalent amount of supply expansion. In practice this would mean that the cost of saved energy should be less than the difference between the cost of energy from new sources and the cost of energy from existing sources. This is clearly a more restrictive criterion than investing in saved energy until the cost of the last increment of saved energy equals the replacement cost. Even if a "no-losers test" is adopted, however, utility rate-basing of *a portion* of a particular housing conservation investment could be an important incentive. That part of the investment which is not rate based could be charged to participants via a long-term loan paid off on the utility bill.

The Oregon plan or some variant upon it appears to us to be a promising arrangement for achieving a high level of space-heating fuel savings in a relatively short time. This new role for utilities will not come about without new policy initiatives, however. Many utilities still see effective conservation efforts as contrary to their interests, and existing regulations at both the state and federal levels would have to be modified to allow utilities to include certain investments "on the other side of the meter" in their rate bases.[55]

The proposal for utility rate basing must be modified for oil-heated housing, since oil suppliers are not utilities. However, the scheme of providing interest-free loans, with the principal to be paid off at the time a building is sold, could still be applied to oil-heated homes if corporations were established which could finance retrofits and receive their earnings via a tax on oil.[56]

Roles for Regulation: Another major factor inhibiting investment decisions based on lifecycle costs is that first costs and operating costs are often paid for by different people. For example, in many rental housing units the owner of the building does not pay fuel bills and is thus not strongly motivated to invest in conservation. This is one of the few areas where we believe that direct regulations, requiring that buildings be thermally tight, should be attempted. But if such regulations are to be effective they should be based on actual energy performance of buildings and not on detailed specifications. A really successful regulatory policy in this area would be contingent on the availability of good performance indices and the associated information feedback technologies.

A closely related barrier to lifecycle costing is the fact that builders are very sensitive to first costs and thus tend not to make the necessary additional investments in conservation needed to minimize lifecycle costs. Here again regulation is called for. But it might be satisfactory that the regulation require for owner-occupied buildings only that prospective buyers be given adequate information regarding energy performance of a building. If the other policies we have recommended here are adopted (especially the energy tax shift), then, because economic forces are strong, the market should work effectively to improve the energy performance of new buildings.

Still another area where regulation may be desirable is in the promotion of better energy efficiency in energy-intensive appliances of a universal type, such as refrigerators. For such appliances, as for automobiles, standards should be relatively easy to implement. Appliance manufacturers may fear loss of sales if they emphasize improved energy efficiency as individual manufacturers, because consumers might not be willing to make the added initial investment. (Recall that to cut the electricity used by a refrigerator in half requires an added first cost of about $75.) An efficiency standard should allay these concerns because all manufacturers would be treated equitably. In this situation the regulation should be in the form of a long-term efficiency standard, with a target several years in the future based on analysis of what could be achieved and what is required to retool manufacturing facilities.[57] Short-term standards are undesirable because they tend to have the effect of a ceiling rather than a floor on energy efficiency. In Figure 8.4 we show the 1979 California standard for refrigeration-freezers, which was set near the high end of refrigerator-freezer electricity use, owing to the successful lobbying effort of manufacturers. With such short-term standards there is little pressure on manufacturers to retool and innovate.

HOUSING-ENERGY POLICY AND THE POOR

Particular attention must be given in housing-energy policy to the plight of the poor, who spend a much higher than average fraction of income directly on housing energy (see Figure 8.7).[58] Efforts should be directed toward assuring that comfortable space conditions and adequate hot-water supplies, refrigeration, and cooking capabilities are available to low-income households at reasonable cost. There is more than one way to meet this objective, however.

One way to protect the poor would be to reduce the price of energy for poor people up to some predetermined subsistence level of energy consumption. Such a direct energy subsidy has been proposed for metered utilities in the form of "life-line rates," which involve a low charge for the initial block of energy consumed each month.[59] For individual purchases of fuel, such as fuel oil for the single-family residence, coupons, possibly with transferable rights, have been proposed as a means to assure consumers access to minimal quantities of low-cost fuel.[60]

Creating an energy price subsidy would be a very inefficient way to protect the poor, however. If the price of energy were kept low up to the subsistence level of consumption, then either *most* energy for direct use by *all* households would have to be offered at a low price, or low-priced energy would have to be offered only to poor people—simply because the direct consumption of energy by the poor is substantially higher than suggested by income (Figure 8.7). The first of these options is clearly undesirable because it would promote excessive consumption, while the second would require a new bureaucratic structure for its administration. A bureaucratic organization which serves the poor in connection with a necessity like energy creates a dependency relationship between itself and the poor. This is a highly undesirable side effect of many policies of "doing good."[61]

We propose instead to take advantage of the fact that energy price increases, even with the imposition of an energy tax, would be more than offset by reductions in fuel bills following the appropriate conservation investments. The task therefore becomes one of assuring that appropriate investments are indeed made in conservation in low-income housing.

The financing scheme we have proposed for all housing—interest-

---

FIG. 8.7

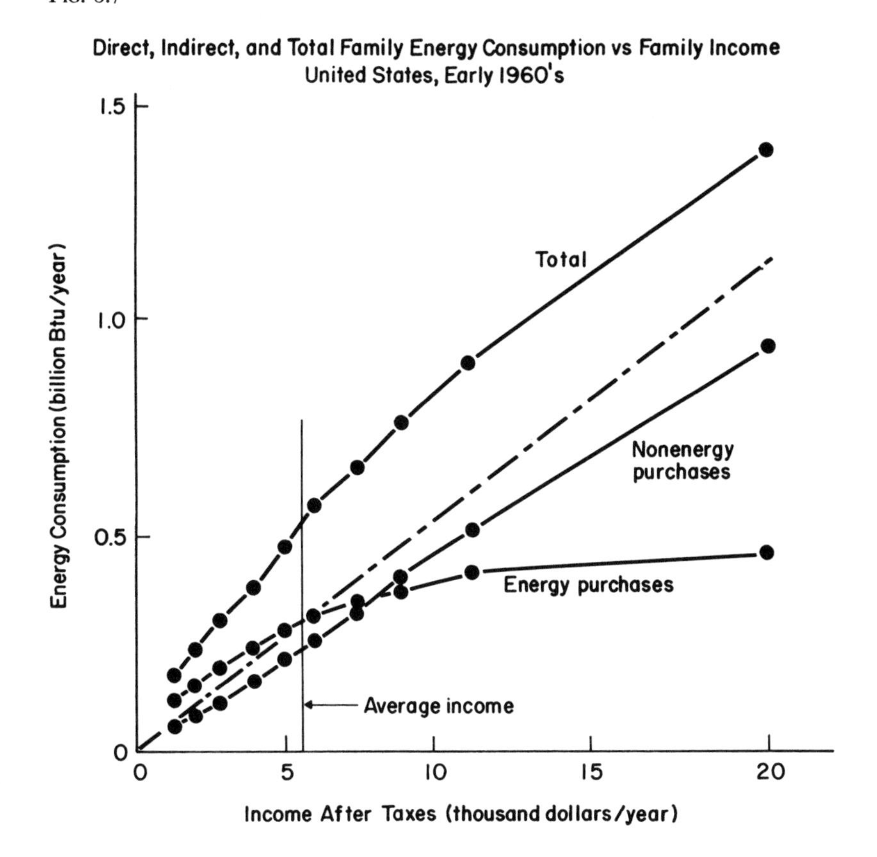

NOTE: The dashed line indicates direct energy use if it were allocated according to income. Direct energy use by the poor lies above this line, showing that a tax on direct energy use would regressive, i.e., it would be larger in percentage terms for poor people than for rich.

SOURCE: R. Herendeen, "Affluence and Energy Demand," *American Society of Mechanical Engineers Technical Digest*, 73-WA/Ener-8, Oct. 1974.

free loans offered by utilities (or an equivalent institution for oil-heated houses), the principal of which would be paid off if and when the house were sold—would make capital available to low-income homeowners who might otherwise have great difficulty in securing financing for conservation investments. Would the program reach all building owners and obtain their collaboration? The financial rewards to the corporations that would renovate housing with conservation investments would be great, so an effective sales effort would be expected. Nevertheless, where low-income housing is involved, special oversight and assistance, e.g., loan guarantees, may be needed to expedite the work and to avoid missing certain types of housing.

By itself, the utility-financing proposal may be inadequate to assure that conservation investments are made in rental properties where landlords do not pay fuel bills. Some landlords may choose not to take advantage of the interest-free loans, fearing that when selling the building, they would not be able to recover the full investment that must be paid for at the time of sale. However, the mandatory energy-performance standard we have proposed for rental properties would help overcome this obstacle.

* * *

THE POLICIES WE HAVE RECOMMENDED involve providing consumers with information based on actual energy performance, government procurement, an energy tax shift, and the phasing out of energy price controls, energy supply subsidies, and most regulations. Of these, price decontrol, the elimination of energy supply subsidies, the energy tax shift, and strategies that provide better information are far more important than the others. These measures are aimed at making the market work better—which is preferable to relying on detailed regulations. Regulatory policies often spawn large bureaucracies, and regulations can be subverted. We would advise that regulations be used sparingly—only in areas where the market mechanism cannot be made to work by other means.

CHAPTER 9

TOWARD THE 60-MPG CAR

RECENTLY ONE OF THE AUTHORS, who lives in Princeton, New Jersey, was invited to give a talk on energy conservation at a conference center some 60 miles west of Washington, D.C. Because of the remote location of the conference and a schedule that required him to return home immediately to prepare for another trip, attending this conference involved intensive air and auto travel. The resulting use of transportation energy for this 12-hour trip was[1]

- 15 times the amount of energy consumed for all purposes (residential, commercial, industrial, transportation) by the average American citizen in 12 hours
- Enough energy to provide 12 hours of electricity for 75 of the energy-conserving all-electric houses described in the last chapter
- 85 times the amount of energy consumed for all purposes by the average citizen of the world in 12 hours

While this experience is somewhat unusual, it underscores the shameless profligacy that has come to characterize travel habits in the U.S., especially in business travel. Given the certain prospect that liquid fuels will be much more costly and in much shorter supply in a matter of decades, such patterns of transportation-energy use cannot be sustained.

Table 9.1 Passenger Transportation Statistics for 1975

	ENERGY USE[a] (QUADS)	PASSENGER MILES PER CAPITA[a]	% TRANSPORTATION BUDGET[a]	TRAVEL TIME PER DAY, CAPITA[a] (MINUTES)	PASSENGER OCCUPANCY[a] (LOAD FACTOR)	VEHICLES[b] (MILLIONS)	VEHICLE MILES[b] (10^{12} MILES)	ENERGY PERFORMANCE[a] (M.P.G.)
Passenger								
Auto	9.1	10,500	79%	53	2.2 (1.9)[d]	98 (110)[c]	1.03	14
Light truck*	1.1	720	9%	4	1.4	11	0.1	12
Subtotal for private passenger vehicles	10.2	11,220	88%	57				
Air	1.2	745	7%					
Mass transit	0.1	720	2%					
Other	0.3	267	3%					
Total passenger	11.8	12,950	100%					

* Used as a passenger vehicle.

[a] Report of the Demand and Conservation Panel of the Committee on Nuclear and Alternative Energy Systems, ref. 2

[b] Transportation Energy Conservation Data Book: Edition 2 (data for 1976), ref. 3.

[c] The discrepancy between Federal Highway Administration and R. L. Polk figures (the former in parentheses) is discussed in reference 3.

[d] The figure in parentheses is from the Dept. of Transportation's nationwide personal transportation study, reference 4.

THE AUTOMOBILE

The future of the automobile, which accounts for over half of energy use for transportation and about ⅓ of total petroleum consumption in the U.S. (see Table 9.1), is the central concern in the transportation-energy problem. The dominant role of the automobile in the United States is illustrated by the facts that the average person spends an hour a day in a car and the average budget associated with each car in 1975 (well before the large increases in the price of gasoline in late 1979) was about $1700 per year, or 16% of personal income.[5] It is noteworthy that only about ¼ of the total cost is for fuel (see Table 9.2). Because of the long-term decline in the real price of gasoline, the oil price hike in 1973-74 merely restored the real price of gasoline to its level in 1960. Even the $1.00 a gallon gasoline of late 1979 was in real terms less than 30% more expensive than gasoline in 1960 (see Figure 2.3).

The history of the price of gasoline offers to the economist a ready explanation of the history of automobile performance with respect to fuel consumption. As shown in Figure 9.1, the fuel economy of cars (miles per gallon) was steadily deteriorating through the 1974 model. Then it began to improve.

The rapid improvement in fuel economy in recent years was stimulated in part by consumer concern with the price of fuel and possible shortages in fuel supply. However, the major thrust for improvement has been federally mandated fuel-economy improvements. The Energy Policy and Conservation Act of 1975 mandated that the fuel economy

Table 9.2 Cost of Operating an Automobile, 1975 (¢/mile)*

Depreciation	4.9
Maintenance	3.1
Garage, parking, tolls	2.0
Insurance	1.6
Taxes and fees	0.5
Interest lost	1.1
Fuel	4.1
Total operating cost	17.3
Gasoline price	57¢/gallon
Fuel cost %	24%

SOURCE: Ref. 2

* For an "average family vehicle of conventional design."

FIG. 9.1

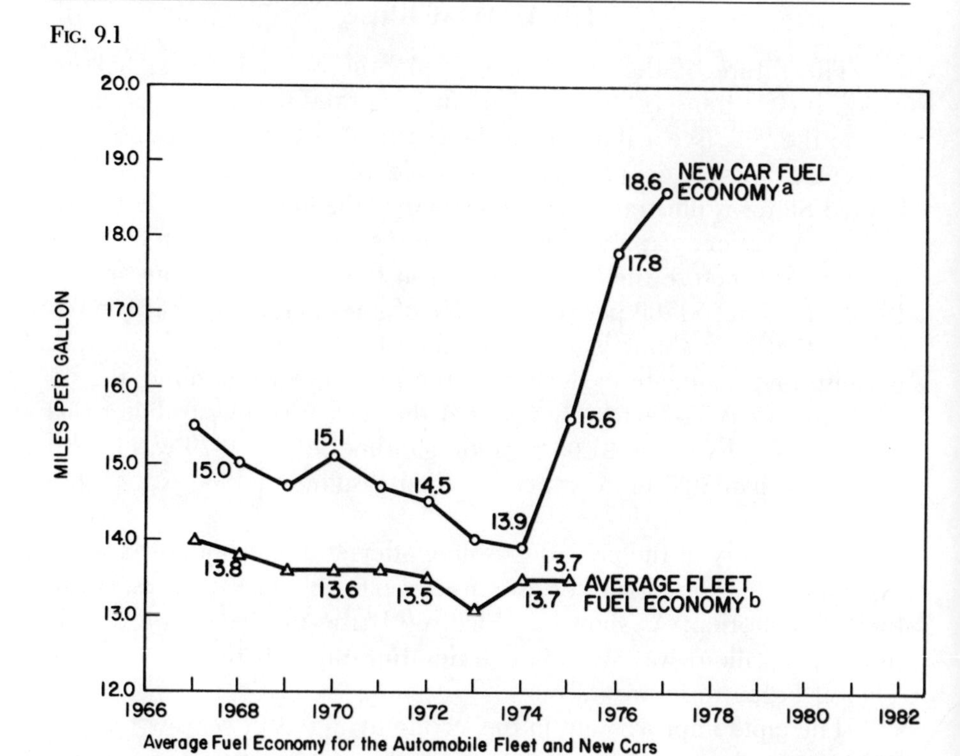

Due to automobile mix, it will take several years before new car fuel economy improvements significantly affect the average fuel economy for the total domestic fleet of cars.

[a] Sales-weighted city/highway combined fuel economy by model year based on post-1974 test procedures.

[b] Average miles traveled per gallon of fuel consumed as reported annually by the Office of Highway Planning, FHWA.

SOURCE: J. D. Murrell et al., *Light Duty Automotive Fuel Economy—Trends Through 1977*, SAE paper 760795, presented at the Automobile Engineering Meeting, Dearborn, Mich., October 18–22, 1976; U.S. Department of Transportation, Federal Highway Administration, *Highway Statistics*, annual.

of new cars rise from an average of 18.0 mpg for the 1978 model year to at least 27.5 mpg in 1985. (The fuel economy standard applies to the average of each manufacturer's cars as measured by EPA for the combined city/highway driving cycle.) This federal regulation, made possible by the availability of a meaningful, measurable performance index, has been the driving force for improvement of cars from an energy standpoint.

Considering the scale of production involved, the ability of auto manufacturers to improve their products' fuel performance in a very short time (see Figure 9.1) has been impressive. And yet it seems that a great deal of luck will be involved if the nation is to achieve its 1985 fuel economy goals. For many years it had been the practice of U.S. manufacturers to produce minor variations of a single type of large vehicle and through attention to style and advertising to attempt to persuade consumers that they had been provided an adequate choice. During the late 1970s U.S. manufacturers devoted the lion's share of their efforts to improving the fuel economy of these large cars. But further fuel economy improvements would come only slowly with this focus on large cars; meeting the 27.5 mpg fuel economy goal on time would be difficult with this strategy.

Detroit learned that the pace of improvements was indeed too slow when large car sales dropped precipitously as a response to the gasoline shortages and sharp gasoline price increases in 1979. But the shift in consumer demand to small cars, while painful to the industry in the short run, provides Detroit the opportunity to meet the 1985 fuel economy goals much more readily than with its large car improvement strategy.

While Detroit has not emphasized small cars, it can rapidly shift production to meet this demand, because it is thoroughly familiar with small car technology, as a consequence of the high penetration of the U.S. market by a variety of foreign lightweight cars. Had the "standard size" car, the beloved behemoth of U.S. manufacturers in the late '50s and '60s, been the only car in domestic production, Detroit would not be able to quickly adjust to this shift in consumer demand toward small cars.

OPPORTUNITIES FOR IMPROVED FUEL ECONOMY

One can better understand the energy performance of an automobile by examining two physical aspects of the automobile separately. The first aspect is the conversion of fuel energy into rotational energy and delivery of the rotational energy to the drive wheels. The second aspect is the task performed by the energy reaching the drive wheels.

The engine converts fuel energy into rotational energy and the transmission delivers it to the drive wheels. The ratio of the energy at the drive wheels to the fuel energy is the *efficiency* of the vehicle. The average efficiency of automobiles is about 12%. Thus in the process of converting and delivering energy to the drive wheels 88% of the energy is lost (see Figure 9.2).

The task to be performed by the energy delivered to the drive wheels is to propel the car through certain distances at certain times. As indicated in Figure 9.3, this energy, which we shall call the task energy,

Fig. 9.2

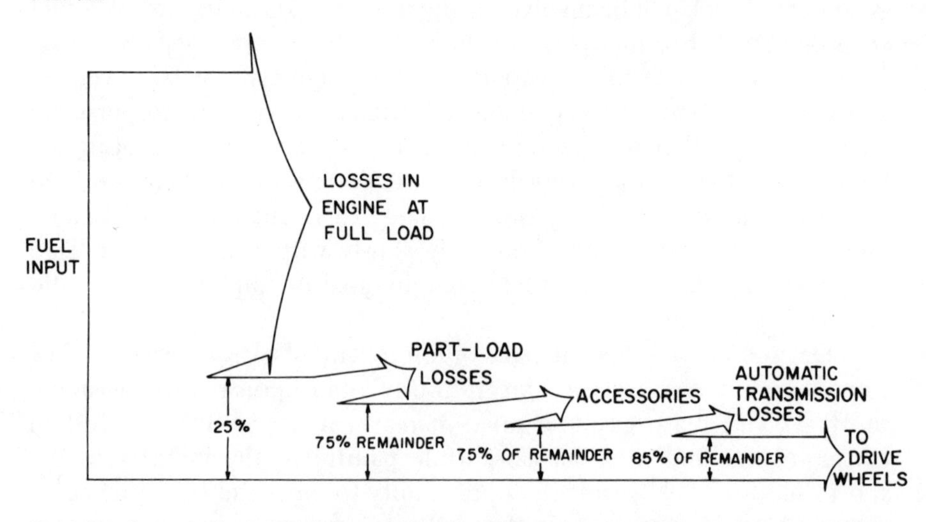

Contributions to the inefficiency of the engine-transmission system of a typical automobile. An engine operating at the "full load" for which it was designed has an efficiency of about 25% for converting fuel energy to energy of rotational motion. The full load, or the level of power for which the engine is designed, corresponds to conditions of high acceleration at high speeds. In most driving much less power is required. Unfortunately, an engine is significantly less efficient when operating at a fraction of full load—as shown by the "part load losses" arrow. Accessories and transmission also take substantial bites out of what is left of the energy, as shown, with the result that the average efficiency of an automobile is only 12%.

SOURCE: *Efficient Use of Energy*, a study by the American Physical Society, 1974, ref. 6.

is dissipated in three ways: (1) air resistance, (2) rolling resistance of tires, and (3) brake losses. Because all these losses increase roughly in proportion to the weight of the car, other factors being equal, it is not surprising that automotive fuel consumption is approximately proportional to automobile weight.

Automotive fuel consumption can be reduced either by increasing the efficiency or by reducing the task energy, since

$$\text{fuel consumption} = \frac{\text{task energy}}{\text{efficiency}}$$

A reduction of task energy by ½ would result in a reduction of fuel consumption by ½, as would a doubling of efficiency. In practice, it is as easy to reduce the task energy as to increase the efficiency.

FIG. 9.3

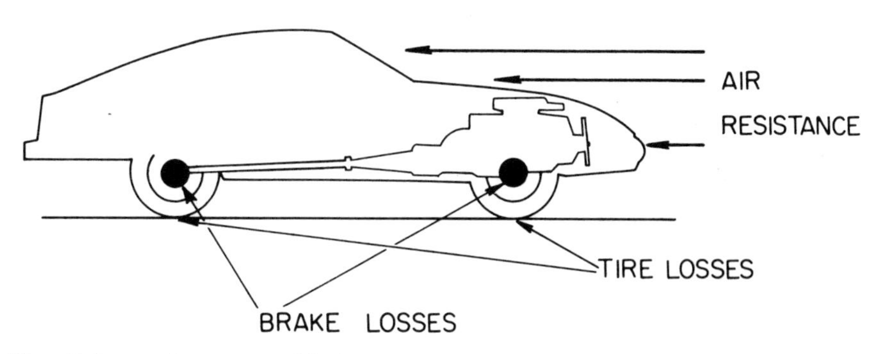

The efficiency of an automobile is a measure of the performance of the engine-transmission system:

fuel consumption × efficiency = energy delivered to drive wheels

The output of the driving wheels, or task energy, is used up to overcome energy losses:

energy delivered to drive wheels = energy losses specified by task

As illustrated, these losses consist of air resistance, the rolling resistance of tires and the brakes:

energy losses = air resistance + tire loss + brake loss

On the average all three of these energy losses are comparable, although air resistance dominates in high speed driving and brake losses are very important in stop and go driving.

REFERENCE: *Efficient Use of Energy*, a study by the American Physical Society, 1974, ref. 6.

What was the principal source of the fuel-economy gains between 1973 and 1977? Not weight reduction. The average weight of new automobiles remained within a few percent of 4000 pounds throughout this period. The reduction in fuel use shown in Figure 9.4 at fixed weight accounts for the 1973–1977 improvement in fuel economy shown in Figure 9.1. Average fuel-economy gains achieved in this period resulted from improvements like radial tires, improved engine-transmission efficiency, and reduction of engine power so as to reduce losses arising from part-load operation. Further modifications exclusive of weight reduction that were being pursued by auto manufacturers in the late 1970s to improve fuel performance included better tires, more streamlining, transmissions with more gears, a switch from gasoline to diesel engines, and the use of turbochargers instead of larger engines to provide occasional extra power.

Since 1977 U.S. manufacturers, with General Motors in the lead, have been demonstrating that substantial improvements are possible

FIG. 9.4

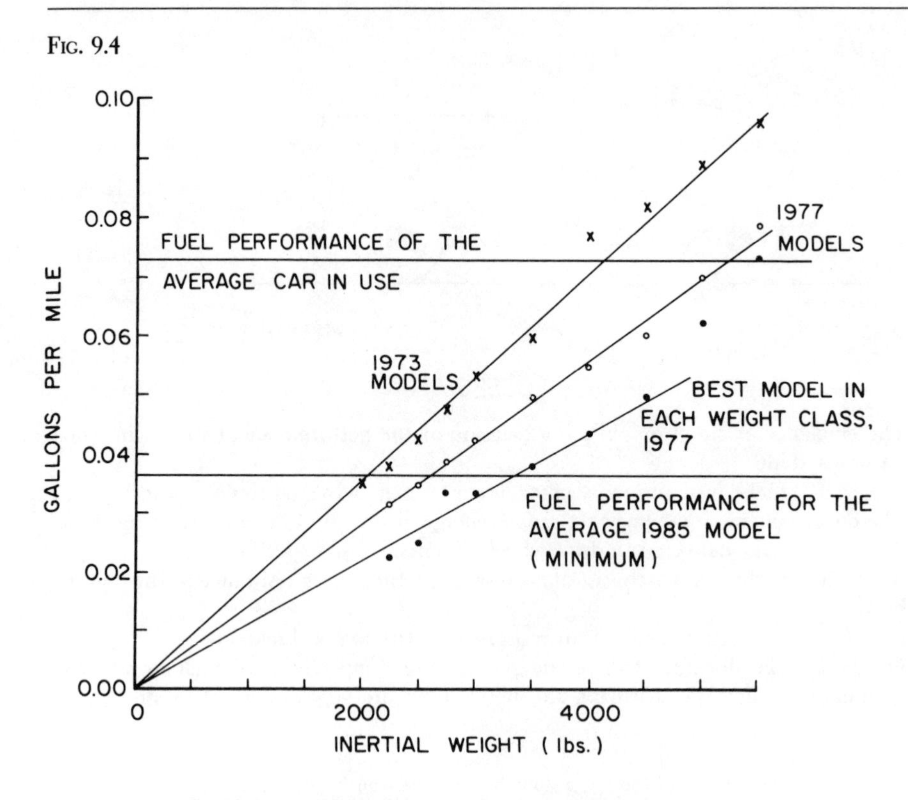

ENERGY INTENSITY vs. VEHICLE WEIGHT

(based on the combined city/highway fuel economy, as measured by EPA)

NOTE: The inertial weight is curb weight plus 300 lbs.

SOURCE: Marc Ross, *Energy and Innovation: The Impact of Technical Change on Energy Demand*, Resources for the Future, 1980.

through weight reduction without reduction of interior space. Most of the fuel-economy improvement mandated for the 1985 models will be achieved without sacrifice of interior space.

The possibilities for efficiency improvements without downsizing would by no means be exhausted with the achievement of a 27.5 mpg average fuel economy. Indeed a 1977 TRW report[7] (see Tables 9.3a and 9.3b) claims that with appropriate R&D over the next decade or so, automotive fuel economy could be improved to over 45 mpg (about 3 times the fuel economy of 1975-model cars) without any size *or* weight

Table 9.3a Effect of Technical Changes on Automotive Fuel Consumption

AREA	IMPROVEMENT	SPECIFIC IMPROVEMENT	REDUCTION IN AUTO FUEL CONSUMPTION
Aerodynamics	1. Low-Air-Drag Design	33% drag reduction	8%
Tribology (friction and lubrication)	2. Piston-Ring-Cylinder Friction reduction	25% reduction	2.4%
	3. Adiabatic diesel engine	60% efficiency increase	40%
	4. CVT—Conventional engine	Operation of engine at its most efficient range	20%
	5. CVT—Regenerative braking and energy storage	Operation of engine at its most efficient range	35%
	6. Improved engine and drive-line lubricants	25% reduction in engine friction	5%
Rolling Resistance	7. Low-Drag Tires	50% reduction in tire rolling resistance	13%
Traffic Management	8. Reducing idling operation in delays, reducing accelerations and decelerations	Up to 30% increase in urban gas mileage; 10% in highway	20%
Combustion Research	9. Diesel Engines, Stirling Cycle, Gas Turbine	30% efficiency improvement	30%
	10. Improved Otto Cycle	25% efficiency improvement	25%
Advanced System Research	11. Valve Resizing	60% reduction in pumping loss	13%
	12. Xylan Coating	Halves engine friction	15%
	13. Idle-off System	Cuts idle loss	12%

SOURCE: Ref. 7.

Table 9.3b Effect of Combinations of Changes

IMPROVEMENTS	AUTO CONSUMPTION IMPROVEMENT
1+7	20%
9+5	54
9+5+1+7	67
3+1+7	55
9+1+7+12	53
10+1+7+12	49.5
10+1+7+8	55
10+4+1+7	52
11+12+1+7	37
2+1+7+11+12	39.4
1+2+7+6+11+12	41.2
1+7+9+11+12+13	43

SOURCE: Ref. 7

reduction. In addition, roughly a 50% weight reduction is practical by reducing weight to the 2000 lbs characteristic of the lighter vehicles being sold today. Thus, weight reductions carried out with respect to 1975-model cars could achieve almost another doubling of fuel economy.

An ambitious combination of fuel-performance improvements is exemplified by a new Volkswagen car under advanced development. A prototype has been delivered to the Department of Transportation for testing, and a version embodying most of the improvements may be available for sale in the early 1980s.[8] This car is a modification of the Rabbit Diesel. The present Rabbit Diesel has a combined fuel-economy rating of 44 mpg; that is, *it uses about ⅓ as much fuel per mile as the average car on the road* in the U.S. in 1975. The modified car has a turbocharged diesel engine, an added forward gear, and important safety and other improvements. It is believed it will achieve a fuel-economy rating near 60 mpg; that is, *it might use as little as ¼ as much fuel as the average car on the road.* While the present Rabbit Diesel is relatively expensive, the purchase-price penalty for such cars should be much less than the savings after domestic competitors are established and the novelty wears off.

Could small (roughly 2000 pound) cars come into wide use in the United States? A number of considerations limit the appeal of small cars with high fuel economy:

safety, relative to large cars
the capability to pull trailers
ride quality, e.g., for long-distance trips
acceleration
ease of entry and egress
capability to carry more than four passengers

At present most small cars perform too inadequately in these areas to suit the tastes of many car buyers. However, substantial technical improvements could be made in the first four areas. We discuss possibilities for improving small-car safety in Chapter 12. The use of turbochargers and changes in suspension could lead to significantly improved performance in the next three areas listed above. In the last two areas listed above, technical improvements will be difficult. But just how important are the problems posed by small cars in these areas? In our view the need for cars large enough to serve some very large people and the need for cars large enough to hold many people are exceptional[4] and need not prevent nearly universal use of small cars. The shift of consumer preferences to small cars could be facilitated by the levy of a gasoline tax that is rebated or used to offset some existing tax. The sensitivity of consumer preferences to price was indicated by the precipitous drop in the demand for Detroit's big cars, following the rapid gasoline price escalation of 1979.

SOLAR-POWERED CARS

Provision of fuel for cars would not be a formidable task if the nation shifted to small cars with high fuel economy. Suppose, for example, that the average car on the road had a fuel economy of 50 mpg—as might characterize the modified VW Rabbit Diesel in actual use. A variety of fuel sources would be practical with such a fuel economy. Because fuel supplies would be stretched so far with such cars, even solar sources could be adequate to fuel cars in the U.S. without confronting any of the serious "limits" problems for solar-energy sources described in Chapter 6.

This could be accomplished, for example, by making liquid automotive fuel from biomass, using solar energy as a heat source to drive the chemical reactions in which the raw biomass is converted to a useful fluid-fuel form. When fluid fuel is produced this way, the efficiency of converting biomass to useful chemical fuel can be 70% or higher[9]—on the order of twice the biomass-utilization efficiency with conventional conversion technology.[10] With the combination of high automotive fuel economy and efficient conversion technology the biomass resource in the U.S. becomes impressive: recoverable crop residues alone could support the entire U.S. automobile population.[11]

It is useful to view the match between the biomass resource and automotive demand another way. A car getting 50 mpg would require about 2½ tons of raw biomass per year.[11] This is about the average yield of residues on 2 acres of cropland[12]—which in turn is the per capita amount of land committed to crops in the U.S. This shows that in an energy-efficient economy significant quantities of energy for transportation could be provided simply as a by-product of food production. This prospect stands in sharp contrast to the fact that today the energy content of the food people consume in the U.S. is equivalent to less than 1% of total energy consumption here.

The biomass option becomes economically interesting as well if the nation strongly emphasizes fuel conservation. For example, if automotive fuel economy would rise to 50 mpg, the average consumer could pay over three times the late 1979 price of $1.00 per gallon without increasing his total fuel bill. This fuel price of about $3.00 per gallon (of gasoline equivalent energy) should be substantially greater than the cost of biomass-derived fuels.

Thus in an economy where energy conservation is stressed, biomass could become an important, renewable, and affordable energy resource. In Chapter 11 we describe another example of this finding, that fuel conservation is a key to the successful development of solar energy. This we believe is a general result.

ALTERNATIVES FOR PERSONAL TRANSPORTATION

Improved automotive fuel economy is just one possibility for conserving passenger-transportation fuel. More far-reaching solutions involve moving away from the automobile based on the internal-combustion engine. Here we consider four possibilities—the electric car, mass transportation, substitution of telecommunications for transportation, and urban redesign so as to reduce the need for transportation.

ELECTRIC CARS

Since the automotive internal-combustion engine accounts for such a large fraction of American petroleum usage, much attention has been given to electricity as an alternative to chemical fuel for automobiles. Unfortunately the prospects are not good for substitution of the electric car for the internal-combustion engine—largely because batteries are so heavy and costly. The battery, or "electricity tank," for a 2500-pound car (3500 pounds if battery weight is included) would probably cost 100 times as much, weigh 10 times as much, and carry the car only 1/12 as far between "refuelings"* as the gasoline tank of a fuel-driven car of comparable weight (see Table 9.4). Even with advanced battery technology, electric vehicles, if they come into wide use at all, will be used primarily for short trips and especially "second car" purposes.

MASS TRANSPORTATION

Many people are very enthusiastic about improving mass transit as a means of reducing the use of automobiles in order to increase urban amenities and reduce energy use in transportation. Mass transit now accounts for only 5 to 6% of total passenger miles in the U.S. (see Table 9.1). But can mass transit take over substantially more of passenger transportation in the next decades?

Unfortunately the prospects for mass transit are not good unless population densities are high. In very high-density areas and very high-density corridors we are familiar with the fact that mass transit can function well. In moderate-density (suburban) areas, however, there aren't enough passengers *outside of rush hours.* At non-rush-hour times

*After the battery is discharged, either a lengthy charging period (e.g. overnight) or an exchange of the 1000-pound discharged battery for a charged one is required.

Table 9.4 Gasoline vs. Electricity Storage for Automobiles

	GASOLINE TANK[a]	"ELECTRICITY" TANK[b]
Capital Cost	$15	$1500
Capacity (Range)	300 miles	25 miles
Weight	100 lb (when full)	1000 lb (when full or empty)
Cost of Fuel (1977 average)	2¢/mile	1.5¢/mile[c]

NOTES:

[a] A 10-gallon gasoline tank for a 2500-lb car with a fuel economy of 30 mpg.

[b] This "electricity tank" (i.e., lead acid battery) would hold 10 kilowatt-hours of electricity for a car weighing 3500 lb (including 1000 lb for the battery). The car would have a fuel economy of 0.4 kwh per mile on a simulated metropolitan driving cycle. Because a great deal of structural support would be needed for the battery, this car would have much less interior space than a 2500-lb gasoline-powered car. This design is obtained from Figure 8-6 in "Electric Vehicles," Chapter 8 of *Should We Have a New Engine? An Automobile Power Systems Evaluation. Volume II. Technical Papers,* Jet Propulsion Laboratory Report JPL SP 43-17, August 1975.

[c] This cost of electricity does not include road taxes or retail charges if the battery is charged away from home. Furthermore the cost of the battery or its replacement is not included here.

vehicles run essentially empty. During rush hours there are enough passengers in suburban areas to fill buses and trains. But the cost is surprising. Typically *time doesn't permit vehicles to make more than one rush-hour run.* Until time for the return trip in the afternoon, the vehicle is idle or runs essentially empty.

These vehicles generally must be high-cost, heavy-duty units. Labor costs (assuming full-time employment is required) are also extremely high; a full-time driver will typically make only one well-loaded trip in an 8-hour day, or two such trips in a 12-hour day. Energy use by heavy buses is also high, and compares favorably on a per passenger mile basis with a *small* car only if the bus has a large load. Typically, however, a bus will be only 20% full. The costs implied by such a low average loading are enormous. Typical bus trips in a small city cost the transportation agencies about $2 per rider in 1977. These costs are increasing rapidly. Deficits increased 800% nationwide from 1970 to 1977 and are expected to continue increasing.[14] The nation's rejuvenated buses, trains, and other mass-transit systems cannot sustain continued growth.

Rapid-rail mass-transit systems (e.g., the Metro in Washington, D.C. or BART in San Francisco) are an attractive option for the very long term because such systems can help foster the construction of communities with high population density as an alternative to suburban sprawl. With high population density the need for short-distance travel would be diminished. Until such population redistribution could be achieved, however, this strategy would probably lead to more overall fuel use than a strategy based on small efficient cars. The VW Rabbit Diesel, for example, would require substantially less fuel per passenger mile than modern rapid rail systems (see Figure 9.5). Moreover, a rapid rail system requires an enormous capital investment and many years for construction. (The buildup of the associated high-density community would take much longer.) Thus rapid-rail transit systems are not likely to provide a much larger share of passenger-transportation in the foreseeable future.

A much more promising option represents only a slight departure from the private automobile: jitney service, which employs cars or vans designed to meet the needs of local commuters. If any new type of short-haul public transportation is to become nationally important in the next few decades, this seems to be it. Company cars or vans, vans owned by other organizations, or personal cars can be used to transport small numbers of people from home to work. The driver can do this job on a part-time paid or volunteer basis. Such jitney service reduces people's need for second cars, reduces street congestion and parking problems, and reduces fuel use. It could be encouraged by assignment of

FIG. 9.5

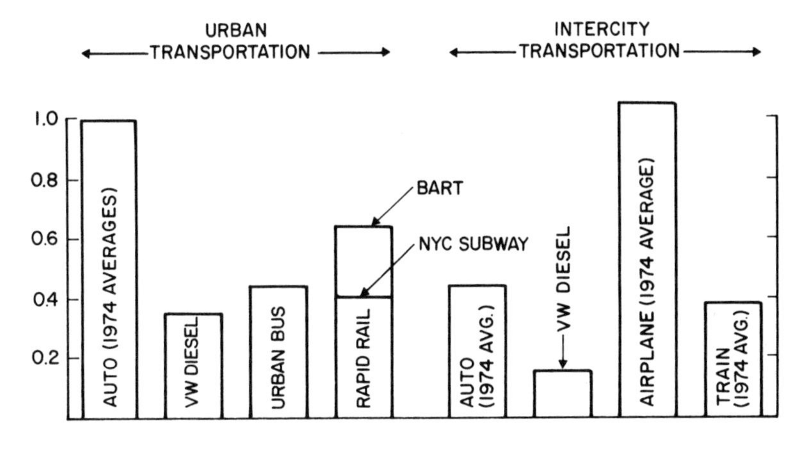

SOURCE: See Note 15

expressway lanes, special parking privileges, information services to put prospective riders in touch with a driver, special insurance provisions, etc.

A striking example of the economics of such service has been given by Frank Davis.[14] Consider a child with disabilities who needs to be transported 15 miles to a special school approximately 200 days a year. It would cost a social service agency roughly $10,000 per year for transportation: operating costs, especially driver's wages, plus paying for the van. If a neighbor who works near the school were to *donate* the 6,000 miles of transportation, the neighbor would be allowed a charitable tax deduction, the social service agency would save $10,000 a year, and congestion and fuel use would be reduced.

SUBSTITUTION OF TELECOMMUNICATIONS FOR PASSENGER TRANSPORT

An electronics-based telecommunications-intensive future could dramatically alter transportation patterns. If mobility is considered "being where you want to be, doing what you want to do," telecommunications may actually increase mobility while reducing vehicle travel and the use of energy for transportation by substituting more efficient and possibly even more effective "electronic" mobility.

The extent to which telecommunications could substitute for passenger transportation is highly speculative. The National Academy of Engineering, in a 1969 study, suggested that *theoretically* telecommunications could substitute for 84% of all transportation. Other estimates range from 16% substitutability for urban travel to well over 40% for business trips. In 1979 a National Academy of Sciences panel concluded its analysis of the substitution of communications for travel with a specific estimate that at most, passenger travel in the long-term future could be reduced about ¼.[2]

COMMUNITY DESIGN TO REDUCE TRANSPORT NEEDS

In the highly industrialized countries of Western Europe the automobile is used only about 40% as much per capita as in the U.S. (see Table 9.5). Much of this difference is probably associated with a much lower demand for short-distance travel. In these countries many people live in relatively small urban centers that provide relatively complete services. One of these services is good public transportation. Thus mass transit may have an important role in fostering desirable patterns of urban development, even if it doesn't account for many of the passenger miles.

This European experience suggests that over the long term, transportation needs in the U.S. could be reduced through urban redesign,

Table 9.5 Auto-Transportation Data for Industrialized Nations, 1972

COUNTRY	PASSENGER MILES/CAPITA	ENERGY/ CAPITA[a]	AUTO PERCENTAGE OF TOTAL PM	PRICE OF GASOLINE[b]
U.S.	10,400	349	92	37.1
Canada	5,800	237	88	27.3
Japan	1,300	35	34	74.6
Six Western	3,900			
European Nations:		83	80	
France	3,100	80	77	87.4
West Germany	4,800	90	82	64.7
Italy	3,300	80	80	116.6
Netherlands	3,800	83	81	85.9
U.K.	4,000	74	80	85.2
Sweden	5,300	128	84	74.6

SOURCE: Tables 5-2 and 5.3 in J. Darmstadter, J. Dunkerley, J. Alterman, *How Industrial Societies Use Energy,* a Resources for the Future report, The Johns Hopkins University Press, 1977.

[a] Gallons of gasoline per capita per year for auto transportation.

[b] Cents per U.S. gallon, regular gasoline.

with appropriate incentives to foster the local enterprises needed to provide a broad range of services.

* * *

FUEL CONSUMPTION BY AUTOMOBILES can be reduced by improving the fuel economy of the automobile, by shifting to other transport modes, or by reducing the need for transportation. In the long term both telecommunications and urban redesign may be effective in reducing transport demand. In our discussion of urban transport alternatives we showed that improving the fuel economy of the automobile based on the internal-combustion engine is far more promising than switching to electric cars or to mass transit. And as shown in Figure 9.5, reliance on small efficient automobiles can be far more effective in curbing transport-energy demand than shifting to alternative transport modes for intercity transport as well. Moreover, improving automotive fuel economy should be far easier than bringing about a modal shift in passenger transportation. One must not lose sight of the value of the service provided by the personal car, whether large or small. At moment's notice one can travel from here to there quickly and at modest cost (except in areas of high population density). No other mode of transport can offer nearly as good a service (again except in areas of high population density). We conclude from this discussion that if the nation pursued increased fuel economy, the chemical-fueled automobile could continue to dominate passenger transportation and that this could be achieved without fuel shortages and at acceptable cost.

PUBLIC POLICY ISSUES

In our discussion of energy conservation in housing we urged that the primary thrust of public policy should be to make market mechanisms work better. In contrast, regulation has an important role in promoting automotive fuel economy.

The substantial improvements in automotive fuel economy now underway are "economic" from the perspectives of both the nation and the consumer. The investment program of U.S. auto manufacturers for making fuel-economy improvements has been estimated to have a *cumulative cost* over 7 years of roughly $35 billion (1978 dollars).[16] The fuel-cost savings *each year* excluding taxes saved are comparable.[17] The average driver will save about $350 per year in fuel costs if his car is improved from 14 mpg to 27.5 mpg.[18] Thus the economics are overwhelmingly favorable.

Despite such favorable economics, it is unlikely that the average consumer would be induced by fuel-cost considerations alone to seek substantial improvements in automotive fuel economy. The cost of gasoline accounts for only about 1/4 of the cost of owning and operating a car. Moreover, an American's car is an important form of personal expression, and only a fraction of consumers have so far been attracted to small cars.

While the market may not be effective in bringing about automotive fuel conservation unless fuel prices increase *much* more, the automobile is rather well suited for regulation. Manufacturing is concentrated in a few firms, and there is a simple miles-per-gallon performance index for automobiles which is measurable, relatively accurate, and easily understood. Moreover, it is quite rational in our view for many individuals to choose cars with relatively poor fuel economy while they (i.e., their representatives) support collective measures to mandate sharp improvements in fuel economy. The contexts for the individual and the collective decisions are quite different. The automotive fuel economy standards set forth in the Energy Policy and Conservation Act of 1975 thus represented both an appropriate policy and politically a much easier course than the alternative, a stiff gasoline tax.

We recommend that fuel-economy standards be strengthened to bring about a much higher average fuel economy than the 27.5-mpg target set for 1985.[19] However, we believe that regulation must be supplemented by a fuel-tax increase because of inherent weaknesses in a pure regulatory approach. One problem with regulation is that standards only simulate actual experience. The manufacturers, required by law only to meet the performance goals as defined by the EPA test, may orient automotive improvements toward passing the test rather than toward making improvements in real on-the-road performance. This shortcoming of the regulations has been apparent to many a disappointed new car buyer who has learned that on-the-road performance is not as good as predicted by the EPA rating. The discrepancy has been significant. For model year 1978 actual on-the-road performance averaged about 2 mpg below the standard, even though the EPA tests indicated that on the average the standard was exceeded by 1.6 mpg.

Consumers will also seek ways to evade the standard, as shown by the "light truck loophole." Lightweight trucks (pickups, panel trucks, and vans of up to 5 tons weight) comprise 3/4 of all trucks. *Most of the use for these trucks is personal transportation rather than operation as trucks.* In a 1972 survey 53% of light-truck use was for personal transpor-

Fig. 9.6

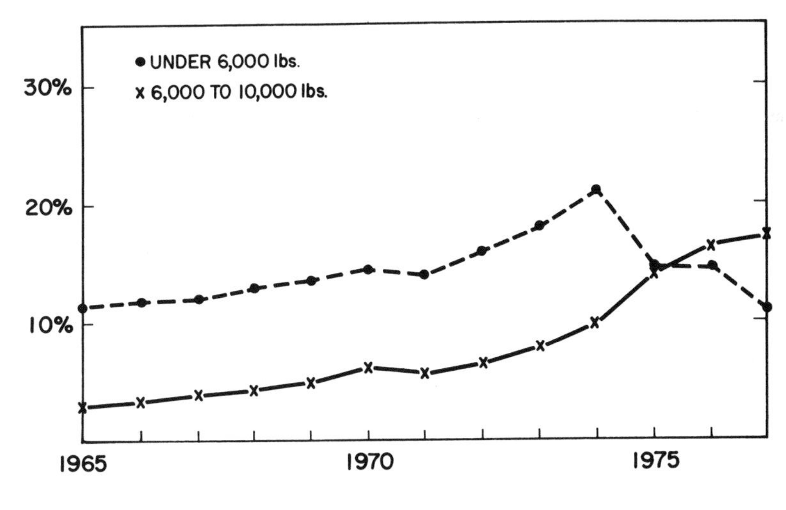

source: Ref. 3

tation, including recreation, and the fraction of this type of use has been rapidly increasing.[3]

The loophole involves the heavier light trucks. The fuel economy of trucks weighing between 3 and 5 tons was not subject to regulation in the mid-1970s. As shown in Figure 9.6, sales of these trucks were increasing very rapidly in this period. This boom is evident to the casual observer in many American towns.

While regulations are being modified to limit such loopholes, modified regulations may be unwieldy, e.g., as one tries to include multipurpose vehicles for which complicated performance indices may be needed. A stiff gasoline tax would be a straightforward means to help close loopholes.

There are many precedents for a high gasoline tax. As the data in Table 9.5 show, the price of gasoline in Western Europe and Japan in 1972, before the "energy crisis," was 2 to 3 times that in the U.S.—mostly because of taxation. The U.S. is abrogating its leadership in the world community today by not having a stiff gasoline tax to demonstrate American commitment to helping solve the world oil problem. A gasoline tax should be levied and gradually increased as fuel economies increase. An eventual tax on the order of $1.00 to $2.00 per gallon (1978 dollars) would not be unreasonable.[20] While many Americans would balk at the notion of gasoline costing $3.00 a gallon, the cost of fuel per mile would be no more at this price for a car getting 42 mpg than it would be for a typical car now on the road using $1.00-per-gallon gasoline. But the "cheater" would be faced with a 200% increase in his gasoline bill!

CHAPTER 10

INDUSTRIAL COGENERATION: MAKING ELECTRICITY WITH HALF THE FUEL

ABOUT 40% OF ALL FUEL CONSUMED in the U.S. is used to produce relatively low-quality energy—heat at temperatures less than a few hundred degrees Fahrenheit. Second-law efficiencies of systems that involve direct fuel combustion to serve such energy needs are very low (see Table 7.1). As we have shown in Chapter 8, one way to significantly improve efficiencies in very low-temperature applications such as space conditioning and domestic water heating is to employ heat pumps. Another way to provide low-quality energy is to exploit the potential for "energy cascading."

When energy is "consumed," it is not destroyed—it is merely reduced in quality. Since there are many situations in which only relatively low-quality energy is needed, there is a large potential for energy cascading, in which initially high-quality energy is used for one purpose and the energy discharged at a lower quality from this activity is used for a second purpose. In principle several different activities can be tied together in an energy cascade to make use of the energy available at lower and lower levels of quality as the energy is used and reused in sequence.

One simple example of an energy cascade was described in Chapter 8—recovery of the waste heat from the dishwasher and clothes dryer in winter for space heating.[1] In the case of the automobile a potentially important energy cascade would involve recovering waste heat from the automobile engine to drive a heat-powered automotive air conditioner.[2] This would substantially improve automotive fuel economy because fuel-economy losses associated with conventional air conditioners are estimated to be about 9% (up to 20% in city traffic on hot days)[3] and because about ¾ of new domestic cars are equipped with air conditioning.[4]

By far the most important application of this energy-cascading principle involves the combined production of electricity and heat, or *cogeneration* as it is called. The cogeneration arrangement which offers the greatest potential for power generation would be to use the very high-temperature heat available in fuel combustion (flame temperatures can be as high as 3000°F) to make electricity first and to recover the exhaust heat from electricity production for applications where low-temperature heat (typically below 400°F) is needed. This could be done in apartment buildings, office buildings, or shopping centers where the heat would be recovered for water heating, space heating, or even air conditioning. Or this could be done in industries which use steam for process applications.

In this chapter we shall focus attention on cogeneration in steam-using industries where the potential savings are especially large. The economics of industrial cogeneration are most favorable for large industrial plants in capital-intensive industries that require a great deal of steam for the production of chemicals, paper, refined petroleum, and the like. Such plants usually operate around the clock and throughout the year, so that the cogeneration equipment can be used most of the time. Because it is often economical in such situations to produce electricity on a steady basis, building cogeneration facilities is an alternative to building baseload* nuclear or coal-fired power plants. In fact if all large industrial steam boilers today were involved in cogeneration, the electricity produced would be equivalent to that generated by 30 to 200 large nuclear power plants, depending on the technology deployed,[5] while energy would be saved through the displacement of central-station power plants at a rate of up to 2 million barrels of oil per day (or 6% of all U.S. energy use).

There are two alternative configurations where the steam needs of industry could be provided via cogeneration—a centralized and a decentralized option. With the centralized option, process steam would be

*Baseload power plants are plants characterized by relatively high capital costs and relatively low operating costs. For economical operation they must be operated near full capacity most of the time. Thus baseload plants produce electricity on a fairly steady basis.

generated as a by-product of power generation at a central-station power plant and delivered to steam-using firms. With the decentralized option, electricity would be produced at the industrial site as a by-product of process-steam production. The decentralized option is far more promising.

One factor favoring decentralization is that there are relatively small fuel savings to be gained from modifying the designs of central-station plants so that plants produce both steam and electricity, at least with the steam-turbine technology that dominates the electric utility industry today. This technology has been optimized to produce electricity alone. Redesign of steam-turbine systems to produce both power and process steam yields very little thermodynamic advantage.*

Not only are the potential gains of thus generating steam as a by-product small; there are serious implementation difficulties as well. Because it is uneconomic to transport steam long distances, steam-using industries would have to be near the power plants supplying their heat, and this is a condition often difficult to fulfill. There is also a serious mismatch in time: large central-station power plants require 6 to 12 years for construction and are designed for a quarter century or more of service. Thus unless there comes to be much more central planning in the U.S. economy than there is today, onsite cogeneration will be favored over central-station configurations.

Onsite industrial cogeneration is not a new concept. In 1920 onsite generation accounted for 22% of American electricity. But over the years this percentage declined—to 18% in 1940, to 9% in 1960, and to only 4% in 1976. In part the declining role of cogeneration is due to the fact that the demand for electricity has grown much faster than the demand for industrial process steam. But also, institutional and economic factors have worked together over time to foster the dominance of central-station power plants.

The principal institutional and economic factors behind the trend away from industrial power generation have been (1) resistance to the concept on the part of the utilities, manifest in extraordinarily high rates charged by utilities for backup power and extraordinarily low rates offered to industrial firms for the purchase of power generated in excess of onsite needs, and (2) a long-term trend toward lower and lower prices

*This is apparent from an examination of second law efficiencies. If a quantity of steam and/or electricity is produced by a system, the second efficiency is the ratio of the theoretical minimum amount of fuel required to the actual fuel consumed by the device (see Chapter 7). While the second-law efficiency is greater for all cogeneration systems than for central station power generation, it is only slightly improved for steam-turbine systems (see row 3 of Table 10.1).

Table 10.1 Performance Characteristics of Alternative Electricity- and Steam-Production Systems

	CENTRAL-STATION ELECTRICITY GENERATION	PROCESS-STEAM[a] PRODUCTION	COGENERATION SYSTEMS[b]			
			STEAM TURBINE	GAS TURBINE	GAS/STEAM TURBINE	DIESEL
1. E/S Ratio[c]	—	—	0.24	0.65	1.1	1.3
2. First-Law Efficiency	0.33	0.88	0.85	0.76	0.69	0.62
3. Second-Law Efficiency	0.35	0.28	0.40	0.47	0.49	0.46
4. Fuel Savings Rate[d]	—	—	0.38	0.86	1.31	1.24
5. Value of Products (electricity plus process steam) in Dollars per Million Btu of Fuel Consumed[e]						
low E price/low S price	2.1	1.3	1.9	2.5	2.6	2.5
low E price/high S price	2.1	2.2	2.6	2.9	2.9	2.7
high E price/low S price	3.2	1.3	2.4	3.3	3.7	3.5
high E price/high S price	3.2	2.2	3.1	3.8	4.0	3.8

NOTES:

[a] For process steam at a gauge pressure of 50 lb per square inch.

[b] Based on the characteristics of systems producing steam at 50 lb per square inch of gauge pressure described in Table 2 of R. H. Williams, "Industrial Cogeneration," *Annual Review of Energy*, volume 3, p. 313, 1978.

[c] The ratio of the produced electrical energy to the produced steam energy.

[d] The fuel savings rate is the fuel saved per unit of process steam through the displacement of central station electricity generation.

[e] High-E/S-ratio cogeneration technologies produce higher-value products than low-E/S-ratio technologies because electricity is a much higher-quality (and hence more valuable) form of energy than low-pressure steam. In this tabulation the low electricity price is 2¢ per kwh and the low steam price is $1.50 per million Btu; the high prices are 3¢ per kwh and $2.50 per million Btu for electricity and steam respectively.

for electricity generated at central-station power plants. If an industrial firm has wanted to cogenerate its own electricity and rely on the local utility for its backup supply, the price charged by the utility for this backup service has usually been at least double the average price industrial firms pay for "full service" electricity. In addition, utilities have usually been unwilling to pay a fair price for electricity generated in excess of onsite needs. An underlying reason for these utility policies is that as a regulated monopoly a utility is allowed to make a profit only on equipment "in the rate base," that is, on property it owns. At least as important as utility resistance to cogeneration is the fact that until recently, declining electricity prices made central-station electricity increasingly attractive over time. As Figure 2.4 shows, the average price of electricity to industrial customers (in constant dollars) was cut in half between 1940 and 1950 and again between 1950 and 1970. But after 1970 this historical trend reversed. And as pointed out in Chapter 4 the price of electricity (in constant dollars) can be expected to increase gradually for the rest of this century. This implies that the economic attractiveness of cogeneration relative to central-station power generation will tend to improve.

TECHNOLOGIES FOR INDUSTRIAL COGENERATION

In industry today fuel is burned to raise steam in a boiler for process use, while in a separate utility installation a boiler is fired to raise steam to drive a steam turbine for generation of electric power (see Figure 10.1). To produce one unit of electrical energy this way, three units of fuel energy are required. The other two units are lost to the environment as waste heat—up the smokestack and in the cooling system. In cogeneration these industry and utility operations are combined (Figure 10.1).

The steam turbine system, while less energy efficient than alternative technologies, is the most common cogeneration technology in use today. With this system high-pressure steam is used to drive a steam turbine for generation of electricity, and low-pressure steam is exhausted from the turbine at the temperature and pressure appropriate for industrial process applications. Alternatively, with a system based on a gas-turbine engine or a diesel engine, combustion gases are used to drive the engine for power generation, and the hot engine exhaust gases

Fig. 10.1

Separate Steam and Electric Power Plants

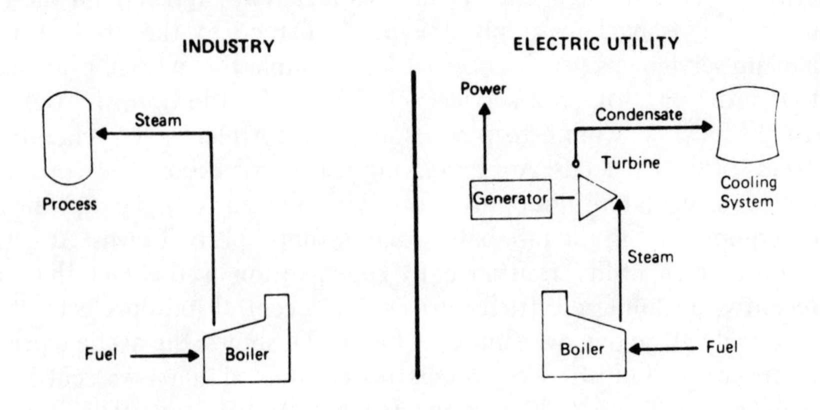

Cogeneration Plants

Back-Pressure Steam Turbine

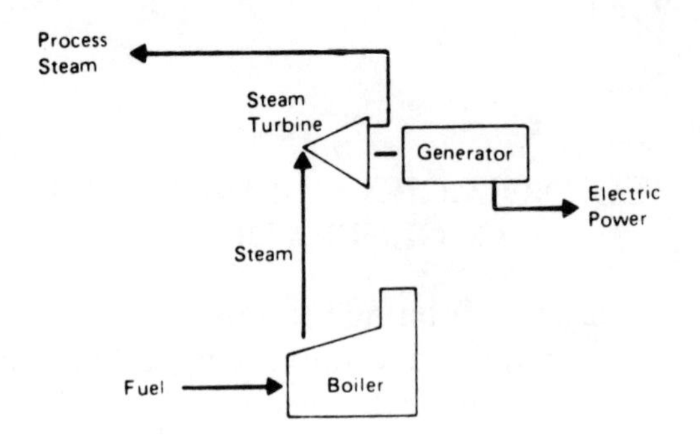

Gas Turbine—Waste Heat Boiler

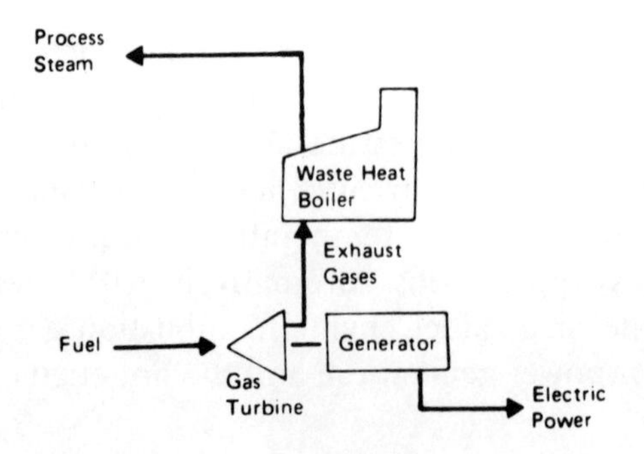

The top figures show how steam and electricity are produced separately at industrial and utility sites today. At the industrial site fuel is burned to generate steam in a boiler, and the produced steam is delivered to industrial processes where the steam heat is needed. At the utility the hot, high pressure steam generated in the utility boiler is used to turn the blades of a steam turbine, thus converting the heat energy of the steam into rotational energy. This rotational energy is then converted into electrical energy by the generator. The spent steam exhausted from the turbine is then condensed into water in a cooling system.

The middle figure shows how steam turbine technology can be used to generate both electricity and steam for industrial processes. In this cogeneration configuration the steam turbine is modified so that the steam exhausted from the turbine is not condensed but is delivered as useful heat to industrial processes. When the turbine is modified this way, the amount of electricity produced is substantially reduced, since the pressure and temperature of the steam exhausted from the turbine are much higher than when only electricity is produced.

The bottom figure shows an alternative cogeneration scheme. In this case the hot, high pressure gaseous products of fuel combustion are used directly to drive the blades of a gas turbine, to convert heat energy into rotational energy. The gases exhausted from the turbine are still hot enough to generate steam for industrial processes in a "waste heat boiler." In a hybrid "gas/steam turbine" system (not shown) the steam generated in the waste heat boiler is used to drive a steam turbine and the exhaust steam is delivered to industrial processes.

---

are used to raise steam for process use in a waste heat boiler. In all these cases the amount of fuel needed to produce a kilowatt-hour of electricity, beyond what would be needed to produce steam anyway, is only about half of what would be needed at a central-station power plant.

Perhaps the most important distinguishing feature of alternative cogeneration technologies is the fact that for each unit of steam energy generated, the amount of by-product electricity produced varies from a low value of about 0.24 units with a steam-turbine system to a high value of 1.3 units with a diesel engine (see row 1 of Table 10.1). This wide variation in the ratio of electricity to steam (E/S ratio) shows that the potential for production of electricity and hence for fuel savings (see row 4 of Table 10.1) varies markedly with the cogeneration technology deployed.

The fuel-saving advantage of high-E/S-ratio technologies is reflected in higher second-law efficiencies for these than for low-E/S-ratio cogeneration technologies, as shown in Table 10.1.* Because the second-law efficiency takes into account the quality of energy in various forms, the technologies with higher second-law efficiencies would tend to produce the more valuable products from a given quantity of fuel. And as the data in Table 10.1 show, high-E/S-ratio cogeneration technologies do indeed produce products with higher dollar value than do low-E/S-ratio cogeneration technologies. These differences in the dollar value of cogeneration products simply reflect the fact that electricity is a much higher-quality, and hence more valuable, form of energy than low-pressure steam.[6]

Not only do high-E/S-ratio cogeneration technologies make more efficient use of fuel than low-E/S-ratio technologies; they would also allow cogeneration to become a major source of electricity for all consumers. This is because with such technologies the amount of electricity produced at industrial sites would often be in excess of onsite needs; in fact very often the amount available for export from the industrial site would exceed that consumed onsite.[7]

THE ECONOMICS OF INDUSTRIAL COGENERATION

The fuel savings offered by cogeneration are often accompanied by capital savings. Since practical cogeneration systems tend to be small—typically only 1 to 10% of the size of central-station power plants—the possibility of capital savings in cogeneration is at first glance surprising. One would think that "economies of scale" would favor large-scale power generation. However, other unique features of cogeneration systems compensate for diseconomies of scale. In a cogeneration system capital costs for power generation may be reduced through cost sharing with process-steam generation: only the capital costs in excess of what would otherwise be needed for steam production in a separate facility

*The data in Table 10.1 show a dramatic difference between first- and second-law efficiencies: the first law efficiency tends to be high when the second-law efficiency is low and vice versa. This curious result arises because the first law efficiency is a poor measure of thermodynamic performance—it does not take into account the fact that electricity is a much higher quality energy form than steam. For example, while the steam turbine has an 85% first law efficiency (i.e., it converts 85% of the fuel energy into electrical energy and steam energy), only 16% of the fuel energy ends up as electricity. (The other 69% is steam energy.) In contrast, for the diesel, with a first law efficiency of only 62%, 35% of the fuel energy is converted into electricity. The second law efficiency of the diesel is higher (46% vs. 40%) because the second law efficiency index appropriately gives more weight to electricity than to steam.

need be charged to power generation in many instances. Also systems with inherently low capital costs, e.g., gas turbines, may be efficiently employed in cogeneration.

The fuel- and capital-savings advantages of cogeneration mean that in a wide range of circumstances cogeneration would provide electricity to industrial customers at less cost than the cost of electricity from a new central-station power plant (the replacement cost for central-station electricity). But at the same time this cogenerated electricity would often be more costly than electricity sold by the utility, the price of which is based on average costs (see Figure 10.2). Unfortunately industrial firms are more influenced in their decisions relating to cogeneration by what they pay the utilities today than by a consideration of what alternative new power sources would cost. Industrial cogeneration provides a classic example of the energy-pricing problem: when new power plants are added to the utility grid, customers are insulated from the full impact of their higher costs because these higher costs are "rolled in" with the costs of the much cheaper old plants to obtain an average price. Society would reap great economic benefits if, instead, decisions relating to new energy sources were made on the basis of a comparison of replacement costs.

A major challenge relating to cogeneration therefore is to bring about institutional changes that cause potential cogeneration projects to be evaluated on a replacement-cost basis. More of this later.

A FUEL STRATEGY FOR INDUSTRIAL COGENERATION

We have shown that cogeneration technology would have the greatest fuel-savings impact if technologies characterized by high E/S ratios (e.g., gas turbines and diesels) were emphasized (Table 10.1). But these technologies today must be fired by oil or natural gas, and it has been the U.S. energy policy to encourage industry to shift from these fuels to coal. Thus present U.S. energy policy is not conducive to fostering the development of high-E/S-ratio cogeneration technologies. In fact only the low-E/S-ratio steam-turbine cogeneration technology is compatible with the "coal for industry" policy.

In the long run coal could fuel high-E/S-ratio technologies. Consider first the gas turbine. With today's coal-combustion technology direct firing of gas turbines with coal would cause serious fouling, corrosion, and erosion problems for the gas-turbine blades. But if ongoing

FIG. 10.2

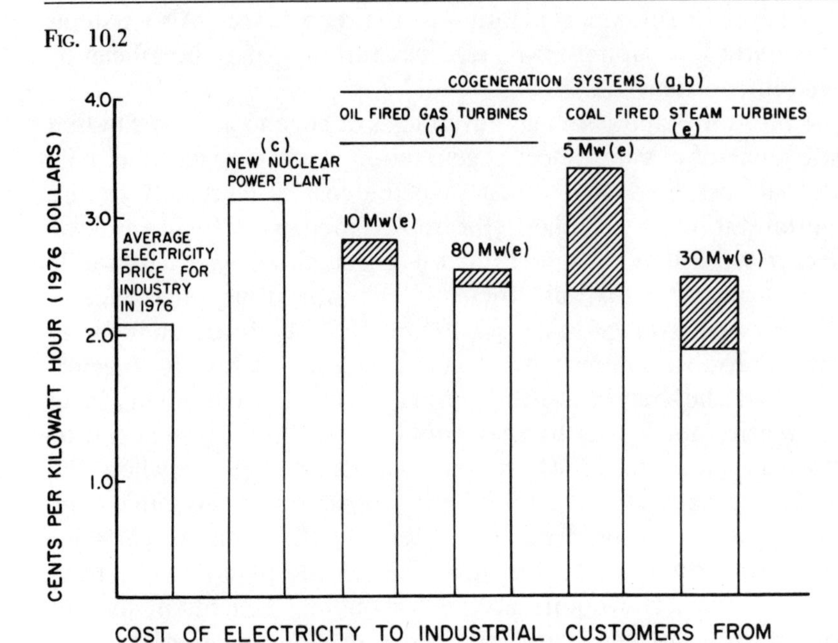

Cost of electricity to industrial customers for alternative power-generating systems (in 1976 dollars). From R.H. Williams, "Industrial Cogeneration," ***Annual Review of Energy,*** **vol. 3, p. 313, 1978.**

NOTES:

(a) To facilitate a comparison with central station power generation we assume utility ownership of cogeneration facilities at industrial sites.

(b) The cost of cogenerated electricity depends on whether or not the capital cost of steam generation can be shared with process steam. If cost sharing is allowed (e.g., in a new facility) then the appropriate costs would be those at the bottom of the shaded bars. If all capital costs must be allocated to electricity (e.g., in an existing facility where an operating boiler must be replaced for no other reason than to install a cogeneration unit), then the appropriate costs for cogeneration would be those at the tops of the shaded bars.

(c) For a nuclear power plant costing $800/kw(e).

(d) For oil costing $16 per barrel.

(e) For coal costing $24 per ton.

R&D efforts on fluidized-bed combustion are successful, within 5 to 10 years coal-fired gas turbines should be commercially available for industrial cogeneration applications. And it may even be possible to develop coal-fired diesel engines. Moreover, for the longer term there are other possible high-E/S-ratio cogeneration technologies such as the Stirling engine for which the burning of low-quality fuels like coal would not be problematic.[8]

Nevertheless the present "coal for industry" policy is in conflict with the objective of maximizing the potential benefits from cogeneration. If industry is forced to shift to coal before advanced coal-burning cogeneration technologies are available, then much of the industrial process-steam resources base may become "locked up" for a quarter of a century or more, and the demand for the advanced technology may be inadequate to spur its commercialization. Thus consideration of the long-term prospects for cogeneration suggests that the "coal for industry" policy should be reevaluated.

It is in fact ironic that through 1979 the government's attempts to discourage oil and gas use and promote coal use by industry has had the opposite of the desired effect. The government can affect sales of natural gas more easily than oil sales, since sales of natural gas are closely regulated. Thus between 1975 and 1979 industrial consumption of natural gas went down 2% per year, but the demand for oil went up 7% per year. And because of supply-availability problems and difficulties relating to pollution control, the industrial use of coal actually declined in this period at an average annual rate of 2%. Thus the principal impact of the "coal for industry" policy in this period was to increase the demand for oil imports and thereby contribute to the erosion of the dollar!

This "coal for industry" policy is foolish. Not only has it had the wrong impact; also it may well be the wrong policy for the long run. As we pointed out in Chapter 4, there is growing evidence suggesting that the natural-gas supplies available at deregulated gas prices may well be far greater than current official estimates.[9] It would of course be foolish to base policies on the assumption that gas is indeed abundant, because these gas-supply estimates are very uncertain. However, this uncertainty should be factored into public policy. Ways should be sought to forestall the industrial conversion to coal until the long-run natural-gas-supply situation is clarified.

Fortunately it is possible to forestall conversion to coal via a fuel-conservation policy. A serious conservation effort would free up for industrial uses large quantities of natural gas that are now used wastefully in buildings. About 40% of natural gas is used in residential and commercial buildings—more than half of which could be saved with a serious conservation effort, along the lines outlined for housing in Chapter 8. By freeing up gas supplies this way for efficient industrial applications like cogeneration, a conservation strategy would diminish the urgency of converting industry to coal and buy time in which to better evaluate the long-run natural-gas-supply situation. It would also buy time to bring to commercialization advanced combustion technologies that permit the use of high-E/S-ratio cogeneration technologies based on coal. However the gas-supply evaluation comes out, this gas-conservation strategy provides a means by which industrial cogeneration can become a major source of electricity through the deployment of high-E/S-ratio cogeneration technologies.

BRINGING ABOUT INSTITUTIONAL CHANGE

The deployment of high E/S-ratio cogeneration technologies often requires largely unfamiliar institutional considerations. To realize the full potential of these technologies, steam-using industries must be motivated to participate in cogeneration projects where substantial quantities of electricity are produced in excess of onsite needs, and utilities must fairly evaluate onsite cogeneration as an alternative to conventional central station power generation.

One important issue concerns the value of cogenerated electricity to the utility.

If more electricity is produced than is needed onsite, then there must be a market for the excess electricity. It would be logical for cogenerating industries to sell electricity to utilities. Unfortunately utilities have tended to discourage such activity; they have usually been willing to pay industrial firms only the "dump price" for this electricity, which is much less than what the electricity is really worth to the utility and is much less than what these firms would need to realize an adequate return on their investments. Partly for this reason industry has focused its attention mainly on those cogeneration technologies that avoid the excess-electricity problem.

In early 1980 the Federal Energy Regulatory Commission announced a new regulation addressing the issue of how cogenerated electricity should be valued by utilities.[10] This regulation, based on the Public Utility Regulatory Policy Act of 1978 (PURPA), requires that electric utilities purchase electricity made available by qualifying cogenerators at a rate reflecting the cost that the purchasing utility can avoid as a result of obtaining this electricity. By requiring the utility to value cogenerated electricity at its avoided cost (which would be the cost of electricity from a new central station power plant if the alternative to the purchase of cogenerated electricity were the construction of such a plant), this regulation overcomes one of the major historical obstacles to cogeneration.

This milestone regulation by itself may be inadequate to foster widespread deployment of high E/S-ratio cogeneration technologies, however. Because steam-using industries are usually more interested in investments relating to their product lines than in investments that would enable them to sell electricity, they will tend to focus their attention on meeting onsite electricity needs and thus will not seek to maximize the potential for onsite power generation. It appears that ownership arrangements for onsite cogeneration other than ownership by the steam-using firm would be more conducive to maximizing the deployment of high E/S-ratio cogeneration.

UTILITY OWNERSHIP

Utility ownership would be one such arrangement. Even without the PURPA regulation relating to the valuation of cogenerated electricity that is sold to the utility, this arrangement would facilitate an evaluation of cogeneration projects on the basis of replacement costs because the utility would have to decide between building cogeneration plants and building new central-station power plants. Utility ownership would also lead to more projects being judged economical than would industrial ownership, since utilities usually require a lower rate of return on their investments than industrial firms. This latter advantage would also mean lower electricity prices for consumers generally.

The option of utility ownership should allay many concerns of utilities about cogeneration. Utility ownership would bring cogeneration into the rate base, thereby providing the utility an incentive to cogenerate. Utility ownership also would provide a straightforward means by which utilities could maintain dispatching control over the electric power entering the grid, to ensure system stability and security.

One concern that many industrial firms might have about utility ownership is that the rate charged by the utility for power might change in the future in unpredictable ways. For example, the implementation of life-line rates for low-income residential customers might be offset by a general increase in the price of electricity for industrial customers. No such problem would arise with industrial ownership. It may be desirable to protect the industrial firm against this prospect.

One modified utility ownership arrangement for cogeneration that should be attractive to industrial firms has been suggested by Albert Smiley, who proposes that utilities be allowed to form profit-making subsidiaries that would deal exclusively in cogenerated electricity and steam. These subsidiaries would own and operate cogeneration facilities at industrial sites. All the electricity produced via cogeneration would be sold to the utility at a price equal to the least costly source of new central-station electricity. The industrial firm would buy electricity from the utility according to the established rate schedule, but the price paid by the industrial firm for steam would be negotiated; i.e., the steam price would not be subject to regulation. Because the subsidiary would often be selling electricity to the parent utility at a price higher than the cost of cogenerated electricity, the subsidiary could offer the industrial firm discount prices for steam and still make a handsome profit in many cases. The state commission which regulates utilities could be attracted to this arrangement by a provision that the profits of the subsidiary be divided between utility stockholders and electricity consumers; e.g., consumers could be given periodic lump-sum rebates. Thus all interested parties would share in the social gain afforded by cogeneration.[11]

SHOULD ELECTRIC-POWER GENERATION BE DEREGULATED?

Cogeneration could be encouraged by creating competition in electric power generation as an alternative to fostering ownership of cogeneration facilities by regulated utilities. Today there is no competition in electric power generation; utilities are regulated monopolies whose profits are determined by state regulatory commissions. While this regulatory activity may have been justifiable in the past, recent technological developments suggest that some degree of deregulation of power generation is probably called for.[12] Policy changes in the direction of deregulation would probably have a much greater impact on the development of cogeneration than the investment tax credits and other incentives usually proposed to foster cogeneration.

Why are electric utilities regulated in the first place? Basically because of the belief that the provision of electricity represents a natural monopoly. The natural-monopoly concept is that whenever unit costs decrease as the plant size is made larger, there are inexorable cost pressures forcing dominance by a single firm—a natural monopoly. Under these circumstances the consumer is served best by granting a single firm a geographic monopoly and by regulating prices so that

- consumers are protected from monopolistic pricing
- the natural monopoly is allowed a fair return on investment

Competition would not be in the consumers' interest because it would lead to smaller plants having needlessly higher costs and to costly duplication of facilities.

Technological developments have changed this situation, so that the natural-monopoly concept no longer applies to all aspects of the power business. Consider the three parts of the electric power industry—generation, transmission, and local distribution. The rationale for regulation of transmission and distribution is still strong. Transmission technology has evolved so that a single system can serve wide geographic regions economically with ever decreasing electrical losses. Duplication of transmission facilities would be very wasteful. Similarly distribution technology represents a natural monopoly in a particular community.

But generation, which involves over half the costs of providing electricity, no longer represents a natural monopoly. One reason is that successes in the development of transmission technology have made it possible to serve any particular need for power by generation facilities dispersed over a wide region. Furthermore, as we pointed out in Chapter 4, there no longer appear to be economies of scale to be captured by building generating plants larger. In fact the "point of diminishing returns" to increasing scale may already have been exceeded with the largest power plants. Finally, cogeneration power plants can produce electricity competitively with central-station power plants 10 to 100 times as large (see Figure 10.2).

In light of these developments it is natural to propose that the power-generation business be separated from transmission and distribution activities and be deregulated.* Making electricity competitive not only would provide a more favorable environment for cogeneration but also would help limit electricity price increases in the new era of rising costs and uncertain future demand.

Deregulation would have to be carried out carefully and gradually so as to minimize problems, such as the possibility that unregulated generation monopolies might develop in spite of theoretical arguments to the contrary and the difficulties of assuring system reliability when generation facilities are not commonly owned.

The PURPA regulation described above[10] represents an important first step toward deregulation of power generation, because it helps foster the creation of a cogeneration service industry. Under this regulation, cogeneration service firms would be able to sell all the electricity they produce to the electric utility at a rate equal to the utility's avoided cost, and they could sell the steam they produce to steam-using firms. Where such steam sales are not regulated the cogeneration service firm

*The organization of transmission and distribution activities might be modified as well. One possibility would be to establish a regionally based natural monopoly for transmission, regulated at the federal level, with community-scale distribution companies regulated at the local level.

would often be able to offer attractive steam sales contracts to steam-using firms, with the result that the cogeneration service industry could grow rapidly. Unfortunately the sale of steam is a regulated activity in many states, so that such negotiated contracts would often not be possible. An important second step toward deregulation of power generation would thus be policies aimed at the deregulation of steam sales to industry.

RESEARCH AND DEVELOPMENT

These considerations point to new directions for research and development. We usually think of R&D in terms of hardware. And indeed hardware R&D will continue to be important for cogeneration, because the possibilities for innovation are great. However, an important new dimension of the overall R&D effort should be to carry out (with present technology) experiments with pricing policies, various degrees of deregulation, and alternative ownership arrangements in "institutional demonstration projects" as we wait for new technology. In these demonstration projects emphasis should be given to high-E/S-ratio cogeneration technologies, which would usually involve exporting considerable quantities of electricity to the utility grid. Perhaps the highest priority for energy policy research relating to cogeneration should be to examine carefully the benefits and drawbacks of various degrees of deregulation of power generation.

If these activities are successful, then cogeneration could become the major source of new baseload electric generating capacity in the U.S. in the last 10 to 15 years of this century[13] as the transition is made to an industrial energy economy where process heat is provided in the much more efficient manner that is required for the new era of high-priced energy.

CHAPTER 11

SOLAR ENERGY AT THE COMMUNITY LEVEL

WHILE IT WOULD NOT BE PRACTICAL to meet most of our energy needs with solar energy in an energy economy that is a larger version of today's, the constraints on solar-energy supply described in Chapter 6 might not be formidable in a world where energy-conservation technologies such as we have described in the last three chapters were in wide use. Moreover, a combined solar and conservation strategy *could* make solar energy affordable for a wide range of options. In Chapter 9 we illustrated this thesis by showing how conservation could promote biomass to the status of a major fuel for automobiles. In this chapter we show how generation of solar electricity could become competitive by exploiting the economic advantages inherent in cogeneration technologies.

The generation of solar electricity is usually regarded as one of the greatest challenges to the solar-energy community. In fact there are many serious analysts who believe that solar electricity will never make a significant contribution to America's energy needs. In a 1976 review of technologies for the generation of electricity from solar-energy sources a prominent scientist concluded:[1]

> It is unfortunate that so many people continue to entertain high hopes for satisfying all our needs for electricity through direct or indirect means of generating it from the sun. . . . For any appreciable contribution to future national requirements for central station electricity, neither direct nor in-

> direct solar energy (other than hydroelectric) is really suitable. There is practically no chance of realizing such a contribution regardless of how vigorously it is promoted and funded by the Congress in response to public aspirations.

Since the author of this statement has long been associated with the development of nuclear power, some solar advocates might tend to dismiss his judgments as biased. Yet similar views have been expressed even by officials responsible for solar energy R&D. In late 1976 Robert Hirsch, assistant administrator for solar, geothermal, and advanced energy systems in the Energy Research and Development Administration, in remarks to the General Advisory Committee of that agency, said:[2]

> What I have learned is that there are a variety of people—in universities, in industry, and in government—who are not at all sure about the future of solar energy. As a matter of fact, a number of them are rather pessimistic about the ultimate economics of concentrator and photovoltaic systems for electric power generation. I guess I have talked to enough people who may have a pro-solar bias and enough who may have a negative bias to balance out, and I can see that there really is a significant question about the ultimate economics of these two approaches.

A simple naive calculation illuminates the concern of these critics of solar-electricity generation. Consider photovoltaic power, which involves the direct conversion of sunlight into electricity (i.e., without first converting the sunlight into heat) in solar cells. These solar cells, familiar to most people as the technology used to produce electricity in outer space missions, are now being developed as a possible terrestrial power source (see Chapter 6). At present solar cells, which convert 10 to 15% of incident sunlight to electricity, cost about $15,000 per kilowatt of electric power produced when the cells are illuminated in full sunlight on a clear day. Because the average amount of sunlight received at a flat surface on the ground is only ⅙ of the peak solar flux, solar electricity costs $90,000 per average kilowatt of produced electricity. This cost must be increased to over $100,000 per average kilowatt delivered because of electrical losses in batteries (electricity must be stored for use at night). This cost may be compared with about $1200 per average kilowatt for a nuclear plant.[3] Since solar energy is "free" whereas nuclear fuel is not, this capital-cost comparison is slightly misleading—but not very misleading because nuclear fuel is relatively cheap if not free. When the costs of electricity produced are compared, the results are 130¢/kwh for solar electricity (neglecting the costs of batteries and other support equipment) and 2.4¢/kwh for nuclear electricity.[4] Thus it appears that more than a 50-fold reduction in costs is necessary for solar cells to become competitive as a source of central-station power. An ambitious photovoltaic R&D program is underway and a 30-fold cost reduction is hoped for by 1986.[5] But whether a 50-fold cost-cutting goal can be reached is highly uncertain.

But even if the skeptics are right, this analysis does not imply that solar electric-power generation may never become economic, only that *solar central-station power generation* may not be able to compete with other *central-station power sources.*

A consideration of the properties of sunlight suggests that arrangements other than central-station power configurations may be preferable for producing solar electricity. Consider first collectors. Because sunlight is diffuse, large collectors are needed, and these tend to be the dominant part of the capital cost of a solar system. Since both the amount of sunlight collected and the materials required increase roughly in proportion to the collector area, there are no significant scale economies to be exploited by building very large collectors. And in fact a significant *diseconomy* inherent in building highly centralized systems is that so doing would preclude the option of solar cogeneration, where sunlight is converted simultaneously into both electricity and heat. Cogeneration is preferred in configurations where solar-energy production is close to where consumption will take place, because it is very costly to transport heat long distances.

How decentralized should solar cogeneration be? The household level may be too decentralized, for at least two reasons. The sophisticated equipment required would be very costly to operate and maintain in small installations. And there are substantial scale economies to be achieved by centralizing solar-energy storage, which is necessary to compensate for the intermittency of solar energy.* On the basis of preliminary analyses it appears that community-scale systems—intermediate in scale between the central station plant and the individual household—may be the most promising configuration for solar cogeneration.

We shall consider here a photovoltaic cogeneration system. We shall show that the naive calculation cited above gives an entirely erroneous impression as to how close photovoltaic technology is to commer-

*Consider a hot-water storage tank. The cost of storage tanks increases roughly in proportion to the surface area (since the amount of material required increases as the area), whereas the storage capacity increases as the volume. Unit costs are lower for large tanks than for small tanks because less tank material and less insulation are required per unit tank volume. However, beyond some point it is not practical to build a tank deeper, so that the cost of further volume increases is proportional to the surface area and no further scale economies can be captured.

cial application and that by simply bringing collector technology already developed into mass production (with no cost reduction for the solar cell!) a photovoltaic cogeneration system could well be competitive with new conventional energy systems.

DESIGN

Instead of a system with solar cells laid out flat in the sun we shall examine a system that employs collectors which track the sun, concentrate the sun's rays 100-fold, and focus these rays on tiny solar cells. There are two advantages of using such concentrators: (1) with a 100-fold concentration of the incident sunlight only about 1% as much solar-cell area needs to be used; and (2) the use of a concentrator permits the recovery of useful heat with the same collector that is used to collect sunlight for electricity generation.

Whereas we suggested in Chapter 10 that cogeneration based on the use of fossil fuels appears to be most promising in industrial applications, photovoltaic cogeneration systems may be best suited for buildings applications. There are two reasons for this. First, the efficiency of solar cells declines as the temperature of cell operation increases.[6] This consideration suggests applications that require very low-temperature heat—less than 200°F, say—such as space conditioning and domestic hot water. Second, a typical concentrating system will recover 4 to 5 times as much heat as electricity. Since buildings require both low-temperature heat and electricity, and typically in a ratio of about 4 to 1, it appears that a photovoltaic cogeneration system is well suited for building use.

We shall consider here for specificity a photovoltaic cogeneration system for a community of new energy-conserving houses of the type described in Chapter 8, located in central New Jersey.[7, 8, 9, 10] See Figures 11.1, 11.2, and 11.3.

SOLAR CELLS

Despite the fact that our photovoltaic cogeneration system is based on the use of solar cells as costly as those now available, the cost of these cells represents less than 10% of the capital cost of the total system, because with concentrating collectors only a very tiny solar-cell area is required.[10h] This component of the total system cost is in fact so small that there is no significant incentive to make solar cells cheaper for this application. What is needed for concentrator applications is more efficient, not necessarily cheaper, solar cells. The use of more efficient cells could lead to a more economical system even if these more efficient cells cost more.

FIG. 11.1

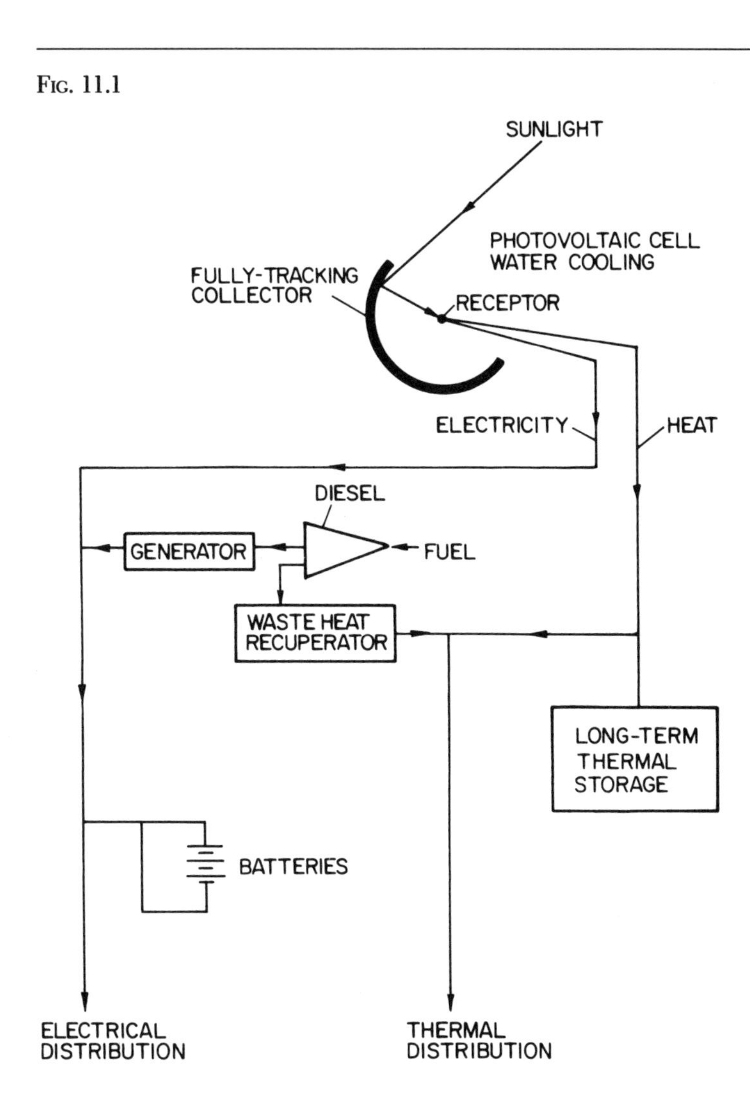

PHOTOVOLTAIC COGENERATION SYSTEM

FIG. 11.2

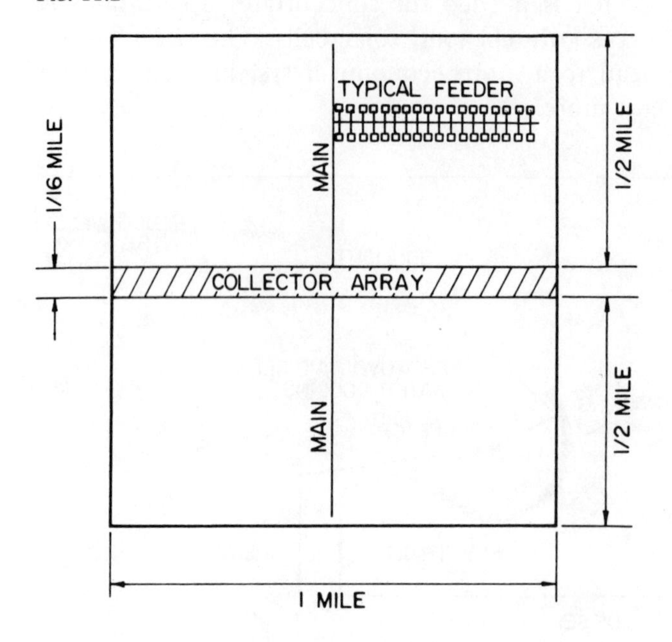

RESIDENTIAL COMMUNITY SERVED BY PHOTOVOLTAIC COGENERATION SYSTEM

The thermal distribution system carries hot water for space and water heating from the central facility to individual houses in the fall, winter, and spring. In the summer the thermal distribution system carries cold water for air conditioning. The "mains" and a typical "feeder" pipe with its connections to individual houses are shown here.

TRACKING, CONCENTRATING COLLECTORS

The single most costly item in our system is the tracking, concentrating collectors and their installation, which account for 35 to 45% of the total capital cost.[10] The fact that the purchase of the collector from the factory represents a large fraction of the total system cost[11] is fortunate, because these collectors are amenable to cost cutting through mass production.[12]

The tracking, concentrating collector is the only significant item in the system we have designed for which an estimated "off-the-shelf" price has not been used for our cost determination. An off-the-shelf price is impossible to give for collectors at this time because these items

Fig. 11.3

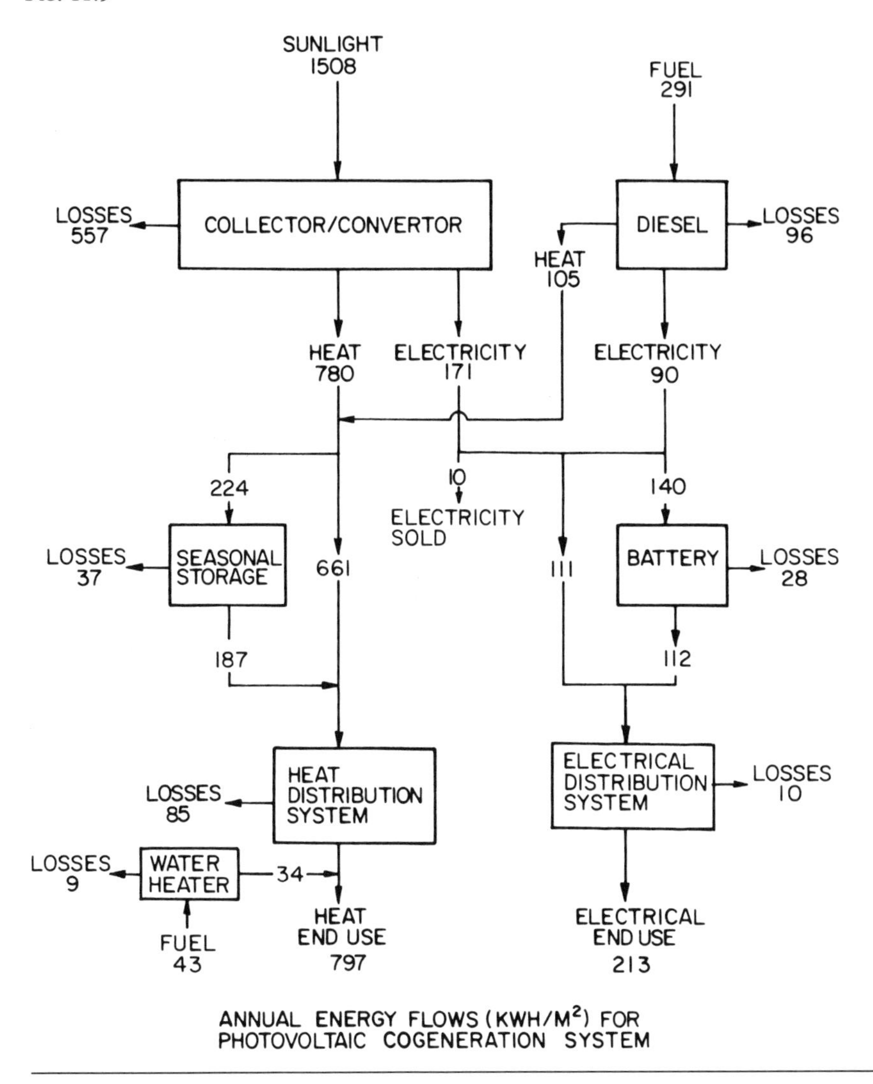

ANNUAL ENERGY FLOWS (KWH/M²) FOR PHOTOVOLTAIC COGENERATION SYSTEM

are custom built today. The collector prices we have used in our analysis,[11] $135 per square meter and $80 per square meter, represent a range of estimates[13] as to what might be achieved in mass production. These prices are comparable to what is often projected for ordinary flat-plate collectors. Is it reasonable to expect that tracking, concentrating collectors could be as cheap as stationary, flat-plate collectors? It is not at all clear that they won't be, because tracking, concentrating collectors do not require the bulky, heat-absorbing material that is needed in flat-plate collectors. Some projections of mass-production prices for tracking, concentrating collectors go as low as about $50 per square meter.[14]

THE FOSSIL-FUEL BACKUP SYSTEM

Providing electricity during long cloudy periods from solar sources is not easy. Without dramatic breakthroughs in battery technology, storing electricity in batteries for long periods would be very costly. One alternative to battery storage would be to recover some heat from the collectors at very high temperatures, store this heat until the cloudy periods, and then use it to generate electricity with some kind of a heat engine. Unfortunately, storing high-temperature heat for long periods is also extremely costly. Still another alternative to the use of batteries would be to rely on the electric utility for backup supplies. But utilities would probably be unwilling to provide such backup with conventional centralized generating capacity, because if there were many such community-scale systems in the grid, their demands for backup capacity would tend to be coincident, presenting a costly peaking problem for the utility from users that would demand few kilowatt-hours throughout the year.

To avoid these problems, we have chosen instead to provide electricity during long cloudy periods with fossil-fuel-fired diesel engines. These units provide about 50% of the required electricity in winter and 15% in summer.[9] Our design would use batteries only for overnight storage of electricity.

Because of this fossil-fuel backup system, solar energy provides only ⅔ of the electricity needs. Despite the critical importance of fossil-fuel backup to the success of this system, annual fossil-fuel consumption per house with this technology would be less than 10% of the average U.S. household's energy consumption in 1975[15] and would account for only about 10% of the annual cost of this system.[10] (It is of course a myth that the use of oil or natural gas must be eliminated from future energy systems. If these fuels are used wisely, oil and gas resources can be stretched into the long-term future.)

ANNUAL HEAT STORAGE

In a photovoltaic cogeneration system low temperature (180°F) heat is recovered in the water that cools the solar cells. The most apparent applications for this heat are space heating and water heating. Unfortunately, however, space heating is done in the winter, while the best sun is available in summer.

One way to take advantage of the summer sun is to store some of it for use in winter—much as ice produced in winter is stored for summer use with the ACES described in Chapter 8. This storage can actually be quite economical if carried out on a large enough scale. In our analysis here we consider a centralized storage system serving 2500 single-family homes.

How much annual storage is needed? So little sunshine is available in winter that only 52% of the winter heat demand is provided by the winter sun with our design.[9] Interseasonal storage accounts for another 36% of winter heat demand. The remaining 12% is provided by waste heat recovered from the diesel engines. As long as the diesel engines are required as a backup electricity source, it is worthwhile to cogenerate with them—that is, to recover the waste heat for heating purposes. With diesel cogeneration the needed seasonal storage capacity is less than it would otherwise be.

The storage capacity required in our proposed system is nevertheless considerable—about 14,000 gallons of hot water per household. If this storage tank were constructed at individual houses, it could cost roughly $4200 installed.[16] But the cost of seasonal storage would be only 25% as large at a centralized facility.[10l] (The cost advantage of centralization would be even greater with wet-soil conditions, because of the large amount of insulation that would then be required.[16])

In this analysis we assume that storage is in a large insulated underground steel tank. After some development, it will be possible to store heat more cheaply at the community level in ponds, lakes, aquifers, or other natural storage reservoirs.

THE THERMAL DISTRIBUTION SYSTEM

The advantages of community-scale storage are partially offset by the costs of distributing to individual houses the heat produced at the centralized facility. District heating can be costly because digging trenches and installing insulated piping underground is expensive. The costs vary significantly with the density of housing. For clustered, multifamily apartment houses or commercial buildings the costs of the thermal distribution system can be reasonable. For low-density single-family dwellings the cost can be high. For our analysis we choose a moderately unfavorable arrangement, a community with 2500 single-family houses per square mile (see Figure 11.2). But even with such low-density housing the economies of community-scale operations are substantial. We estimate[10n] that the thermal distribution system would cost $1100 per house—so that the total cost of storage plus distribution would still cost much less than storage at the household level.

SOLAR AIR CONDITIONING

Even with annual heat storage excess heat is still available from the solar collectors in summer. To make use of this excess heat, we assume that the air-conditioning system is heat- rather than electricity-driven in our residential community. While electric air conditioning is an amenity which causes a serious peaking problem for many utilities (and is therefore a very costly luxury!), heat-driven air conditioning tends to improve the overall economies of solar-energy systems, by putting excess summer heat to work and by making fuller use of the thermal distribution systems through circulation of cold water to houses in summer as well as hot water in winter.

One complication which arises when the thermal distribution system is used to circulate cold water for air conditioning in summer is that the provision of domestic hot water becomes difficult in a system with our design, which has only one delivery pipe and one return pipe. One way to resolve this problem would be to introduce a third pipe into the distribution system to deliver domestic hot water. But so doing would be costly and would lead to significantly greater heat losses in distribution. A better approach would be to use the two-pipe system but to interrupt cooling with a pulse of hot water to each house in the middle of the night. The pipes in our system are sufficiently large that enough hot water could be delivered to meet a day's needs in only 2 hours. While this approach would lead to a significant increase in heat losses during the cooling season, arising because of the repeated heating and cooling of pipes, the overall annual heat losses would be only slightly increased. We have used a third approach in our design: during the relatively short cooling season domestic hot water is provided by burning fuel (oil, say) at the household level. About 20 gallons of oil (or its equivalent) would be needed for this purpose per year. The fuel-burning alternative has the advantage that it can also serve as an emergency backup to the centralized hot-water-supply system.

ECONOMICS

Perhaps the most important lesson that can be learned from the analysis of this system is that the economics of a particular piece of hardware like a solar cell can vary enormously with the type of energy system in which the hardware is deployed. Laying today's solar cells flat out in the sun would result in a cost of produced (not delivered, just produced) electricity in excess of 130¢/kwh. If instead, tracking, concentrating collectors are used to produce electricity, the total cost of *delivered* electricity would be reduced to 12 to 23¢/kwh.[17] If we go further and use these same collectors to recover heat as well as electricity, the cost of electric-equivalent energy is reduced in half to 7 to 11¢/kwh.[10]

This second reduction is significant because it may make the solar photovoltaic system competitive with the alternative of an all-electric house—whether the all-electric house involves resistive heat, a conventional heat pump, or an ACES.

An instructive way to view the economics of alternative supply systems is to compare consumers' total annual energy bills—which include both fuel and capital charges. Such a comparison, along with a comparison of the associated fossil fuel use, is summarized in Figure 11.4 for alternative ways of providing energy for the model energy-conserving house described in Chapter 8 and for a more conventional house of the same size and with the same amenities, built to FHA standards. So as to assure that the costs shown are comparable, the loan payment on the added investment required to make the house "tighter" is included in the cases involving the model energy-conserving house.[18]

The annual energy bill shown in Figure 11.4 is based not on present prices, but rather on national average direct replacement costs for energy (i.e., the replacement costs exclusive of environmental and other social costs that might be reflected in an energy tax). Our estimates are that these direct replacement costs are about 50%[19] and 40 to 80%[20] higher than the 1975 national average prices for gas and electricity respectively. While individual consumer decisions are based on actual prices, societal decisions relating to the commercialization of new technologies should be based on replacement costs.

The annual costs shown in Figure 11.4 are given for alternative financing arrangements for the photovoltaic cogeneration system (as well as for the ACES—see discussion in Chapter 8 under "The Annual Cycle Energy System"). The reason for this is that the financing one might consider first for this system—public utility financing—may be less practical than the alternative of municipal utility financing.

While we argued in Chapter 10 that industrial cogeneration units should be tightly interconnected with the utility grid, our analysis in this chapter shows that in the case of residential cogeneration based on photovoltaic technology, interconnection with the grid provides no strong advantages. The primary reasons for this judgment are that there is a good match between both heat supply and demand *and* electrical supply and demand with this technology and that an onsite backup

Fig. 11.4

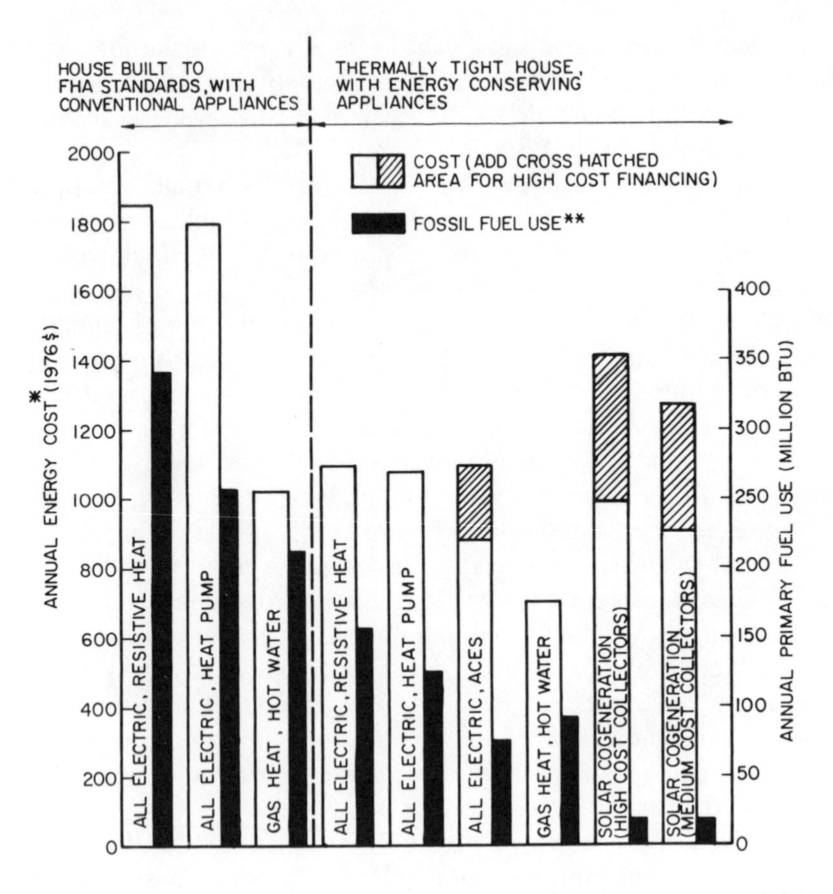

*Includes fuel cost, operation and maintenance cost, and loan payment on capital investment

**The fossil fuel use shown is the fossil fuel use on-site plus the fossil fuel (or its equivalent) consumed to generate all electricity provided by conventional central station power plants.

system can be relatively cheap. Moreover, since the scale of these operations is so small compared with what public utilities are accustomed to managing and since the operation would provide the utility with no excess electricity for other customers, it would probably be more practical to organize these systems on an ad hoc basis as small municipal utilities, financed by the sale of municipal bonds. Financing by tax-free municipal bonds leads to lower costs to consumers. (The high and low cost financing options shown in Figure 11.4 for solar cogeneration refer to public and municipal utility ownership, respectively.)

The comparison of annual energy bills for alternative systems shown in Figure 11.4 reveals several interesting things. First, it appears that the solar cogeneration system would not be competitive with a system based on gas heat.[21] Even under the most favorable conditions the solar cogeneration system would be 30% more costly than the system in which space heat and hot water are provided by gas and other end uses are served by electricity from central-station plants. Economically one simply cannot beat the combination of energy conservation and gas heat at the present replacement cost for gas.* As Figure 11.4 shows, the total monthly energy bill (based on replacement costs) for an energy-conserving gas-heated house would be less than $60 per month. Indeed, the price of gas delivered to this community would have to rise to over $8 per million Btu in 1976 dollars (equivalent to heating oil costing $1.15 per gallon) before the lowest-cost solar cogeneration system shown would be cheaper.

While this solar cogeneration system cannot beat the gas-heat option, Figure 11.4 shows that a solar cogeneration system is roughly competitive with the all-electric option—whether one uses resistive heat,[22] a conventional heat pump,[23] or an ACES.[24] This is a very significant result because of the growing importance of all-electric houses. The number of electrically heated houses nearly doubled between 1970 and 1975,[25] and the major trade publication for the electric utility industry projected in 1978 that the increase in the number of such houses in the period 1975–1995 would be approximately equal to the increase in the total number of households in this period,[26] implying that the percentage of houses heated by electricity would increase from 13% in 1975 to 36% in 1995.

Existing public policies that favor moving away from the use of oil and gas also make the ability to compete with the all-electric option significant. It appears, moreover, that if a relatively modest energy tax were imposed on fossil and nuclear fuels, to reflect social costs of energy production and use not covered by market prices (see Chapter 15), the

*Gas heating has the same economic advantage over any system based on central-station electricity. Nevertheless even today electric heating is being installed in new homes. The way the economic balance has been redressed (in part) is that electrically heated houses have been built thermally tighter than fossil fuel heated houses. In effect the very low cost of saved energy has been utilized to make up in part for the high cost of electric heating.

solar cogeneration option could even have an edge over the all-electric option.

Still another remarkable aspect of the comparisons shown in Figure 11.4 is that the combination of energy conservation at the point of end use (e.g., having thermally tight houses and energy efficient appliances) and solar cogeneration is *substantially more attractive economically* than the all-electric option for a *conventional* house and may cost no more than a system providing gas heat to a conventional house. In this sense fuel conservation makes the solar cogeneration option "affordable" under a wide range of circumstances.

These economic calculations are of course peculiar to a system built in central New Jersey. In this area the heat and electricity provided by the solar cogeneration system would be roughly balanced by the demands for heat and electricity. An attractive feature of the photovoltaic cogeneration system is that such a favorable balance can be achieved in a wide range of circumstances. Further north more seasonal storage but less air conditioning would be needed, whereas to the south the reverse would be true.

Finding space for collectors should not be a serious problem in most areas with this system. Because end-use conservation has been emphasized in these houses, the total collector area required per house is only about 180 square feet, or less than ¼ of the roof area of a typical single-family house.[9] Nevertheless, in some areas, especially already developed areas, there may be practical difficulties in finding enough space for the solar collectors or in establishing the district heating network; in some areas insolation is not adequate; at some sites the topography would be inappropriate. Thus the photovoltaic cogeneration option would certainly not be suitable for all buildings. The most promising applications would be new building developments, both residential and commercial.

Our economic analysis shows that photovoltaic cogeneration could help create initial markets for both solar cells and tracking, concentrating collectors, so as to help bring down prices for these items through large-volume production. But it is unclear whether or not photovoltaic cogeneration will become "the wave of the future." Two considerations suggest that *several decades from now* solar cells for buildings might be used primarily in simple stationary arrays. First, if solar cells do become very cheap, then tracking, concentrating collectors may not be needed to bring down costs. Second, technological innovations aimed at making houses thermally tighter may eliminate most of the need for energy for space heating. With very tight houses it might no longer be economical to have district heating. The fixed cost of the distribution system would decrease only slightly as the houses got tighter and tighter. Thus the cost of delivering each unit of energy would become very large. Also the heat losses from the piping would become comparable to the heat losses from the houses. For an energy-conserving house such as the target house proposed by the American Physical Society (see Chapter 8 under "Space Heating for New Housing"), a small furnace burning an easily storable liquid fuel would be a heating-supply option hard to beat. But until solar cells become truly cheap and until this or a similar energy-conserving design becomes a conventional housing design, photovoltaic cogeneration for housing could be a major application of photovoltaic technology.

PUBLIC POLICIES

We have shown that the fuel-conservation strategy of cogenerating heat and electricity could make solar photovoltaic technology competitive with the alternative of an all-electric energy system for buildings and that the combination of end-use fuel conservation in buildings with solar cogeneration could be no more costly than providing energy from conventional sources to conventional buildings.

Because the economics of solar cogeneration appear to be promising, the major hurdles to the use of this technology are institutional, rather than technological.

We do not mean to imply that there are no technological problems. Perhaps the single most important developmental task is to mass-produce low-cost tracking, concentrating collectors. Beyond this it is desirable to seek innovations that would reduce component costs below the levels we have estimated. Among the more promising avenues for innovation would be the quest for more efficient (rather than cheaper) solar cells and the development of "natural" thermal-storage systems (aquifers, ponds, lakes, etc.).

Among the major institutional obstacles to solar cogeneration are all of the institutional obstacles to fuel conservation. Our estimate that the economics of solar cogeneration are favorable is based on the use of replacement costs (which are much higher than present prices) for conventional energy supplies. Thus, as in the case of fuel conservation, policies are needed that would facilitate economic decision making on the basis of replacement costs. An energy tax shift imposed on conventional energy sources would also give a strong boost to this technology.

Moreover, policies that facilitate lifecycle costing, so important for fuel conservation (see Chapter 8 under "The First-Cost Problem"), are doubly important for solar cogeneration. Because roughly 80% of the cost of energy from a solar cogeneration system is associated with capital costs, policies that promote the availability of capital for applications such as solar cogeneration would be critical.

New institutional arrangements are also needed for managing and creating solar cogeneration systems. We have already suggested that the nature and scale of these systems may be more amenable to municipal than public utility ownership and operation. Since there has been a long-term trend away from municipal utilities toward centralized public utilities, it may be desirable to identify policies that would help stem this trend.

Who would create a solar cogeneration system? At present we probably do not have a suitable industry for building these systems. One potential industry would be the architectural engineering firms which build power plants, refineries, and other large installations. But such firms may be too big for the job. Their projects typically involve billion-dollar investments. The total investment required for our solar cogeneration system for a community of 2500 houses is only $20 million. Such firms may also be an inappropriate match for the tasks involved. We have shown that if solar cogeneration is to be successful, the energy supply technology must be carefully integrated with the technologies for using energy in houses. Perhaps it would be better instead to foster the development of a whole new industry of "mini" architectural engineering firms which could work closely with or as part of the housing industry. The development of such an industry could be nurtured by government procurement actions, such as the purchase by government of a solar cogeneration system for a housing development at a new government installation.

While it is beyond the scope of this analysis to set forth a blueprint for action to promote solar cogeneration, we feel that the primary thrust of public policy should be to make market mechanisms work better and to nurture the development of this technology in the private sector. A government Manhattan Project for Solar Cogeneration is not needed.

CHAPTER 12

IS FUEL CONSERVATION HAZARD FREE?

IDEALLY WE WOULD LIKE to enjoy the benefits of energy without having to suffer the hazards, such as those described for energy-supply technologies in Chapters 5 and 6. Unfortunately, however, most technologies, including fuel-conservation technologies, are not without safety and environmental problems. In this chapter we briefly discuss some representative problems associated with some of the technologies discussed in Chapters 8 to 11.

BUILDINGS

Perhaps the most serious problems arising from improving the energy performance of buildings (including heating systems and appliances) are associated with the reduction of air infiltration. Heating the air that leaks into buildings accounts for a major fraction of a building's fuel consumption for space heating in winter; similarly infiltrating warm air accounts for much of the need for air conditioning in summer. Thus the reduction of air infiltration is a major element in conservation. Several problems can arise as a result:

- Low levels of air infiltration can give rise to odor problems.

• In houses with gas-fired appliances such as gas stoves there can be increased levels of combustion-generated indoor air pollution (such as carbon monoxide and nitrogen dioxide). The levels of such pollutants near gas appliances in many houses are already above the levels set by EPA standards for out-of-doors.[1]

• There can be increased levels of pollution associated with, say, plastics, certain insulating materials, or the materials used in constructing buildings. Among such emanations, formaldehyde is already present at disturbing levels in some types of housing. Exposure to formaldehyde may cause burning of the eyes and irritation of the upper respiratory passages. High concentrations may produce coughing and a sense of pressure in the head. Formaldehyde may also pose serious long-term health hazards that are not yet well understood.[1]

• Radon-222 is a radioactive gas (produced when radium-226 undergoes radioactive decay) present in some houses at significant levels. This gas emanates from building materials such as concrete or brick, from the soil under building foundations, and possibly from tap water taken from wells or underground springs. At low ventilation rates in houses lifetime exposure to radon could result in increasing the incidence of lung cancer by an amount equal to the observed rate for male non-smokers.[1]

• Increased moisture problems, e.g., mildew and fungus, can be present where or when there is excess humidity.

At a high ventilation rate these various pollutants would be "flushed out," but if the air infiltration rate were reduced to or below the level advocated in Chapter 8, these problems would be serious in some houses. There are, of course, specific corrective actions that can be taken for some of these problems, e.g., exhaust fans for the kitchen or bathroom. However, a more interesting general approach to these problems is to restore the flow of fresh air but to control its introduction into the building by using a device which allows for the exchange of heat (and perhaps also humidity) between the flows of outgoing and incoming air (see Figure 12.1). Inexpensive heat exchangers of a wide variety of designs are now becoming available for use in controlled ventilation systems for houses.[2] With a heat exchanger, the load on the space-conditioning system is reduced by a factor of 2 to 4 for the same rate of introduction of fresh air. Thus the presently typical fresh-air infiltration rate of one air exchange per hour (i.e., introduction in 1 hour of a volume of air equal to the house's volume) might be maintained with ¼ of the associated thermal loss, exceeding the conservation goal set in Chapter 8.

FIG. 12.1

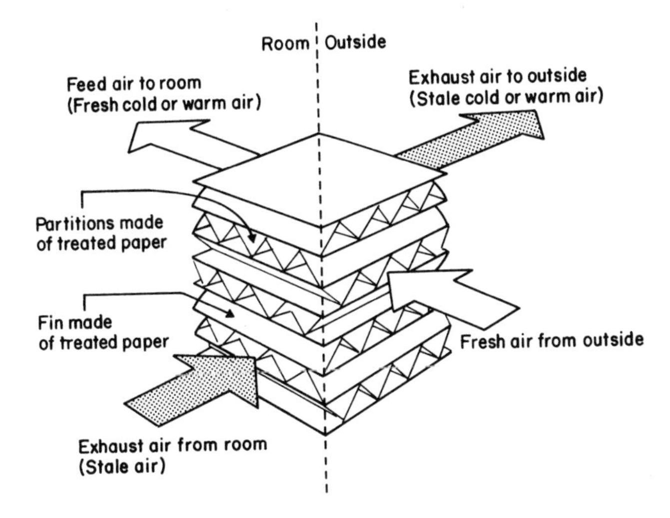

The above device is a heat exchanger which transfers heat between the stale air being exhausted from a building and the fresh air being introduced. The device saves heat in winter and "coolth" in summer. The device is constructed of plates and fins made of treated paper. The fresh air and exhaust air passages are totally separated, allowing the fresh air to be preconditioned to the temperature and humidity levels of the room air without mixing with the exhaust air.

The principle of operation is based on the heat transfer properties and moisture permeability of treated paper. Total heat (sensible heat plus latent heat) is transferred to the fresh air being introduced into the system via the medium of treated paper.

SOURCE: "Mitsubishi Enthalpy Exchanger-Lossnay," a pamphlet prepared by Mitsubishi Electric Corporation.

AUTOMOBILES

Most automotive redesigns aimed at fuel conservation have no apparent safety problems associated with them. The crashworthiness of small cars, however, is one major safety concern which has been identified. Broadly speaking, small cars are as easily designed as large cars for safety in collisions with barriers. Of less but still great importance are head-on or side collisions with heavier vehicles; in these the small car comes off second best. For this reason it is not very surprising that accident surveys show that fatal accidents occur more frequently in small cars. On a per car per year basis, almost twice as many accident fatalities involve occupants of small (subcompact) cars as occupants of large cars.[3] It is probably not at all accurate to extrapolate this experience to indicate what fatality rates will become in case of greatly increased use of small cars, but the problem is clearly troublesome.

Two developments appear promising for mitigating this problem: (1) Smaller vehicles will naturally encounter fewer large vehicles on the road as the relative number of small vehicles increases. Accentuating this trend by separating small vehicles from large vehicles on some roads is a possibility which deserves consideration. (2) Smaller vehicles can be designed with more emphasis on safety. This was the focus of the design of the modified Rabbit Diesel discussed in Chapter 9.

COGENERATION SYSTEMS

The cogeneration systems considered in Chapters 10 and 11 each have as a component a fuel-burning engine, e.g., a gas turbine or a diesel, located, respectively, at an industrial plant and in a residential area. The engine is used in full or in part to generate electricity. In each case the system requires much less fuel to generate a kilowatt-hour of electricity than does central-station electricity generation; so, other things being equal, total emissions of pollutants per kilowatt-hour are less. In actuality, such a simple comparison cannot be made because the fuels and technologies used in the cogeneration systems differ from those used in central-station generation. Moreover, the distribution of emissions would be very different. The sources of emissions would be located near points of use with cogeneration, whereas central power stations tend to be sited remotely from demand centers, often in low-population areas.

No convincing method of analysis has been developed for the assessment of the relative pollution impacts of distributed versus centralized sources. In our opinion, the proper response to this problem is that new fuel-burning devices located in populated areas should meet rigorous emission standards. If coal or other "dirty" fuel is to be used, this poses a serious challenge, but it may well be possible to meet this challenge through development of technologies like fluidized bed combustion.[4] (With this technology sulfur is removed from sulfur-bearing fuel in a relatively simple and effective way during combustion.) If natural gas or high-quality, clean fuel oil were available for these applications, the challenge to technology would be relatively easy to meet. And indeed in an energy-conserving economy these clean-fuel options may well be available for the indefinite future. As we pointed out in Chapter

11, a relatively small amount of liquid backup fuel can go a long way in making viable a community-scale solar cogeneration system. And in Chapter 13 we describe a long-term strategy for using natural gas in energy-conserving industrial applications.

CONCLUSION

These examples show that fuel-conservation technologies, like energy-supply technologies, do indeed pose hazards that cannot be ignored. However, there are crucial differences between fuel-supply and fuel-conservation technologies.

One important difference involves the nature of the risks. As we showed in Chapter 6, a high level of implementation of some solar energy-supply systems would imply severe constraints on land use with broad ecological and style-of-life implications. Even more severe are the irreducible problems of fossil fuel and nuclear fission described in Chapter 5. Irreducible risks like the potential for climatic change associated with the atmospheric build-up of carbon dioxide and the potential for nuclear weapons proliferation inherent in a commitment to nuclear power seriously threaten the stability of society at large. These problems are not at all comparable to pollution from fuel-conservation technologies, which at worst would strongly affect a small part of the population or would somewhat degrade the environment of many.

Another important difference is that problems of standard setting and of enforcement tend to be rather different for energy-supply and energy-conservation technologies. As we pointed out in Chapter 5, it is proving to be very difficult to impose new environmental regulations on the energy-supply industry and to enforce new regulations that have been adopted, owing both to the powerful efforts by the industry to resist regulations and to the complexity of these regulations. In contrast, the setting and enforcement of environmental standards for fuel-conservation technologies to be used in tens of millions of diverse dwellings should be a much easier process. Regulations on the sale of certain products would be required (e.g., regulations on insulation to ensure that chemically unstable materials are kept off the market), as would regulations at the point of installation.

Safety regulations for consumer products, which would usually involve a simple ban on sales of products that could be especially hazardous, would be much simpler to administer than present complex regulations that attempt to cure environmental ills by applying "bandages" to "sores" created by large established technologies. Since there is such a rich diversity of technological options for meeting particular conservation demands, such bans on the more hazardous products should not be a significant constraint on technological innovation.

The viability of establishing and enforcing environmental standards at the point of the installation of conservation technologies is suggested by the largely successful historical experience with the system of local authorities established throughout the nation to impose and enforce fire-hazard and safety standards in buildings.

In summary, while conservation technologies are not hazard free, the risks to society even in the worst cases are far less than would be the case with many energy-supply options, and the regulation of these hazards should be a much easier and more effective process than the regulation of energy-supply technologies.

CHAPTER 13

NEGATIVE ENERGY GROWTH

Instead of posing as prophets
We must become the makers of our fates.

KARL POPPER,
The Open Society and Its Enemies.

A FUEL-CONSERVATION SCENARIO

In Chapter 3 we derived an energy projection to the year 2010 which we call a Business-as-Usual (BAU) scenario. While energy demand in that scenario would grow only about ⅓ as fast as projected by many official and industrial groups, we are persuaded that a level of energy use higher than the BAU level in 2010 is extremely unlikely. Nevertheless, we showed that providing the energy to meet the BAU demand level in 2010 (about 45% higher than the 1975 level) would require truly heroic efforts and would involve serious public health, environmental, and global-security risks.

On the basis of the analysis of the last several chapters we are now prepared to present an alternative energy scenario that would reflect a substantial commitment to fuel conservation. As in the case of the Business-as-Usual projection, the Fuel-Conservation scenario should not be regarded as a prediction. As we stated earlier, the level of energy use several decades in the future is a matter of realistic societal choice rather than a matter of prediction.

In developing a Fuel-Conservation scenario, we shall assume the same demographic trends, the same GNP growth, and the same trend in the mix of economic output between goods and services as in the BAU scenario. This means that in evaluating growth rates for energy, as expressed below and in equation 3.4

$$\begin{pmatrix}\text{annual energy}\\ \text{growth rate}\end{pmatrix} = \begin{pmatrix}\text{annual growth rate}\\ \text{in functional demands}\end{pmatrix} - \begin{pmatrix}\text{annual rate of decline}\\ \text{of energy intensity}\end{pmatrix} \quad (13.1)$$

we shall assume the growth rates for various functional demands are the same as in the BAU scenario. But while the rates of decline in energy intensity in the BAU scenario were based on historical data on the response of energy demand to changing prices, these rates in the Fuel-Conservation scenario will instead be based on the possibilities for fuel conservation. Energy efficiency improvements estimated on the basis of historical responses to expected changes in energy prices represent much less than the economic potential for fuel conservation. Numerous institutional obstacles prevent evaluation of alternative energy-using technologies in terms of lifecycle costs and the full replacement cost of energy supplies. In the Fuel-Conservation scenario we assume that policies would be enacted to overcome these obstacles. Thus while the scenario we describe here is both technically and economically feasible, it would happen only if energy policies such as those we have set forth in this book were adopted.*

RESIDENTIAL

We have given considerable attention to fuel conservation in housing in this book, partly because we have found that the potential for fuel savings is substantially greater than most people realize. Specifically we have shown that a reduction to ⅓ of the present level of fuel use for heating would be feasible and economically justified for existing housing, with retrofits, and that new housing units with considerably greater amenities (both more floor space and more appliances) than the average housing unit today could be provided with energy services using not more than ⅓ the fuel used today. If effective conservation policies were adopted so as to reduce by ⅔ the average energy intensity of housing in 2010 (corresponding to an average annual rate of decline in energy intensity of about 3% per year), total energy requirements for the substantially increased housing stock of 2010 would be only 60% of what energy requirements for housing are today (see Table 13.1).[2]

* A somewhat similar scenario labeled "aggressive energy conservation policy aimed at maximum efficiency plus minor lifestyle changes" was developed recently by a panel investigating energy demand in a National Academy of Sciences study.[1]

Table 13.1 A Fuel-Conservation Scenario for Energy Consumption in 2010

	1975 ENERGY DEMAND (QUADRILLION BTU)	DEMAND GROWTH: PARAMETER	DEMAND GROWTH: AVERAGE ANNUAL GROWTH RATE (%)[a]	AVERAGE ANNUAL RATE OF DECLINE OF ENERGY INTENSITY (%)[b]	AVERAGE ANNUAL GROWTH RATE OF ENERGY DEMAND (%) (1975–2010)	2010 ENERGY DEMAND (QUADRILLION BTU)
Residential Buildings	16.2	Households, Household Income Effect	1.3, 0.4	3.2	−1.5	9.6
Commercial Buildings	9.9	Service-Sector Employment	2.5	3.2	−0.7	7.8
Transportation						
Autos and Light Trucks	10.2	Adult Population	0.9	3.5	−2.6	4.2
Other	8.2	GNP	2.7	1.8	+0.9	11.1
Industry	26.1	Goods Production	2.5	2.0	+0.5	31.3
TOTALS	70.6				−0.3	64

NOTES:

[a] These are the same demand growth rates as in the Business-as-Usual scenario.

[b] See text.

COMMERCIAL

While we have not examined energy use in commercial buildings in detail in this book, the energy waste so characteristic of residential buildings is apparent in commercial buildings as well. One would expect, for example, that fuel consumption for space heating would be less per unit of floor space for commercial buildings than for residential buildings because there is less "building skin" per unit of floor area. But in fact commercial buildings in 1975 consumed 40% more space-heating energy per unit of floor area than did residential buildings![3] In the last several decades energy has represented such a small fraction of the total cost of operating buildings that commercial-building design has been mindlessly oblivious to fuel-efficiency considerations. For example, between 1950 and 1970 energy use per square foot of typical New York City office buildings roughly doubled as emphasis was given to more lighting, more glass, and more air conditioning in the era of low-cost energy (see Table 13.2).

Much can be done to reverse the trend toward energy-intensive commercial buildings, however. On the basis of the energy saving possibilities demonstrated in an experimental General Services Administration office building in Manchester, New Hampshire, the federal government issued in 1975 target design guidelines for new federal office buildings that energy use per square foot should not exceed a level of about ⅓ of that for typical existing office buildings (Table 13.2).[4] Moreover, work carried out in conjunction with the Department of Energy's Building Energy Performance Standards (BEPS) program showed that if office buildings of recent vintage had been built to reflect conservation objectives, energy use per square foot would have been reduced to about ⅓ the level of energy use in typical existing buildings (Table 13.2). This redesign effort did not represent the economic optimum: in this exercise the cost of saved energy was estimated to be about 25¢ per gallon of oil equivalent energy. A far greater level of investment in conservation is economically justified, since the average price of energy for commercial buildings was 40¢ per gallon of oil equivalent in 1978, while the replacement cost was then perhaps 65¢ per gallon of oil equivalent.[5]

Much can also be achieved with existing buildings. For example, at Ohio State University energy use in one campus building was reduced by ⅔ at a saved-energy cost of only 3¢ per gallon (Table 13.2).[6]

Because of such possibilities we shall assume that a ⅔ fuel reduction per unit of building space by 2010 holds for commercial as well as residential buildings.[7] The net effect of this efficiency improvement and

Table 13.2 Total Annual Energy Use By Commercial Buildings (Thousand Btu/ft²)

	FUEL CONSUMPTION[a]
NYC Office Buildings, by Vintage[b]	
1950–54	264
1955–59	320
1960–64	391
1965–69	493
U.S. Average (1975)[c]	
All Commercial Buildings	325
Office and Public Buildings	344
Target for New GSA Buildings[d]	100
Redesign of Recent Vintage (1975–76)	
Large Officc Buildings[e]	
Before Redesign	190
After Redesign	110
McCampbell Hall, Ohio State University[f]	
Before Retrofit	481
After Retrofit	147

NOTES:

[a] Here electricity consumption includes generation, transmission, and distribution losses, by associating 11,300 Btu of fuel with each kwh of delivered electricity.

[b] Charles W. Lawrence, "Energy Conservation: Its Implications for Building Design and Operation." *Proceedings of the Conference on Energy Conservation: Implications for Building Design and Operation*, Bloomington, Minnesota, May 23, 1973.

[c] J.R. Jackson and W.S. Johnson, "Commercial Energy Use: A Disaggregation by Fuel, Building Type and End Use," Oak Ridge National Laboratory Report ORNL/CON-14, February 1978.

[d] F. Dubin, "Energy Conservation Studies," *Energy and Buildings*, vol. 1, No. 1, p. 31, May 1977.

[e] See note 5 and "Economic Analysis of Proposed Building Energy Performance Standards," prepared for the Division of Building and Community Systems, Assistant Secretary for Conservation and Solar Application, U.S. Department of Energy, by Pacific Northwest Laboratory, Lawrence Berkeley Laboratory, Oak Ridge National Laboratory, September 1979.

[f] R.H. Fuller and D. Sullivan, "Energy Efficiency at the Ohio State University," *Energy and Buildings*, vol. 1, No. 4, p. 393, June 1978.

our projection for growth in commercial-building space is that commercial energy use would be reduced to 80% of the 1975 level by 2010 (see Table 13.1).

AUTOS AND LIGHT TRUCKS

A 1977 TRW report (see Table 9.3) claims that with the appropriate R&D it would be possible to increase automotive fuel economy to over 45 mpg without reducing either interior car space *or* auto weight below the 1975 average of about 2 tons. Beyond such purely technical fixes, fuel economy could be doubled again by cutting auto weight in half.

We shall assume here that fuel-economy standards are tightened and that an energy tax shift is enacted so that the average fuel economy in 2010 is 45 mpg. This would be achieved with some mix of technical fixes and downsizing. Fuel consumption by 2010 would be reduced dramatically in this sector—to 40% of the 1975 level (Table 13.1).

OTHER TRANSPORTATION

We have not done an analysis of the fuel-savings possibilities for "other transportation," which consists mainly of freight transport and air travel. Therefore, instead of making an independent projection in this area, we adopt a projection made by a recent National Academy of Sciences (NAS) panel, adjusted to reflect the fact that we assume a higher GNP growth rate.[8] The NAS panel's projection is described as involving an "aggressive energy conservation policy aimed at maximum efficiency plus minor lifestyle changes." As shown in Table 13.1, energy demand would grow faster in this sector than in any other in the period to the year 2010.

INDUSTRY

We have emphasized only one energy activity in the industrial sector, industrial cogeneration. One reason we did not emphasize others is that energy use is much more diversified and the details are less well understood in industry than in other sectors. But also, for most industrial activities we believe that the details of improving energy efficiency can be left to industry, which has demonstrated that it is quite responsive to energy prices. Public policies are needed to bring energy prices in line with total replacement costs (including social costs not normally included in corporate accounts) via an energy tax shift. This tax shift would give industry a strong incentive to make investments in energy-conserving technologies. It would do so not by specifying that particular technologies be used but rather by creating a new economic climate in which energy would be more costly in relation to other factors of production. In this new economic climate industry should be free to make its own technological choices, as long as such choices are consistent with general social constraints to protect workers and the environment.

The efforts to conserve fuel in industry in recent years have been impressive. Between 1973 and 1978, the energy required to produce a dollar of industrial output declined at a rate of about 1.6% per year.[9]

Much of the improvement in the energy efficiency of the overall economy which has taken place since energy prices began rising has come from improvements in the industrial sector. In this period there was practically zero growth in industrial energy use because the improvement in energy efficiency approximately offset the increase in output. This effect was particularly dramatic between 1976 and 1978, when the energy required to produce a dollar of industrial output declined 7%.

Several complementary considerations suggest that the recent downward trend in the energy intensity of industry will continue or even accelerate slightly. A continuing decline in energy intensity can be anticipated both because of expected shifts in the mix of industrial output and because of expected efficiency improvements.

As we pointed out in Chapter 2, the trend in manufacturing in recent years has been toward greater emphasis on advanced processing and fabrication and less emphasis on processing basic materials. We expect this trend to continue. This means we should expect more product types which are not material intensive, such as electronic devices; weight reduction and materials substitution, for example in automobiles; increased product durability, as proposed, for example, by Porsche;[10] design changes to enhance repair and remanufacture, so that, for example, a worn-out appliance could be returned to the factory to be remanufactured and sold again as a "like new" appliance to another customer;[11] and design changes to enhance recycling, for example, by minimizing the content of deleterious impurities that would otherwise be incorporated in materials produced from scrap.[12] This trend has far-reaching implications for energy use[13] because the energy required per dollar of output in the six major basic-materials-processing industries (food, paper, chemicals, petroleum, primary metals and stone, clay and glass) is 7½ times as great as for the rest of manufacturing.[14]

The opportunities for energy-efficiency improvements in manufacturing are very rich. In the short run much can be accomplished with better "energy management" and relatively little investment—through use of sophisticated automatic control systems; retrofitting of energy-conversion equipment, such as placing air-fuel mixture controls on burners; operational changes, e.g., better scheduling so that energy-consuming equipment is turned off when it is not needed; and better maintenance. The literature contains hundreds of significant suggestions,[15] and companies are commonly achieving cost-effective reductions in fuel use of 10 to 20% through improved energy management. Minor operation and maintenance changes known as "good housekeeping"—including such measures as formation of energy-management committees, unannounced inspection by management, and repair of windows and leaking steam-distribution systems—have accounted for much of the very recent decline in industrial energy use per unit of production cited above.

Another type of change is improved energy-conversion technology. Industrial cogeneration, discussed in Chapter 10, is an example. Other examples of major targets of opportunity include waste heat recovery,[16] retention of infrared radiation,[17] and various possibilities for improving the efficiencies of mechanical drive systems. In British case studies it has been found that typically less than half the electricity used in electromechanical drive systems is transmitted as useful work to the material being processed.[18] One possibility for improvement is to power machine tools via hydraulic instead of electromechanical transmission systems, which can result in a 20% saving.[19] Another possibility involves modification of electric motors. Exxon has developed an "alternating current synthesizer" which could reduce electricity requirements for industrial motors by up to 50%.[20]

Perhaps the most important opportunities for improvements in energy efficiency in the long run involve fundamental changes in processes. There is strong evidence that major technical improvements—such as the recently introduced float glass process, the Alcoa chloride process for aluminum manufacture,[14] the dry process for making cement,[21] and various opportunities for fundamental process improvement in steel manufacture[22] result in savings in many factors of production, including energy and materials.[23] Charles Berg has suggested many new avenues for technological developments that could lead to significant reductions in both energy requirements and total production costs.[24]

The wealth of technological opportunities for energy efficiency improvements in industry provides the basis for our judgment that the energy intensity of industry could decline at an average annual rate of 2%/year (slightly faster than the average for the period 1973-78) through the year 2010—if the appropriate public policies were adopted. The major challenge to public policy would be to make capital available. Unfortunately, capital has been in short supply in most energy-intensive industries in recent years. At the same time much of the available capital has been absorbed by the energy-supply industries—especially by the oil companies and the electric utilities (see Chapter 4). Freeing up some of the capital that goes into supplying energy would go a long way toward overcoming the capital-scarcity problem in industry. Since new investments in expanding our energy supply amount to some $50,000 or more per barrel per day of oil equivalent, while much indus-

Table 13.3 Industrial Energy Use-Output Relations (1972)

	GDP ORIGINATING IN INDUSTRY AS A PERCENTAGE OF GDP	INDUSTRIAL ENERGY USE PER $ OF INDUSTRIAL GDP (U.S. = 1.00)
U.S.	31.1	1.00
Canada	29.0	1.24
Six Western European Countries	41.8	0.57
France	48.1	0.40
W. Germany	50.7	0.52
Italy	33.3	0.69
Netherlands	35.6	0.70
U.K.	34.8	0.80
Sweden	25.7	0.84
Japan	40.9	0.65

SOURCE: See J. Darmstadter et al., note 26.

trial conservation requires a net investment of $5000 or less per barrel per day of saved energy, a shift away from energy-supply expansion could probably free up much capital for general industrial modernization as well as for more narrowly defined conservation projects. A general policy aimed at reducing overall energy demand and thereby the need for energy-supply expansion so as to foster this capital-investment shift is outlined in Chapters 14 and 15.

If the energy intensity of industry declined, as we assume it would in this scenario, at an average rate of 2% per year till the year 2010, the energy required to produce a real dollar of industrial output at that time would be only half of what it was in 1975. This is roughly the level of improvement suggested by recent econometric analysis as a long-term response to a two- to threefold energy price increase.[25] With such an improvement U.S. industry would be using slightly less energy per dollar of output than was used in Western Europe in 1972 (see Table 13.3). Since we project that industrial output will increase about 140% by 2010, the net effect would be that industrial energy use would increase about 20% (1975-2010) (see Table 13.1).

SUPPLY IMPLICATIONS

The net effect of increased demands and efficiency improvements for our Fuel-Conservation scenario is that fuel use in the U.S. would be about 10% less in 2010 than in 1975. This comparison points to a major benefit of the Fuel-Conservation scenario: it would make the task of providing supplies much easier. Indeed reliance on conventional fossil and nuclear fuels could be cut back substantially in this scenario. In Table 13.4 we show what the energy supplies would have to be in 2010 for this scenario *if nuclear power were gradually phased out and if fossil-fuel use were cut back to ½ its level in 1975.*

Let us suppose that essentially all the rest of the energy supply were provided by solar energy. How much would be needed in the various forms? Analyzing in detail what the solar-energy forms might be is beyond the scope of this book. However, we expect that two types of solar-energy resources will certainly be developed: organic wastes and hydroelectric power. The major sources of new hydroelectric capacity would come from expanding existing facilities and from building new small-scale facilities (at which the amount of power produced would typically be less than 1% of the power produced at a modern central station electric power plant). Organic wastes and hydropower would be adequate to provide nearly ¼ of total energy needs.

With a successful development effort much of the remaining ⅕ of

Table 13.4 Energy Supplied for the Fuel-Conservation Scenario in 2010 (Quads)[a]

Nuclear[b]	2.9
Fossil Fuel[c]	32.8
Organic Wastes[d]	9.5
Hydro[e]	5.5
Other Solar[f]	13.3
	64

NOTES:

[a] The numbers refer to present fossil-fuel-equivalent inputs to obtain the energy carriers created from these various sources.

[b] If no more nuclear plants were built beyond those now operating or under construction, then about 1/2 of the 1985 nuclear capacity would be retired by 2010, leaving 50 Mwe of nuclear capacity, which is approximately the same as the 1978 level of installed nuclear generating capacity.

[c] The number shown is 1/2 the level of fossil-fuel use in the U.S. in 1975.

[d] The estimates of fuels from organic wastes are based on Poole and Williams (ref. 27), except that the estimates for crop residues and manure given for 1975 by Poole and Williams are increased according to the expected growth in the population and the estimates for pulp waste are increased in proportion to the level of goods production. See note 28.

[e] The hydroelectricity produced in 1975 was (in equivalent fossil-fuel terms) 3.2 quads. We assume this is supplemented by new hydropower capacity (mainly small-scale hydro) amounting to 50 Gw (see ref. 29). At 50% capacity factor and an effective heat rate of 10^4 Btu/kwh this corresponds to 2.3 quads of fuel-equivalent energy.

[f] This is the residual, measured as fossil-fuel-equivalent fuel displaced.

the energy supply, equivalent to about 80% of the present level of imports, could be provided by a wide variety of other solar options—such as solar-energy systems for buildings, photovoltaic converters, wind, solar process heat, and biomass plantations. The solar-energy sources would be distributed so as to exploit the advantages of particular geographic regions. To provide this much solar energy by 2010 would be an ambitious undertaking; to provide even more solar energy in this time frame, in our view, would be unrealistic, because of the intense requirements for capital that would be involved.

In Chapter 3 we pointed out that in the long run at least 40% of the total energy use should be provided by fluid fuels (liquids and gases—e.g., petroleum, natural gas, or synthetic fluid fuels derived from solid chemical fuels). This constraint could be met even with the fossil-fuel cutbacks we hypothesize here, since the total quantity of chemical fuels (fossil fuel plus organic wastes) would amount to ⅔ of the total energy supply. The end uses of chemical fuels that appear most reasonable on the basis of the analysis presented in this book are:

transportation	—	liquid fuels as at present
space heating	—	sharply reduced oil and gas use; but storable liquid fuel for backup on solarized buildings
industrial heat	—	natural gas
industrial feedstock	—	coal, biomass

This list shows that some of the conventional wisdom of today's energy-supply orientation would be invalid. Very little oil and gas would be used in buildings because of the substantial opportunities for conservation and "solarization." But natural gas could well become the preferred fuel for industrial heating applications, because of the inherent difficulties involved in providing industrial heat with solar energy, because of the environmental advantages of natural gas relative to other fossil fuels,* and because natural gas will probably be much more abundant, and thus less costly, than oil (see note 5, Chapter 4). In a Fuel-Conservation strategy the present policy to try to force industry to stop using gas would be converted into a policy to encourage industry, through the mechanism of price, to use gas more efficiently. Indeed it is very likely that a Fuel-Conservation strategy such as that outlined here would stretch out supplies of natural gas so that it could play a major role in the economy for a long time to come.

*Not only does the combustion of natural gas generate less of the conventional air pollutants but also burning natural gas produces less CO_2 than the burning of other fossil fuels—half as much as coal, per Btu of released energy.

While total energy demand would decline in this scenario, electricity demand might continue to grow. In a maximum-electrification variation of the Fuel-Conservation scenario, electricity demand would increase about 35% above the 1978 level by 2010, at which time about 50% of all energy would be used to generate electricity. The electric-power generating capacity which would need to be added for this would be about 150 million kilowatts, which is roughly the amount of new capacity scheduled to be brought into operation in the period 1979–1984.

In 1979 nearly all of this capacity was already under construction, so that much of this new generating capacity will almost certainly be built. Thus in this maximum-electrification variation of a Fuel-Conservation scenario *no more electric generating capacity than what is already under construction would be needed.* The industrial cogeneration and solar electric generating capacity that would be built therefore would be replacement rather than expansion capacity.

SOCIOECONOMIC IMPLICATIONS

The Fuel-Conservation scenario represents a sharp departure from our present energy course, and it is important to discuss the socioeconomic impacts of such a radical change.

CONSUMER EXPECTATIONS

In many people's minds fuel conservation is linked with 55-mpg speed limits on 70-mph freeways, with turned-down thermosats in winter, and with gasoline rationing. To characterize fuel conservation as a strategy entailing consumer sacrifices is wrong, although such talk has the gloss of truth because any sudden reduction in energy supply would require the curtailment of some activities. The fuel conservation we have discussed in this book is conservation of a different sort: the use of more efficient technology. Conservation technology provides Americans the opportunity to realize their economic goals in the face of more costly energy.

After several decades of a fuel conservation program similar to that proposed here, the typical individual would be well fed, live in a large dwelling area with a comfortable controlled environment, own a car, and have substantial leisure time. The differences from the present expectations of consumers are insignificant: frivolous uses of natural resources, as in packaging, would be much reduced, and more effort to recycle certain materials would be justified as a market response to high energy prices; the average car would be smaller than today's; air travel would be relatively more expensive. Energy use per capita would be at the level of the early 1950s, and the technology would be much more efficient.

THE ROLE OF TECHNOLOGY

Fuel conservation is usually regarded as a short-term energy strategy. A recent utility ad reads, "Save till we're able to build more power plants." But to talk of fuel conservation as only a stopgap endeavor is wrong. We have shown that many important conservation measures involve long-term investments, and that the advantages of conservation will be reinforced by continuing energy price increases. And although much can be accomplished with on-the-shelf technology, the long-term prospects for conservation would be greatly enhanced through technological innovation.

This country needs a wide range of products—better windows, indoor shutters, heat exchangers for household ventilation systems, better gas and electric heat pumps, economical thermal-storage systems, small-scale fluidized-bed combustors, superchargers for automotive engines, new basic-materials processes in industry, and generally improved designs that would lead to higher energy efficiency. The innovations needed for a Fuel-Conservation strategy are very different from the kinds of new technology one usually associates with government R&D—a space station, a 100,000-barrel-per-day solvent refined coal plant, a fast-breeder reactor, or a subway between New York and Los Angeles with a travel time of 30 minutes.[30] Such projects involve such large investments and high risks that the private sector would not be interested without major government subsidies. In contrast, because of the relatively small scale and great diversity of conservation technology, the innovation needed for a conservation strategy could be carried out in the private sector. The major R&D challenge to government, therefore, is to create a climate conducive to innovation in the private sector. An energy R&D policy along these lines is described in the next chapter.

THE ROLE OF GOVERNMENT

In many people's minds energy conservation means a bigger role for government. This impression arises not just because conservation is usually identified with emergency rationing but also because even energy-efficiency improvements are usually discussed in terms of automotive fuel-economy standards, appliance energy-performance standards, and the like. Throughout this book, however, we have shown by example that market strategies tend to be more effective in promoting conservation than regulatory approaches. Indeed the comprehensive energy-conservation effort represented by the scenario set forth in this chapter probably could not be carried out except by a market-oriented strategy, because most technological opportunities can be realized only through highly diverse decisions which account for the special attributes of each building, each production facility, and the people who operate them. An energy policy to promote such a market strategy is described in the final chapter of this book.

PART III

REINVENTING AMERICA

CHAPTER 14

ENERGY AND INNOVATION

Let the feelings of society cease
to stigmatize independent thinking.

JOHN STUART MILL,
"On Genius," 1832

MOST PEOPLE BELIEVE that without growth there would be technological stagnation. But the negative-energy-growth scenario we set forth in the last chapter would be viable only in an economy in which there was continuous technical change. While much fuel conservation can be accomplished with technology that is already commercially available, continuing improvements in fuel conservation can be sustained only if innovations are continuously introduced into the marketplace. Moreover, the successful development of solar energy will come about only if new approaches are encouraged. As we have shown by example, some of the most promising strategies for solar-energy development are different from those being emphasized in current programs. In both these areas the opportunities for technical change are great.

INNOVATION ON THE WANE IN THE U.S.

Unfortunately the present economic climate in the U.S. is not conducive to innovation. While the United States leads the world in total research and development (R & D) effort (see Figure 14.1), it actually lags behind other nations in R & D oriented toward the improvement of industrial productivity (see Table 14.1). It is shocking, in fact, that after more than a decade of slow productivity growth, R & D on economic growth and productivity in 1978 accounted for less than ½% of the total federal R & D effort (see Table 14.2).

The state of private sector support of R & D is even more troubling. As pointed out in 1978 by President Carter:[1]

> In recent years private sector research and development has concentrated on low risk, short term projects directed at improving existing products. . . . Emphasis on the longer-term research that could lead to new products and processes has decreased.

While nonfederal funding for research and development increased (in constant dollars) at an average rate of 7.6% per year in the period 1953-1967, this rate fell to 2.0% per year in the period 1967-1977.[2]

Fig. 14.1

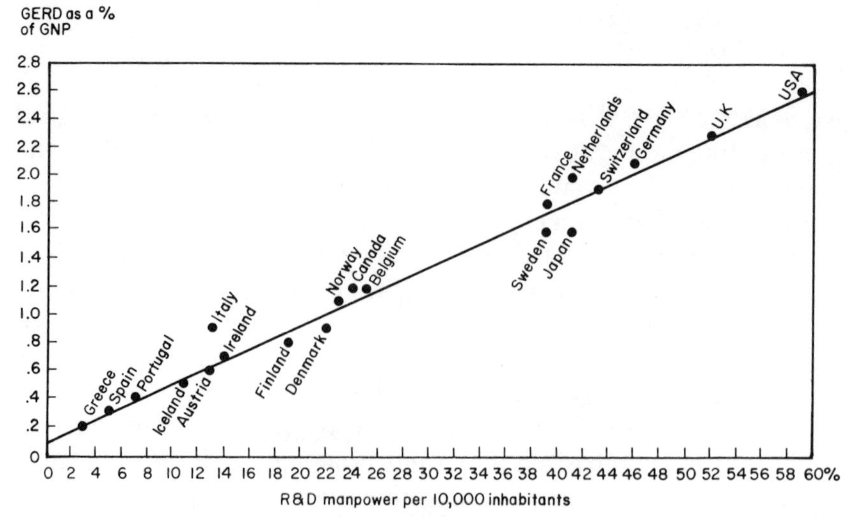

SOURCE: OECD, "Patterns of Resources Devoted to Research and Experimental Development in the OECD Area, 1963-1971" (Mimeo), May 17, 1974, p. 4.

NOTE: GERD = Gross Expenditures on R and D.

Table 14.1 Percentage Shares of Public R & D Expenditures
1968–69

COUNTRY	MILITARY, SPACE, NUCLEAR	ECONOMIC, AGRICULTURE, MANUFACTURING SERVICES	WELFARE, HEALTH, ENVIRONMENT	OTHER, INCLUDING UNIVERSITY
United States	79.3	6.0	12.7	1.9
United Kingdom	59.4	22.1	3.7	14.8
Japan	8.7	25.0	4.0	62.2
Sweden	52.2	13.1	8.2	26.3
France	55.2	16.5	2.7	25.6

SOURCE: OECD statistics, 1971.

Understanding the reasons for this general failure of private R&D to advance even as rapidly as production will help in formulating policies to foster innovations in fuel conservation and solar energy.

One reason often cited for the decline in private-sector R&D is that the regulatory climate in the U.S. is stifling innovation. This point has been articulately set forth by physicist Freeman Dyson[3] in an article entitled "The Hidden Cost of Saying No." Dyson states that so much time and money must now be committed to proving certain types of products safe that few corporations are willing to make the required investments. Carl Djerassi points out, for example, that to bring a new chemical birth-control agent to market requires an investment of $18 million and 17 years to conduct the required long-term studies.[4] As another example of the inhospitable regulatory climate for industry, Philip Abelson, the editor of *Science* magazine, has pointed out that in the making of steel, companies must comply with more than 5000 regulations issued by 27 different federal agencies.[5]

While these examples show that increasing public concerns about the side effects of technology are hampering innovations that involve potentially hazardous materials, it does not follow that the trend toward tighter regulations is hampering innovation in general. Some products are hampered by tightening regulations, but others may be encouraged, as we shall see below.

Another reason sometimes cited for declining innovation is that the U.S. is becoming a mature industrial power and is losing the incentive to innovate. There is evidence that a mature industrial society operating at high levels of production is not strongly inclined to innovate, perhaps because it is satisfied to maintain established areas of produc-

Table 14.2 Federal Research and Development funds, according to function, for fiscal year 1978. Percent.

Function	Percent
National Defense	49.0
Space	11.9
Energy conversion and development	10.6
Health	10.2
Environment	4.2
Science and technology base	4.0
Transportation and communication	3.1
Natural resources	2.3
Food, fiber, other agriculture	1.9
Education	1.0
Income security, social services	0.6
Area and community development	0.4
→ Economic growth and productivity	0.4
International development	0.3
Crime	0.2

SOURCE: National Science Foundation, "Fiscal Years 1969–1978, Analysis of Federal R&D Funding by Function," NSF77-326, 1977.

tion and is wealthy enough to become preoccupied with largely nonindustrial endeavors. According to E. J. Hobsbawm,[6] the seeds of British industry's downfall in the mid-20th century were sown in the late 19th century, with the failure to support education, research, and fundamentally innovative investments relevant to industrial change, in contrast to efforts made in Germany and the United States. Even the great industrial inventions made in Britain in the 1860s were adopted abroad much more quickly than at home.[7]

But the argument that the U.S. has lost the incentive to innovate overlooks the pressing technological needs faced by the U.S. today. In the area of energy alone much new technology is needed to reduce our dependence on insecure sources of foreign oil, to minimize the impacts of high energy prices, and to cope with increasingly severe environmental problems. Thus while some mature industries may be ineffective theaters for innovation, there are many areas where pressing social needs cry out for innovation.

Attributing the decline in innovation to regulation or to the maturity of U.S. industry addresses the problem too narrowly. New strategies for innovation are not being undertaken because the nation does not know where it is headed. The future can no longer be regarded as an extension of the past, and the nation does not have a public policy which defines a new course. Because of this uncertainty new energy-investment strategies are not being pursued—they are perceived as being too risky. Industry operates in a very uncertain regulatory environment: on the one hand, tightening regulations restrict the scope of industrial activity; on the other, many of these regulations are coming under strong attack in the political process. Future demand for energy is also very uncertain—because future energy prices are very uncertain, because the impacts of future prices on demand are poorly understood, and because it is unclear how new markets will evolve.

What is needed to create a climate conducive to a new course for investment is an energy policy that reduces uncertainty. Long-run regulatory objectives must be more clearly formulated, a long-run pricing policy is needed, and mechanisms are needed to help develop new markets in socially desirable areas.

There is no unique formula for fostering innovation by eliminating uncertainty. Here we consider two alternative approaches: the creation of a new government-industry partnership and the development of a market-oriented strategy in a controlled economic climate.

The development of a new government-industry partnership would be a logical culmination of our present energy policy, which is preoccupied with expanding energy supplies. Such a partnership would probably be necessary to bring about the Business-as-Usual energy future set forth in Chapter 3. However, this approach would probably be incapable of dealing with the increasingly serious side effects of new technologies, and it would be grossly ineffective and inefficient as a means of fostering technological change.

The alternative we recommend is to foster innovation in the private sector by making markets work better. In this approach social goals would be achieved through general policies that control the economic climate in which private firms operate, but within this controlled economic climate the course of innovation would be directed primarily by market mechanisms. Government's direct involvement with the R&D process would be restricted to areas that tend to be neglected by the private sector—areas that today are largely neglected by government as well.

A NEW GOVERNMENT-INDUSTRY PARTNERSHIP?

A proposal we hear more and more often these days from those with a vested interest in sustaining the present energy course is that a new government-industry partnership should be created. Such an arrangement would indeed reduce uncertainty relating to new investments: regulatory conflicts could be more easily resolved if government and industry were partners instead of partial adversaries in the development of new technologies, and some of the financial risks could be shifted from the shoulders of industry to government. In what follows we describe what we believe are fundamental flaws underlying the government-industry-partnership proposal.

SHOULD WE BREAK THE REGULATORY STRANGLEHOLD?

A major motivation for a government-industry partnership is to ease the regulatory burden on industry. But while present regulations are indeed making some kinds of innovation difficult, changes in the regulatory system should be made cautiously. It is important to distinguish between *economic* and *social* regulations.

The high level of public concern about the adverse side effects of modern economic activity has resulted in creation of many new regulatory agencies, which include the Environmental Protection Agency, the Occupational Safety and Health Administration, the National Highway Transportation Safety Board, the Nuclear Regulatory Commission, and the Consumer Product Safety Commission. These new regulatory agencies differ significantly from the previous generation of regulatory agencies. The purpose of most of the old agencies is economic regulation—the control of prices, product allocation, and market entry. ("Market entry" refers to new firms engaging in the sale of a particular product.) The purpose of most of the new agencies is social regulation, the protection of workers and the public from noneconomic side effects of economic activity. (One important new agency has also been created for further economic regulation in the energy area—the Economic Regulatory Administration of the Department of Energy.)

In the area of economic regulation a good case can be made that many policies have outlived their usefulness. For example, we feel that new policies aimed at partial or even total deregulation of the generation of electric power would be effective in promoting economic efficiency and would, in particular, be conducive to the implementation of industrial cogeneration technology (see Chapter 10). More generally we have stressed that new economic regulations should be adopted only where the market cannot be made to work effectively, and in these cases regulations should be formulated in terms of performance criteria rather than product specifications, so as to allow industry the maximum flexibility (see, for example, Chapter 8).

Social regulation is inhibiting the introduction of many new products, particularly those involving toxic chemicals. Many potential new pesticides, herbicides, pharmaceuticals, and other chemical products are being kept off the market by regulations. But at the same time great risks to public health and safety would be involved in blindly moving ahead with new products in these areas. Should the nation continue the present trend toward requiring that products be proven safe before being marketed, or should it return to the "safe until proven harmful" philosophy of control? This question has no easy answer because neither the costs nor the benefits of proposed products can be adequately quantified before they are used. While there is considerable room for improving particular regulatory procedures in the relatively new endeavor of social regulation, we are led to conclude that a strong regulatory policy should be pursued for products where significant public hazards might be involved—for two reasons.

First, we believe that technology has reached a turning point in its history. Industrial society has achieved such a high level of affluence and consumes natural resources at such an intensive rate that the adverse social impacts of new technologies have become quite apparent—in terms of air and water quality, toxic substances in foods, and the like. The intensity of many of these problems is so great that it is unlikely that present concerns will prove to be transient. In fact the nation is entering a new "era of ecological management" in the exploitation of natural resources, in which increasing attention will be given to the adverse side effects of new technology (see Chapter 15), and tough regulations will be called for to bring modern technology to heel. Second, we are also persuaded that the opportunities for technological innovation are so great that technological development could move ahead vigorously even within the framework of strong social regulations. But *technological innovation must be pursued in new directions.*

We have presented a glimpse in the preceding chapters of the far-reaching possibilities for innovation in fuel conservation and solar technologies. While none of these technologies are risk free (see Chapter 12), there is such a wide range of possibilities that many relatively safe and clean technologies can be developed and deployed. Industry should view today's environmental concerns, not as a constraint to progress, but as a spur to innovation in new directions. Here are some examples of this potential:

- Strict enforcement of federal air-pollution standards for sulfur oxide emissions would help create markets for fluidized-bed-combustion technology (a promising advanced technology for coal utilization—see Chapter 10), simply because sulfur oxide control via fluidized-bed combustion may well prove to be less costly than sulfur oxide control via stack-gas cleanup. (In a 1976 Exxon report[9] it is estimated that the potential market for coal-fired fluidized-bed-combustion boilers would be reduced by ⅓ if states that now have air-pollution regulations that are more permissive than federal standards were able to maintain their own standards.)

- Concern over automotive air pollution was a major factor prompting a critical reexamination of alternatives to the present gasoline-fueled internal-combustion engine in the early 1970s. One result has been the recent introduction of the automotive diesel engine in the U.S. The diesel has low hydrocarbon and carbon monoxide emissions (although it does pose other air-pollution problems) and offers the added advantage of high fuel economy (see Chapter 9).
- Water pollution control standards have led to new developments in the washing and bleaching of wood pulp for papermaking. Higher consistency pulp (i.e., a higher ratio of fiber to water), higher levels of recycling of water, new bleaching chemicals, and new bleaching processes are all leading to sharp reductions in water discharge and substantial reductions in fuel use.[10]

These are not isolated examples, but rather are hints of the widespread possibilities for technical change that is compatible with public concerns about the environment.

SHOULD THE GOVERNMENT ASSUME A GREATER ENTREPRENEURIAL ROLE?

The proposal for a government-industry partnership to help bring new technologies to the marketplace is often advanced by advocates of breeder reactors, coal-gasification projects, and other capital-intensive technologies that entail substantial financial risk. This point of view was articulated in 1978 by Harold Finger of the General Electric Company:[11]

> Any energy policy that does not *emphasize* increased supply—not simply treat it on the back burner as a byproduct—is totally inadequate to face the needs we are going to have in the years ahead.
>
> Therefore, it is really in the area of energy supply that major effort is required, but, unfortunately, it is the area in which the nation seems to be most hamstrung. What is disturbing about that is that we are blessed with an abundance of domestic resources and all the capabilities that we need, but they have not been brought to bear on the solution of our energy problem.
>
> . . . I am convinced that only by a collaborative effort, in which the government works closely with industry to try to provide an incentive for industry to do what it is capable of doing, can we provide the energy that the nation needs for continued economic growth and providing an adequate job base.

. . . I believe it is essential that the DOE [Department of Energy] work closely with the existing industrial capability and capacity, not as adversaries but as partners in trying to solve this very tough national problem. . . .

It is truly remarkable that the same voices which abhor government red tape and bureaucracy would call for government-industry collaboration in solving the energy problem. In practice a collaborative effort between government and industry is one where the government becomes an entrepreneur, assuming a major share of the financial risks involved. This entrepreneurial role for government would probably be necessary if the nation were to continue the present course of energy growth, because industry has been reluctant to bear the high costs and financial risks of expanding energy supplies to meet a (very uncertain) growing energy demand. Because the private sector has been unwilling to meet these challenges, President Ford, at the urging of Vice President Rockefeller, in 1975 proposed a 100-billion-dollar Energy Independence Authority to provide loans and loan guarantees to the private sector for large-scale projects and the speed-up of development of emerging technologies.[12] This proposal was followed up in July 1979 by President Carter, who proposed the formation of an Energy Security Corporation to carry out a 12-year massive energy-supply development program (emphasizing synthetic fuels from coal and oil shale) with $88 billion in federal subsidies—subsidies in the form of price guarantees, federal purchase agreements, loan guarantees, and direct government ownership of new facilities.[13]

We find such proposals to be seriously flawed in several respects. One problem is that industrial and public interests tend not to coincide for these activities. The industrial objectives of expansion and profit making are often in conflict with the public objectives of environmental protection and public health and safety, so that a greater government entrepreneurial role would tend to compromise the government's responsibilities to protect the greater public interest.

One might think that if technologies involving serious potential hazards were developed under government auspices, this development would in fact be carried out in accord with sufficiently stringent regulations to assure the protection of the public against potential risks. Unfortunately, however, this has not happened in actual practice, as one can readily learn by studying the course of development of light-water-reactor technology, a model for the government-industry-partnership proposal.

Consider the issue of light-water reactor safety, which has been an important issue in the public debates over nuclear power since the early 1970s and which became the focus of broad public attention with the Three Mile Island accident in March of 1979. Long before this accident the independent technical community recognized the needs for improvements in reactor safety. In 1974 a panel of physicists conducted a study under the auspices of the American Physical Society and recommended modest changes in reactor designs that could lead to substantially improved safety.[14] The measures proposed for serious consideration included several reactor design modifications that would greatly reduce the chances that there would be massive releases of radioactivity to the environment in the event of a fuel meltdown accident—an event which government and industry have always regarded as improbable but which was only narrowly averted at Three Mile Island. By the end of 1979 none of these proposals had advanced beyond the research level in federal programs.

The root of the problem with nuclear-energy technology and more generally with any hazardous capital-intensive technology lies in the fact that it is very difficult to bring about needed changes in such technology when it is in an advanced state of development. Our knowledge of the risks involved with hazardous technologies is imperfect. As we learn more and public attitudes change, modifications of regulations and designs are sought. But because of the very large investments involved, changes in the design of technology made late in the development-commercialization process are typically very costly. Moreover, even where changes might not be so costly, both industry and the government bureaucracy responsible for the technology tend to resist making changes, perhaps for fear that an admission that the technology is not what it should be might undermine the whole effort. Because of this problem it is desirable to keep separate the entrepreneurial and regulatory responsibilities for new technologies—a separation that is difficult to maintain when the government is entrepreneur.

Another fundamental problem with government-industry collaboration is that it would be a major step toward central planning. We have just noted that the institutions responsible for developing large-scale, hazardous technologies tend to resist innovations designed to make the technologies less hazardous. Indeed central planning may inhibit innovation generally. Charles E. Lindblom, in examining the problem of technological lag in European Communist nations,* asks:[15]

*These observations apply only to the civilian sector. The Soviet military has proven to be an efficient innovator.

> Why does technological innovation lag? The simplest answer is that communist enterprises are without the incentives to innovate that propel enterprises in market-oriented systems. Innovation is dangerous to an enterprise manager in a system in which he is required to meet output quotas. If he experiments with new technology, he may fail to achieve his quota. If he succeeds, he fears that new quotas will be imposed on him.
>
> But that answer is incomplete. Presumably, in recognition of the weakness of innovative motives at the enterprise level, leadership will locate responsibility for innovation elsewhere. They do. Pressure for innovation comes down from above, from party and from specialized institutions for encouraging innovation, such as special plans (as a component of annual economic plans) drawn up for innovation, directives ordering production of new kinds of products, engineering design bureaus, technical publications, and canvassing of technical publications abroad. But these methods have not been generally successful.
>
> Formal planning for technological innovation is subject to all the defects of planning already identified above, and plan achievement in this sector is among the lowest of all sectors of the economy. It is also Soviet practice to skimp on allocations and funds for capital replacement. And for various institutional reasons, even after a decision is made to build a new plant or convert to a new technique of production, it takes two or three times as long to accomplish the task as it does in the market-oriented polyarchies.*

It will be very difficult to persuade most people that organizing large-scale, centrally planned R&D projects is not a promising way to bring about technological change. One often hears people bemoaning the inability of government to solve the energy problem in such words as "If we can put a man on the moon, we ought to be able to. . . ." Efforts such as the Manhattan Project and the Apollo Project have a hold on peoples' imagination and conform to the widely held paradigm that planned change is desirable. Such projects are the norm for U.S. R&D in military, space, nuclear-energy, and other large-scale energy-supply ventures and are responsible for the very large federal R&D budget (see Table 14.1). But while government may have been successful in the crash programs to develop the atomic bomb and to put a man on the moon, the "big project" approach is inappropriate for the development of commercial products for the civilian economy, where considerations of cost and market demand are central. This approach does not work for commercial products because the effort is not managed by those who would undertake the risk to produce and sell the product.[16]

* Lindblom calls a political-economic system involving rule by many a polyarchy rather than a democracy.

This general criticism of centrally planned R&D is strongly reinforced in the area of fuel conservation because of the characteristics of the needed technologies. First, a government entrepreneurial role is wholly unnecessary because bringing most new fuel-conservation technologies to the marketplace does not require the large concentrations of capital, the long developmental lead times, and/or the high financial risks that characterize conventional new energy supplies. Furthermore, as stated by James Utterback, government management would be wholly inadequate to the challenge:

> The options for innovation in conservation are so numerous and so clearly linked to perceived need that centralized evaluation is at best infeasible and at worst counter productive.[17]

TOWARD A MARKET STRATEGY FOR INNOVATION

As an alternative to a new government-industry partnership we propose that innovation be fostered in the private sector by creating a controlled economic climate that gives industry clear signals as to the areas where innovation would lead to new markets. In addition, the government's direct role in R&D should be restricted to those areas where market mechanisms clearly would not be adequate.

To facilitate discussion of the specific measures that would carry out this general policy, it is useful to distinguish between the various stages of the innovative process:

Basic Research: Scientific investigation aimed at expanding the store of fundamental knowledge.

Applied Research: The application of scientific methods and principles to the solution of technical problems. (This research is generic, i.e., not aimed at specified products.)

Development: Investigation, design, and creation of specific protoproducts and production processes.

Demonstration: Testing of these systems in near-commercial applications.

Commercialization: Introduction of these systems into the marketplace.

The most important distinction to be made among these activities is that research is a generic activity while development, demonstration, and commercialization are product-specific activities.

At the present time the federal government has a role in all these activities. It is our view that the government is much too involved in product-specific activities and not involved enough in generic activities. And we would add to this list another critical activity relating to innovation where the government role needs to be strengthened: education. Specifically we propose that:

- The government's role relating to development, demonstration, and commercialization should be to create an economic climate conducive to these activities in the private sector.
- Government support of research and education relating to innovation should be bolstered.

In effect we are calling for a reduction in the direct government role in the final stages of the innovative process and an increase in government support in the earlier stages. Government support for development, demonstration, and commercialization is an inefficient means of spurring innovation and poses serious conflicts between the governmental roles as entrepreneur and as guardian of the public interest. On the other hand, strong government support is needed for research and education because the risks are usually too high, the benefits too diffuse, and the payback period too long to justify strong private-sector support for these activities.

We shall now discuss in detail how this transformation of the R&D effort might be accomplished. While we shall be referring to energy-related technology, much of what we have to say is relevant to other technological areas as well.

TRANSFERRING DEVELOPMENT, DEMONSTRATION AND COMMERCIALIZATION ACTIVITIES TO THE PRIVATE SECTOR

The invisible hand of the market is clearly preferable to government planning for guiding the myriad technological changes that would lead to improvements in the efficiency of energy use throughout the economy. Then, if it is granted that most development and demonstration work should be carried out by the private sector, the question is, What conditions would stimulate a much larger private-sector effort?

The most important policy initiatives to support private innovation concern energy prices. The subsidies supporting the production of conventional energy supplies should be eliminated, and wellhead, minemouth, and import taxes should be imposed on fuels to eliminate differences between energy prices and replacement costs, to cover the social costs of side effects of energy production and use, and to take account of the fact that nonrenewable energy resources belong not just to the present generation but to future generations as well. The taxes should be imposed as a tax shift, offsetting some existing taxes.* The higher

*See Chapter 15, under "An Energy Tax Shift," for discussion of such a shift.

energy prices resulting from this tax shift would create a much more favorable climate for bringing both fuel-conservation technology and solar-energy technology to the marketplace.

We have already pointed out that some regulatory reform, particularly in economic regulation, is also needed to create a climate conducive to innovation. But as we have shown, it is not necessary to relax the social objective of minimizing the adverse impacts of new technology in order to spur innovation. To the extent possible, regulations should be based on *performance* rather than *specification(s)*, so as to allow industry the maximum flexibility for innovation.

Industry must be prepared to innovate in a climate of changing social regulations, however. Perhaps the single most important lesson to be learned from recent trends in social regulation is that knowledge of the adverse impacts of new technology and public concerns about these risks will continue to change, resulting in some cases in progressively tighter regulations. Changing regulations will inevitably force fundamental changes in some technologies and perhaps abandonment of some. To minimize the potential dislocations associated with such actions as well as those associated with changing economic conditions, it is important that *technological diversity* be stressed in the innovative process. Most fuel-conservation and solar technologies tend to be small in scale and varied so as to conform to diverse applications; and in some cases they tend to be regionally specific. As we have shown in Chapter 12, adaptation to changing concerns about the adverse impacts of technology is relatively easy for such technologies.

Public policy to establish a climate conducive to private innovation must be sensitive to the different roles of small and large firms. Small *new* firms are responsible for much product innovation (as contrasted with innovation in production processes).[18] J. Herbert Hollomon,[19] John Jewkes and his collaborators,[20] and others[21] have shown that new small firms provide a much larger fraction of innovations in the U.S. than their relatively small R&D expenditures or their total size would suggest. Furthermore, as Hollomon has pointed out,[19] much innovation takes place by invasion, where a firm with a new product displaces an existing firm producing the old. These considerations highlight the importance of the creation of new firms as a spur to innovation. Yet there has been a dearth of new companies in recent years. According to Dr. N. Bruce Hannay, vice president for research and patents at Bell Telephone Laboratories,[1] the number of new venture technology companies coming to the market with stock issues dropped in the late 1970s from several hundred a year to essentially zero.

The solution to this problem lies, we believe, not with a specific bureaucratically controlled policy to make more venture capital available for small firms, but with general policies to establish an economic climate that would stimulate the demand for fuel-conservation and solar technologies: energy-price-deregulation policies, an energy tax shift, fiscal policies to promote the availability of capital for long-term homeowners' loans, and the like. Such policies are called for because capital becomes available when expanding markets give rise to high retained earnings.

Making more venture capital available leads to technological innovation through the formation of new firms. Innovation of a different kind, evolutionary innovation in production processes, tends to be emphasized by existing large firms.[18] Firms must be relatively large to bear the risks of maintaining an ongoing R&D program. Yet the commitment to R&D does not simply increase with the size and maturity of the firm.[8] Some very large firms tend to view innovation conservatively and may in fact use their extensive capital resources and control over the market to suppress innovation and prevent the entry into the market of small firms with radical innovations. A particularly favorable environment for innovation might be one with many middle-sized firms. According to Frederick M. Scherer:[22]

> . . . giant scale has a slight to moderate stultifying effect. . . . All things considered, the most favorable industrial environment for rapid technological progress would appear to be a firm distribution which included a preponderance of companies with sales below $200 million, pressed on one side by a horde of small, technology-oriented enterprises bubbling over with bright new ideas and on the other by a few larger corporations with the capacity to undertake exceptionally ambitious developments.

Medium-sized firms often need to innovate to successfully compete with the truly giant corporations, and they are large enough to support their own R&D effort and to produce and market products on a large scale. Thus general antitrust and tax policies which favor such firms relative to very large firms might lead to greater innovation.

INCREASING GOVERNMENT SUPPORT FOR RESEARCH

While we believe the government should get out of the product development and demonstration business, we also feel that its commitment to research must be greatly expanded if the innovative process is to be sustained for the long term.[23]

Research, both basic and applied, is essential to the innovative process. But often there is complete uncertainty as to whether important research efforts will lead to marketable innovations. Moreover, the lead time between research and market penetration is often much longer than the planning time horizon of the private sector. For these reasons generic research activities should be largely the responsibility of the government. The government's responsibility relating to research arises also because the public interest is broader than private interests, particularly in areas relating to the adverse impacts of new technology.

Applied research in particular, i.e., fundamental research on topics selected for their technological importance, should be given much more support than it has received in the past. Heat transfer in turbulent flow, the thermal response of buildings made of various materials to changing weather conditions, friction and lubrication, and air drag on surface vehicles are examples of problem areas relevant to fuel conservation for which new research could provide valuable funds of knowledge on which improved product design could be based.[24, 25, 26] An excellent example of the importance of this kind of research is the aerodynamic research carried out in the 1930s and 1940s by NACA (National Advisory Committtee for Aeronautics), the predecessor of NASA (National Aeronautics and Space Administration), which proved very useful to aircraft manufacturers in specific design work.[27]

The startling fact is that very little research of this nature is underway. W. C. Reynolds, who reviewed research expenditures for a number of government agencies, concluded in a 1977 report:[28]

> These numbers support two opinions that I have heard from many colleagues. The first is that too much emphasis is being placed on development. . . . The second is that there is a "gap" between the basic research programs and the development programs, that applied research is not receiving sufficient emphasis.

It is no accident that applied research has been neglected in the U.S. This is a direct consequence of the fact that R&D in the U.S. has been dominated by military, space, and nuclear efforts. As Hollomon has observed:[19]

> The great demand for technically trained people in the United States beginning in the early 1950s encouraged the educational establishments which produced them to concentrate on the areas of science and technology which the government was supporting. These fields were primarily those dependent on the immediate application of new capability originating from science: electronics, solid state physics, nuclear physics, aeronautics, astronautics. The federal government supported virtually no activities

> in the universities for the purposes of improving manufacturing technology or for understanding the nature of the distribution and marketing system of the country or the means by which information can better be disseminated for these purposes.
>
> The amount of research and development directed toward space and defense has declined somewhat. But the difficulty is that the emphasis placed upon this activity has introduced a distortion which will take years to disappear. . . .

While there will always be the need for modern specialized science and engineering in areas such as space and nuclear physics, applied research relating to fuel conservation will draw heavily on the older, "classical" disciplines of science and engineering. (This point is argued persuasively in a 1975 American Physical Society report on fuel conservation.)[29]

Another type of research that needs more emphasis is research on the side effects of technology—on public and occupational health and safety, environmental impacts, national security, etc. This research should be of two types: fundamental scientific research aimed at identifying and understanding adverse impacts, and policy research aimed at exploring alternative technological and institutional strategies for coping with these problems. This kind of research should find a more hospitable home in a government-supported R&D program like the one we are outlining here than in the present program, where the government as entrepreneur has repeatedly reacted defensively to criticisms of technologies for whose commercial success it has responsibility.

RENOVATING THE PROCESS OF RESEARCH MANAGEMENT

Not only should government programs emphasize research instead of development and demonstration; also, the government should adopt a looser administrative structure for this activity. Today most R&D projects are designed and conducted through a series of interactions between the investigators and the project administrators in Washington. These interactions are usually initiated by an RFP (request for proposal) sent out to prospective investigators by the funding agency for competitive "bids." Once a project is funded, the field work is continually monitored by bureaucrats who represent "mission oriented"* agencies and who are often familiar only in general terms with the research they are monitoring.

*Political decisions unrelated to progress in the project often lead to a change in the mission objective during the course of a project.

We have already pointed out the failings in such attempts to organize creativity in the economies of Eastern Europe. That the problem is not limited to Communist bureaucracies is suggested by the analysis of research in large corporations by Jewkes and his collaborators:[20]

> The chances of success are . . . reduced where the research group is organized in hierarchical fashion, with ideas and instruction flowing downwards and not upwards, and is held to be at the beck and call of the production and selling side of the firm; where the direction of research is so closely defined that it gains a momentum rendering it impossible for intriguing sidelines and odd phenomena to be followed up; where the allocation of functions is determined in such a way that voluntary and ephemeral groupings among the research workers are impossible or are frowned upon; where men are asked to report at regular intervals upon ideas around which their minds are still anxiously groping; where achievements are constantly being recorded and assessed; where spurious co-operation is enforced by time-wasting committees and paper work and where painstaking efforts to "avoid overlapping" frequently quench originality. The awkward, lonely, enquiring, critical men—the men of "wide-ranging, sniffing, snuffing, undignified, unself-dramatising curiosity" as Sinclair Lewis once described them—may well be a positive nuisance in such surroundings.

The alternative of a looser administrative structure for energy research would be feasible if a large fraction of the total R&D funding were provided for broad-based programs instead of for very specific projects. One reason for emphasizing program as opposed to project support is purely practical. Because of the diversity of opportunities for technological innovation and the relatively small scale of the technologies involved, much of the research on fuel conservation could probably be carried out most effectively in many small research groups instead of in a few large laboratories. As Enrico Fermi has observed:[30]

> There is much to be said for the small group. It can work quite efficiently. Efficiency does not increase proportionately with members. A large group creates complicated administrative problems, and much effort is spent in organization.

If a large number of relatively small research groups were to be supported, it would be impossible for any finite group of "project monitors" in the funding agency to keep abreast of detailed developments in the field. Also an emphasis on program support would probably result in more innovation in the course of the research, because the work could more easily shift in response to continually emerging new knowledge.

The model of federal support for basic research in physics in the

1950s and 1960s, appropriately modified, may be a good one for organizing energy R&D: support could be provided to those groups that "do good work" in broadly defined areas.[31] How would the funding agencies decide what is good work? For basic research the creation of a peer-review system like that used in basic physics research is appropriate. For applied research the review group should be broadened to include, besides peers, technologists who actually build things and could provide feedback to scientists and engineers regarding the needs of industry. Similarly the review group for research on side effects of technology should include, besides scientific peers, representatives of groups that would be affected by the technologies being studied. In any case, however, the review process should be, like that for research in basic physics, largely external to the funding agency.

Bringing potential users into the review process is one way of ensuring that applied research is being directed toward social needs and not merely toward the needs of the researchers. Another way would be for a consortium of potential users to organize and govern an applied-research effort. For example, a research program carried out under the joint auspices of an industrial association and a professional society would provide a close link between the skills of the laboratory scientists and engineers and the market interests of the manufacturers.[26,27] The problem of proprietary interests should not arise as long as product development is not involved. Such arrangements could be especially helpful in spurring innovation by small firms. As Hollomon has pointed out:[19]

> Innovation and technical change also appear to take place more slowly in an industry made up of a large number of very small firms unable to cooperate and share in the cost of innovation. If the cost or risk of innovation is high compared to the firm's assets, the innovation is unlikely to take place. Small firms cannot even employ the kinds of technical people necessary to stimulate the use of new techniques or to be aware of the possibilities. The home-building industry is made up of thousands of firms tied primarily to their localities and unlikely to adopt quickly new techniques. The productivity of the entire construction industry probably has not improved in the last decade.

While almost every other industrialized nation encourages the collaborative work of industrial associations, the U.S. does not. Perhaps antitrust laws or the fear of them has discouraged collective technical work. But this situation could be changed. According to Hollomon:[19]

> Recent discussions with government officials imply that they consider the following four questions in evaluating the legality of associative activities: (1) Is the collective association open? Can anyone be a member? (2) Is support for collective technical work available for everybody in the industry? (3) Does it not deal with the allocation of markets? (4) Does it not deal with the determination of prices? If these conditions are met, then collective technical work is not only possible but should be encouraged, except in certain special industries.

This model of having users organize research could also be applied to research on the adverse impacts of technology and the exploration of technological alternatives. Over the last several years a large number of environmental, consumer, and other citizens' groups have been organized throughout the country to take action against technologies which they believe pose significant public risks. Often such groups do not have access to the technical expertise needed to adequately evaluate the risks involved. Moreover, such groups would usually require outside funding to support such research activity. The National Science Foundation Science for Citizens Fellowship program,[32] which provides fellowships for scientists to work with citizens' groups, is an interesting experiment in support of this kind of activity, which could lead to better-informed public debates over technological risks and alternatives.

REORIENTING TECHNOLOGICAL EDUCATION

A complementary long-range objective of the energy R&D program should be to provide the appropriate technical manpower for the kinds of innovation needed. Our present technical educational system emphasizes fields which grew out of World War II military research: electronic, nuclear, and aerospace technologies. The schools of engineering have been shaped by federal support of research in these areas. Subjects such as production engineering, process engineering (e.g., for basic materials like steel and paper), and more basic topics like combustion and lubrication are moribund at most of our universities. A veritable renaissance of skills development and research related to society's basic technology is required.

It is also desirable to greatly extend the educational responsibilities of universities to more effectively include midcareer reeducation. Over the last couple of decades our institutions have turned out very narrow specialists—some of whom are jobless or want to change careers to work on "problems of society." Such career changes would be facilitated if the universities established technical retraining programs. Demographic trends underscore the urgency of establishing such programs. Figure 3.4

shows that between 1980 and 1990 there will be an explosive 25% increase in the 25-to-44-year-old age group in the population, indicating fierce competition for jobs. Dislocations could be greatly mitigated by retraining programs that stressed broader, marketable technical skills.*

In summary we believe that research and education should be the activities involving the greatest direct government support. This means that universities should have an important role in the energy R&D process. They have a negligible role today—as anyone phoning for an appointment at the Department of Energy might surmise when asked by the secretary at the end of the line, "And which company do you represent?" But greater university involvement in energy R&D will require not only more government funding for such activity but fundamental changes on the part of universities as well. The conversion of engineering schools in the 1960s into schools for certain very highly specialized areas of technology and science may have been appropriate when the problems of defense and space were our dominant concerns. Now we must turn engineering talents to more pragmatic ends: the survival and well-being of postindustrial society.

*If universities do not seize this opportunity to take on a new role, their future is bleak: the traditional college-age population (18 to 24) peaked in 1980. See Figure 3.4.

CHAPTER 15

A FUEL CONSERVATION POLICY

The best laid plans o' mice an' men
Gang aft a-gley
An' lea'e us nought but grief an' pain
For promised joy.

ROBERT BURNS,
"*To a Mouse*,"1785

THE FUEL-CONSERVATION STRATEGY we have outlined in this book is based on technical changes which would be less costly to the economy as a whole than the alternative of increasing energy supplies. Despite favorable economics, however, many of these measures will not be implemented and none will be fully deployed *in the present economic climate*, because under the present rules the energy market functions very badly. The relative costs of the conservation and increase-of-supply alternatives as seen by energy consumers are very different from the relative costs to the economy as a whole—mainly because energy prices are influenced by a pervasive system of subsidies and of price and quantity controls imposed on energy markets by government. Thus the extent to which fuel-conservation measures will be exploited depends on *political decisions* concerning energy-related taxes, subsidies, regulations, etc.

The specific fuel-conservation initiatives we have proposed to overcome the barriers to economically justified conservation are summarized in Table 15.1 for each of the areas explored in this book. These measures are categorized into the six different general types of policies listed in the note to Table 15.1.

Our primary emphasis throughout this book has been on policy measures aimed at making market mechanisms work better. This emphasis is reflected in our proposals to eliminate energy-supply subsidies, to levy energy taxes as substitutes for existing taxes, to provide consumers with better information as a basis for more rational decisions in the marketplace, to decontrol fuel prices, to begin deregulation of electric-power generation, and to shift the responsibility for development and demonstration of new technology from the government to the private sector.

Our principal proposals—to emphasize fuel-conservation technol-

Table 15.1 Public Policy Suggestions

Housing (Chapter 8)

Information Programs
- House-doctor programs
  - public program (6)
  - licensing for private programs (1)
- Programs to generate feedback to consumers and policy makers regarding changing patterns of energy use (5)
- Requirement that household-energy-use information be available when a house is sold (5)
- Research and development (3)

Decontrol of Energy Prices (1)

A General Fuels Tax (2)

Measures Needed to Complement a Fuels Tax
- Elimination of energy-supply subsidies (2)
- Loan guarantees for retrofitting low-income housing (2)
- Thermal performance standards for rental buildings (1)
- Performance standards for energy-intensive appliances (1), (5)

Market-Stimulation Measures
- Utility involvement in conservation investments (1)
- Fiscal policy to enhance the availability of capital for conservation investments (2)
- Government procurement of fuel-conserving buildings (4)

Automobiles (Chapter 9)

Strengthened Fuel-Economy Standards (1)

A Substantial Road-Fuel Tax (2)

Improved Measurement of Fuel Economy (5)

Industrial Cogeneration (Chapter 10)

R&D on Technologies Characterized by a High Ratio of Electricity to Steam Production (3)

Table 15.1 Public Policy Suggestions (Continued)

Modification or Elimination of Industrial Fuel-Allocation Program (1)
A General Fuels Tax (2)
Elimination of Energy-Supply Subsidies (e.g., investment tax credit) (2)
Experimentation with New Institutional Arrangements (3)
- Utility ownership of onsite facilities
- Deregulation of power generation, with a modified institutional structure for power transmission and distribution

Community Solar Cogeneration System (Chapter 11)

A General Fuels Tax (2)
Elimination of Energy-Supply Subsidies (2)
R&D (3)
- System Design
- Components

Land-Use Planning (6)
R&D on Alternative Institutional Arrangements
- Private utility ownership (1)
- Municipal utility ownership (6)
- Private-sector ownership, with deregulation (1)

Conservation Hazards (Chapter 12)

Housing Safety Standards, Locally Administered (1)
Small-Car Safety
- R&D (3)
- Standards (1)

Air Pollution from Cogeneration Systems
- Regulations for dispersed sources (1)
- R&D on air pollution from fluidized-bed combustion (3)

Research and Development (Chapter 14)

General Conditions
- A General Fuels Tax (2)
- Elimination of energy supply subsidies (2)
- Fiscal policy to encourage venture capital (2)
- Appropriate antitrust enforcement (1)

Specific R&D Policies
- A reduced government role in development and demonstration (3)
- Increased government support for research and education (3), (6)
- Government procurement of new high-performance products (4)

NOTE: the numbers in parentheses refer to the following policy types.
(1) Regulation, deregulation
(2) Taxes, subsidies, fiscal policy
(3) Research and development
(4) Government procurement
(5) Information and exhortation
(6) Government planning and organization, government enterprise

ogy and to adopt a market-oriented strategy for implementing this technology—involve marked departures from established trends. Although most spokesmen give lip service to more effective use of energy, public policies promoting fuel-conservation technologies have—with a few exceptions—not been enacted. Instead, a prodigious effort is being made to expand energy supplies. Similarly, the trend in energy policy has been toward more central planning instead of toward greater reliance on market mechanisms in bringing about technical change. Despite these trends, however, the course we propose is a natural next step in the history of natural-resource management in the United States.

A HISTORICAL PERSPECTIVE ON ENERGY POLICY OPTIONS

The U.S. is well endowed with the energy resources and most of the other natural resources needed for economic development. The way people have viewed this bounty and the means taken to manage it have undergone continual change.[1] A brief history of public policy for natural-resource management will provide the context in which to examine the alternative approaches to energy management we shall describe in this chapter: central planning and reliance on markets operating in a controlled economic climate.

THE ERA OF HAPHAZARD RESOURCE EXPLOITATION

Throughout most of the nineteenth century the U.S. was a frontier society in which land, water, and other natural resources appeared to be inexhaustible, with unconstrained opportunities for private developers. In this era the tradition of government subsidy for exploitation of natural resources was established.[2] The government gave land to any private developer for mining, lumbering, farming, grazing, damming, and transport systems, and supported this development by using general tax money to build canals, railroads, and roads. Destruction of land and water resources was tolerated. Taxes on most mining industries were unusually low.*

THE ERA OF MANAGED GROWTH

Toward the end of the nineteenth century, and especially during the administration of Theodore Roosevelt, a new era began, character-

*Nash (ref. 2) recounts how mining interests dominated California government in the early years to such an extent that the only important tax revenue from mining was a head tax on Asian immigrants employed as miners.

ized by the scientific management of natural resources.[3] At this time a new breed of public land managers, opposed to haphazard exploitation of water resources and federal lands, came into power. They trod a path between the preservationists, a new group which opposed development, and the traditional business, or laissez faire, interests, who were interested mainly in quick profits. These "conservationists" of the early twentieth century recognized that through scientific management—that is, through exploitation of innovations in forestry, agriculture, hydrology, etc.—industry could obtain greatly increased yields. They appreciated that the building of a small dam could prevent the building of a dam of optimal size; that if the forest products industry was to survive, it must be concerned about replanting and caring for forests.

The rise of scientific management was accompanied by the growth of the large corporation, which so dominates the U.S. economy today. Among other advantages, large corporations were able to bring together the technical manpower and capital resources needed for scientific management.

In this same period, a new government institution was born—the regulatory agency. The first such agency was the Interstate Commerce Commission, set up in 1887 to regulate railroads. This was followed by the Federal Communications Commission, the Federal Power Commission, the Civil Aeronautics Board, and others. While these agencies were nominally set up to represent the general public interest, they were largely concerned with the elimination of cut-throat competition, which in this era had become a "limit to growth," that is, a constraint on corporate expansion. These agencies thereby helped to provide stability and opportunities for sustained growth to the large established corporations, which welcomed the government's regulatory role.

THE ERA OF ECOLOGICAL MANAGEMENT

By the 1960s, scientific management of natural resource exploitation had begun to be viewed as inadequate. While scientific management made possible great increases in production, the adverse side effects of continued growth were being felt by this time in many areas: the high levels of industrial and consumer activities were leading to substantial environmental disruption.*

Well-publicized analyses, beginning in 1962 with Rachel Carson's indictment of the pesticide DDT in *Silent Spring*, have shown that

*There has also been heightened concern about the exhaustion of nonrenewable resources. However, as we pointed out in Chapters 4 and 5, this is a less serious problem than environmental disruption.

today's industrial activities are threatening the health and welfare of the poulation both directly and through the impact of these activities on natural systems. We have shown in Chapter 5 that today's energy-supply technologies pose threats which could have global catastrophic consequences in a matter of decades.

Environmental problems are not new.* What are new are the pervasiveness of the impacts and the widespread public interest in doing something about them. The public interest is aroused not only because of the intensity of the problems but also because the economic growth which created these problems has brought about a relatively high level of affluence: most people rank amenities such as safety at work, safe consumer products, and a pleasant, healthful environment above increased income—once elementary economic security has been attained.[5] As pointed out in Chapter 14, these concerns have led over the last decade to the creation of many new agencies responsible for "social regulation"—the protection of workers and the public from noneconomic side effects of economic activity.

Social regulation is an embryonic activity that is still experiencing growing pains. While the preponderant point of view both in government and in influential groups outside of government is supportive of this trend, the new regulations are being challenged by an "antiregulatory" group that sees them as posing unjustified, inflationary burdens on production.[6] The principal purpose of this group is not to do away with social regulation, but to insist that the costs and benefits of specific regulations be quantified and that regulations be implemented only if the benefits exceed the costs. But *this cost-benefit approach to social regulation will not work,* either technically or politically.

Technically it will not work because of fundamental problems that prevent meaningful quantification of costs and benefits. Quantification is difficult, first of all, because of substantial technical uncertainties concerning risks. For example, on the basis of present knowledge it is not known whether or not there is a level of exposure to certain air pollutants below which adverse health effects do not occur. Another example is that it is not known what shifts in the global pattern of rainfall would result from a doubling of the amount of CO_2 in the atmosphere as a consequence of the burning of fossil fuels. And even if these technical uncertainties could be resolved, the monetary assessment of damage is

* For example, concerns about air pollution date back at least to the fourteenth century. Under Richard II (1377-1399) and Henry V (1413-1422), England took steps to regulate and restrict the use of coal, apparently because of the smoke and odors produced by its combustion. See ref. 4.

highly uncertain because ethical as well as economic considerations are involved. Thus, the dollar cost of lung disease and of life shortening arising from air pollution and the cost of the shifting patterns of rainfall that could arise from the atmospheric buildup of CO_2 are wildly uncertain and dependent on individual systems of values. As a result, social regulation will continue to be determined politically, largely on the basis of qualitative considerations.

Furthermore, within the political process the trend will be toward more rather than less social regulation, for there is no turning back the clock on efforts to limit the adverse effects of technology; we have entered a new era of ecological management. The problems are real, and they often turn out to be more serious the more we learn about them. Moreover, public concerns about these problems will likely increase through greater awareness of the risks. Both studies of *potential* risks and graphic descriptions of *real* disasters, such as the forced evacuations in 1978 and 1980 of the chemically contaminated Love Canal community near Buffalo, New York, attract wide attention and are heightening the public awareness of the side effects of modern technology.

The antiregulatory critics of social regulation do not present a viable strategy for dealing with the adverse impacts of technology; nevertheless, their concerns do point to a fundamental problem—that our present approach to natural-resource management is becoming increasingly unworkable. What needs to be done, however, is not to remove the burden of social regulation from industry, but to *renovate* the entire process of natural-resource management. Today natural-resource management, which involves, in addition to economic and social regulation, an increasing government role in the development of technology, is moving toward detailed central control of the processes of production. In the following section we discuss the problems posed by this trend toward central planning. We then describe an alternative market-oriented approach to ecological management which we believe is inherently more viable.

THE SIREN SONG OF DETAILED CENTRAL PLANNING

Many voices from different parts of the political spectrum are calling for central management of national and world resources. W. W. Rostow has stated in an interview[7] that the large government planning agencies can do the job if they are supplemented by a central directorate of economists:

> If you don't like the idea of national planning—and I have no inherent liking for it—the simple fact is that right now the big bureaucracies in Washington are planning these sectors. They have their hands on transport, energy, agriculture, air and water, but they just aren't planning in an effective way and coordinating it.
>
> My recommendation is that we set up, as I believe we will have to, a central planning unit. It could be put in a medium-sized townhouse in Lafayette Square and limited to 30 professionals. It is not a question of getting a massive new bureaucracy, but we do need to get a concept of what ought to happen in these sectors. We have got to see the maximum job the private sector can do. We have got to get the public policies that allow the private sector to do the maximum and see what the residual costs are for government, meanwhile making sure we have priorities so the balance of the economy isn't excessively distorted.

The case for central planning has been perhaps most eloquently formulated by J. K. Galbraith,[8] who argues that the development of central planning is the logical next step in the evolution of the U.S. economy. Galbraith has described the present private economy as consisting of a "planning system," made up of the roughly 1000 corporations that provide about half of all privately produced goods and services, and a "market system," made up of the roughly 12 million smaller firms that constitute the rest of the private economy. Galbraith calls that part of the economy controlled by the large corporations the planning system because here corporate planning has to a considerable degree replaced market forces as the guide to decisions relating to production. Today's individual integrated corporation manages production from raw materials to finished products, where, in former times, the series of intermediate products, associated with the transformation of raw materials into finished products, were bought and sold by separate businesses. Government has played a supportive role for the corporate planning system—both through the protective activities of agencies for economic regulation and through massive subsidy programs, such as the highway-construction program that made possible the growth of the powerful road transport industry. But, according to Galbraith, the planning system has become so large that its frequent failings can be acutely disruptive of the overall economy. He believes that the only effective response to these failings is a much more active government role. In 1973 he set forth his argument for central government planning through a discussion of the failure of the auto and oil industries to effectively coordinate their corporate activities relating to the supply and demand for petroleum products:[9]

> The organization by the automotive industry of the increased use of its products was . . . a triumph of the planning system. . . . The full magnitude of the increase in demand [for petroleum products] was not foreseen. Nor was it foreseen that the building of the needed pipelines, refineries and docking and discharging facilities for large tankers would involve difficult and, on occasion, irreconcilable environmental conflicts. In consequence there is now doubt as to whether the petroleum products made necessary by the great expansion in automotive, truck and other use will be available. A new phrase has been born. There is much talk of an "energy crisis."
>
> As with other faults of the planning system the possible shortage of petroleum products is being discussed as though it were sui generis—another purely adventitious failure. We now know that it is nothing of the kind. The planning system involves an intricate coordination as between its several parts in pursuit of their purposes. There is every likelihood that, from time to time, this coordination will fail. And these failures are already fairly common. . . .
>
> When, as in the case of the automobile and the oil industries, planning by one industry imposes requirements on another that it cannot meet, it is, of course, taken for granted that the state will intervene. By public action freight and people will be returned to the more economically fueled railroads. Or there will be subsidy to hitherto unprofitable sources of energy. Or there will be technical support to the development of new energy supplies. The state, in short, will take steps to effect the coordination of which the planning system is incapable. It will impose overall planning on the planning system. This is the next and wholly certain step in economic development—one that is solidly supported by the logic of the planning system. . . .
>
> The solution is to recognize the logic of planning with its resulting imperative of coordination. And government machinery must then be established to anticipate disparity and to ensure that growth in different parts of the economy is compatible. . . .
>
> There will have to be a public planning authority. This, in turn, will have to be under the closest legislative supervision. For here will be encountered the most difficult of all the problems of the public cognizance. That will be to have planning that reflects not the planning but the public purpose. The creation of the planning machinery, which the present structure of the economy makes imperative, is the next major task in economic design.

This argument by Galbraith, like similar arguments presented by many others, seems highly plausible. It fits the twentieth-century paradigm of management—that increasingly pervasive planning and control are good. Despite this plausibility, we here argue that detailed central management of energy technology would be ineffective for the goals we have proposed.

One basic problem with detailed central management is that activities which would be planned and coordinated do not behave like simple machines. This problem is far more difficult for government planners than for corporate planners, whose responsibilities and goals are much narrower. Many important variables that bear on the activities involved cannot be meaningfully incorporated into central management models, and in many instances the variables that cannot be modeled adequately will be more important than those that can. Consider, for example, the track record of the central planning effort for nuclear-power development. The cost competitiveness of nuclear power, the safety of nuclear technology, and the demand for electricity have all been subject to elaborate quantitative evaluation by the government agencies responsible for nuclear power. Yet it can be seen with hindsight that no analysis could have done justice to many of the variables involved. It is difficult to imagine how planners could have anticipated the amazing capital-cost escalations of nuclear power and the rapidly changing political climate regarding the risks of nuclear-fuel reprocessing and radioactive-waste disposal, not to mention the slackening of demand growth for electricity owing to saturation and price increases. Government failure to deal adequately with such variables has far more serious consequences for the economy as a whole than would failure in corporate planning efforts, because the various competing firms usually do not make exactly the same mistakes.

Another problem inherent in detailed central planning is that the many different suborganizations which must work together all have different actual goals; all are to some degree in conflict with each other and with the general intent of the functions they carry out. Lindblom discusses this divergence in East European experience:[10]

> For example, low-level members of the hierarchy misrepresent to their superiors what their capacities are for fear that the truth might lead their superiors to ask for too much from them. . . .
>
> Yet another commonplace problem is failure of internal control. Subordinate agencies pursuing their own special interests escape the authority of their superiors or become engaged in battles with other subordinate agencies, a problem so familiar to the Soviets and the Chinese that each has a name for it: vedomstvennost and pen-wei chu-yi.

Precisely this type of conflict occurs, for example, between the government's entrepreneurial and regulatory roles in the mangagement of new energy technologies. The government's efforts to promote new technology are in basic conflict with its responsibilities to protect the environment and the public health and safety.

A further problem is that the data base is usually inadequate for decisions at the center, for two reasons. First, because of multiple responsibilities and therefore the limited time available for any particular decision, the decision maker can assimilate only highly aggregated, i.e., very general, information. In addition, the information made available is often "filtered" selectively by staff so that it is not really representative of the issue at hand. These problems become more crucial the more centralized the decision making. Large corporations, though viewed as examples of successful centralized management, are in fact usually made up of units possessing a surprising amount of autonomy.

Perhaps the most serious weaknesses of all were pinpointed by Augustine Cournot in 1877:[11]

> . . . one must always fear, in the installation of economic regulations, the influence of men or classes who seek their particular interests against the general interest, for ordinarily the general interest is defended or advanced less effectively than the particular interest. Finally, even if one attributes all requisite knowledge and impartiality to the legislator, it is still necessary to entrust the execution of the regulations to a large number of agents of undistinguished capacity and morality, who discharge their duties without distinction only to the extent necessary to preserve their positions and obtain regular advancement, not the way men act when regulating their own affairs.

Finally, as we pointed out in the preceding chapter, central planning is an inefficient way to promote innovation, especially for such diverse, evolving technologies as those involved with energy use.

For all these reasons we conclude that detailed central planning is not merely inefficient; *it may not even be able to move a society toward its objectives.* It is wasteful, if not also counterproductive, to burden central government with more responsibility for detailed control than is essential to the well-being of the nation. The real question is, then, *To what extent is detailed central control necessary?*

Some activities do require detailed control at the highest levels. The U.S. government has decided that much contemporary energy-supply technology requires such control. We agree with that general assessment. *If* one assumes the necessity of continued growth in resource consumption and therefore continued and increasing reliance on today's large-scale supply technologies, then the need for detailed central control is inevitable. The discussions in Chapters 4, 5, and 6 showed that the investment required for each unit of additional energy *supply* is much higher now than in the past (so that the energy industry will be unwilling to invest in supply operations without extensive government assistance) and that potential side effects of these technologies are substantial (so that detailed centralized regulation is necessary to protect the public interest).

Alternatively, today's energy problems can be attacked at the roots by pursuing a new course for technological development—one emphasizing fuel conservation—in which the need for government assistance and restrictive government regulation is much less.

RELIANCE ON MARKETS IN A CONTROLLED ECONOMIC CLIMATE

Our investigation of the potential for fuel conservation is motivated by the same goals that are spurring the growth of pervasive central planning in the energy area: minimization of the costs and the adverse side effects of energy technology. However, the measures we have proposed for implementing a conservation strategy represent *a fundamentally different approach to natural-resource management.* Present policy fosters the development of increasingly unmanageable types of technology, in tandem with ever greater efforts at central management. We recommend, instead, fostering technologies which require relatively little central management, in tandem with a greater reliance on market mechanisms in a controlled economic climate.

We recommend a market-oriented policy because central planning is neither desirable, necessary, nor adequate for fostering the deployment of fuel-conservation technologies. We have already described the inherent general weaknesses of central planning. Although central planning may be necessary, despite these weaknesses, for the management of large-scale *energy-supply* technologies that involve high public risks, the private sector should be fully capable of managing the more benign, relatively small-scale technologies needed for an *energy-conserving* future. Moreover, while centralized planning may be suitable, in favorable circumstances, for managing a handful of standardized energy-supply technologies, a market strategy is the only effective way of managing the myriad energy-conserving technologies, which must be tailored to meet the changing needs of millions of individual consumers.

But while reliance on the market is the only way to keep down energy and energy-related costs and to manage effectively the evolution of energy technology, we are not talking here about that will-of-the-wisp the absolutely free market described by Adam Smith:[12]

> [Every businessman] neither intends to promote the public interest nor knows how much he is promoting it . . . he intends only his own security, . . . he intends only his own gain, and he is in this . . . led by an invisible hand to promote an end which was no part of his intention. . . . By promoting his own interest he frequently promotes that of society more effectively than when he really intends to promote it.

Instead we are talking about the market being free on a fine scale, in the sense that transactions between individual consumers and producers would not be encumbered by the heavy hand of public policy administered via price controls and allocation rules. Rather, such transactions would be guided by a public policy that controlled the market on a broad scale, via measures that would create the general economic climate in which these transactions would be carried out. The physicist and the artist are familiar with this distinction: one kind of order on a fine scale and quite a different kind of order on a broad or coarse scale.

The major public policy initiatives associated with this market strategy would be a deemphasis of economic regulation, the elimination of energy-supply subsidies, the levy of an energy tax as a substitute for certain existing taxes, measures to encourage the development of an energy-service industry (a new industry which provides not energy per se but energy services—in buildings, for example, this would mean comfortable temperatures, and adequate lighting and ventilation), and measures that protect the poor from the impact of high energy prices. All of these measures must be carried out in concert if there is to be a truly effective market-oriented policy for energy conservation.

THE ROLE OF REGULATION

Social regulations would still be important in our market-oriented strategy. However, such regulations would not be the impediments to economic activity that they are now for conventional energy-supply technologies. As shown in Chapter 12, there is usually such a diversity of technological opportunities for reducing the energy requirements for a particular activity that the most risky approaches can be avoided. Thus industry can minimize its regulatory burden by choosing a course of development that involves technologies posing low public risks.

For most *economic* regulations the situation is different. It has been shown that the indirect costs associated with the inefficiency of economically regulated markets are high.[6] Moreover, economic regulations help create the "need" for central planning. In contrast, movement toward deregulation would strengthen the market system. For such reasons economic regulations should be reexamined with an eye to elimination. We have stressed that policies to deregulate energy prices are needed so that energy prices will rise to the level of replacement costs. Not only should we deregulate oil and gas prices; we also should begin to move toward the deregulation of the price of electric-power generation (see Chapter 10). Deregulation of energy prices should be accompanied by the elimination of the federal fuel-allocation programs. These programs, involving a hierarchy of preferred fuel consumer classes, are made necessary by the pervasive system of price controls, which boost the demand for energy above the level of available supplies. Economic regulations relating to energy have evolved to the point where there is an astonishing degree of involvement of the federal bureaucracy in the marketing and management decisions of individual firms (see Table 15.2).

Table 15.2 Some news items concerning Department of Energy regulatory activity.

. . . Colorado Interstate Gas Co. [applies] for permission and approval to abandon its Little Polecat purchase meter station and a 450-horse power compressor, all located in Park County, Wyo.[41]

Petitioner [Gibson Drilling Co., Kilgore, Tex.] currently receives 35 cents per Mcf and requests a rate of $1.75 per Mcf of said gas (natural gas from a particular well in Rush County, Tex.). Petitioner plans to rework said well, but without special relief abandonment will result.[42]

Jewell Oil Co., Inc. is an independent distributor of motor gasoline and No. 1 and No. 2 fuel oils in and around Cobb, Wis. Prior to and during 1972 Union Oil Co. of California supplied Jewell with these products. . . . [The following issue arises]: If a supplier has arranged to supply a purchaser through a substitute supplier . . . and the substitute supplier terminates its relationship . . . is the original supplier still obligated . . . ?[43]

Take notice that . . . Ernest D. Huggard filed an application . . . to hold the following positions:

Vice President, Control, Atlantic City Electric Co., Public Utility, and Director, Vice President, Deepwater Operating Co., Public Utility.

Any person desiring to be heard or to protest such application should file a petition. . . .[44]

We are not so foolish, however, as to reject all types of economic regulation. We have suggested building energy-performance standards for rental housing because renters need protection from the high cost of operating energy-inefficient buildings (Chapter 8) and tighter fuel-economy standards for cars because, unless fuel prices are much higher, economic forces may not be adequate to continue automobile technology's current forward motion (see Chapter 9). Yet, even in these cases, we are less than optimistic about the potential efficacy of regulation. Because it is mass-produced by just a few firms, the automobile is close to being the ideal subject for regulation; yet even automotive standards have become shaky after a few years as "the system" has exploited both weaknesses in the method of determining performance and loopholes in standards (e.g., for light trucks). Thus, the imposition of performance standards should be a "policy of last resort", and, where possible, performance standards should be complemented by an energy tax or similar market-oriented policy.

With respect to regulation the thrust of our proposal is to ease the burden of control by encouraging technical development in relatively benign directions. This approach does not rest on an assumption of public spiritedness on the part of business, but on the *creation of an economic climate in which business would serve the public interest by acting in its own interest.*

THE ELIMINATION OF SUBSIDIES

Recent studies show that the special subsidies enjoyed by the energy-supply industry are substantial.[13] For investor-owned electric utilities, investment tax credits and rapid depreciation cost taxpayers over $2 billion annually. A similar sum is involved in the tax privileges of publicly owned electric utilities. Electric utilities also benefit from the very large federal R&D budget for new electrical supply technologies and from federal enterprises supporting electricity. In all, the subsidy to electric utilities amounts to about 10% of gross sales. The oil and gas industries likewise receive billions in tax reductions. Some of these privileges have been curbed in recent years, however. Total subsidies for oil and gas at present constitute a smaller percentage of gross sales than for electric utilities, perhaps 2 or 3%.[14]

Strangely, government and the public believe in the effectiveness of such subsidies. The energy industry and its supporters argue that oil companies need "incentives" to drill more exploratory wells and that utilities need tax breaks to help them build new power plants. This sloppy thinking is due partly to the effects of economic regulations which keep the price of energy to consumers below the replacement cost. If the price of energy were equal to the replacement cost, no incentives would be needed because energy demand would be lower and the industry would have adequate profits for bringing forth new supplies as needed. An essential step toward creating an economic climate conducive to fuel conservation is the elimination of all subsidies.

AN ENERGY TAX SHIFT

We propose that the elimination of energy price controls and subsidies be complemented by the levy of an energy tax. Revenues from this tax would be used to offset some existing taxes, so that the new tax would actually be an *energy tax shift*.

An Energy Tax Versus Conventional Incentives: Because higher prices for energy will not be popular, many will ask why we don't instead make the more appealing recommendation that fuel conservation be promoted with conventional incentives such as investment tax credits, accelerated depreciation, grants, and loan subsidies.

We believe an energy tax shift is preferable to such conventional incentives because an energy tax would promote fuel-conservation innovations more effectively than incentives.

By increasing the price of energy, an energy tax says in effect, "Conserve or pay more." Assuming investment funds are available and that a variety of conservation technologies are being marketed, a higher price allows the consumer maximum flexibility in conserving: he can either modify his energy-consuming behavior or choose from among the available technologies. If he elects the latter approach, he has complete freedom in making his choice.

By giving the consumer this freedom of choice, the energy tax creates a climate conducive to innovation. In contrast, conventional incentives, though better than nothing and politically popular (because they are an established means of rewarding special interest groups), stifle innovation by encouraging only a narrowly defined set of technological options for meeting energy needs that are diverse and continually changing.

We illustrate this problem with incentives by considering how a public official might establish an incentive program to promote conservation in industry. Suppose that a tax incentive, e.g., an investment tax credit or accelerated depreciation, is to be offered to firms that make energy-conserving investments. Since the responsible official would in effect be giving public money to firms that save energy, he would tend to be very careful in establishing guidelines for eligible investments, so as to make sure that he could not be accused of misappropriating these funds. Thus he would tend to restrict eligibility to single-purpose, well-understood technologies in applications where energy-savings opportunities could be unambiguously identified.

But, as Charles Berg has pointed out, many of the most promising opportunities for fuel savings arise with multipurpose activities—where innovations are aimed at simultaneously reducing the costs of several factors of production, such as capital, labor, and energy—and with new technologies, where performance is not well understood.[15] In some instances incentives may actually slow down the process of modernizing industry by extending the useful life of existing facilities that would otherwise be retired and replaced by new plants that would offer not only greater energy savings but an overall improvement in total productivity.

An alternative to giving industry incentives for specific equipment would be to base incentives on performance, i.e, on demonstrated energy savings. But this approach is also seriously flawed, because it poses a formidable measurement problem: saved energy relative to what? One might propose that energy consumed per unit of output be used to determine the incentive. One problem with this proposal is that direct energy consumption associated with a particular product represents only part of the total energy consumed. One must also take into account the energy embodied in purchases of inputs to the production process. But measuring this indirect energy use would be administratively impossible. The task of measurement is further complicated by the fact that quality changes in the product and changes in the production process which affect energy use (e.g., the introduction of air pollution control technology) are often taking place at about the same time that energy-conserving equipment is being installed. And even if these conceptual problems with measurement could somehow be overcome, the manufacturer often could not afford the instrumentation for, or for proprietary reasons would not permit, detailed measurement and outside evaluation of energy flows within his plant.

The difficulties with particular incentives are not due to inattention to detail. They are inherent. Incentives require simple, fixed specifications. Improved technology of energy use requires changing and multipurpose approaches in highly diverse situations.

The Rationale for an Energy Tax Shift: An energy tax shift—i.e., a tax on fuels as imported or extracted from the well or mine, with the revenues returned to taxpayers to offset some existing taxes—is the cornerstone of our proposal to create an economic climate conducive to fuel conservation.* The energy tax along with price deregulation would raise the price of energy to a level above the direct replacement cost, so that the price would reflect costs of energy not reflected in completely free market transactions. The tax would be made up of two parts†: one part would cover the social costs arising from energy production and use; the other would arise because nonrenewable resources belong not just to the present generation, but to future generations as well. Thus, the tax would result in reduction of the side effects of energy production and use and the conservation of nonrenewable resources for future use, by encouraging energy sources that are more benign and renewable, such as "saved energy" and solar energy.

The portion of the tax intended to make energy consumers pay the costs of side effects should be viewed as complementary to a regulatory strategy for controlling these problems. For reducible risks, regulations can be partially effective, but they are very imperfect public policy instruments. In the case of pollution, regulations usually specify that emissions or ambient concentrations of pollutants cannot exceed specified threshold levels. Yet in many instances no threshold can be established for ill effects. A fuel tax to cover pollution costs could partially overcome this shortcoming of regulation. Such a fuel tax would be far easier to enforce than a tax on specific pollutants; moreover it could cover a variety of damaging side effects of energy systems, as discussed in Chapter 5. For irreducible risks, a tax may be more practical in many instances than an outright ban on the offending activity; the tax would permit an established industry to selectively wind down the offending activity instead of halting it all at once, so that economic dislocations could be minimized.

The component of the energy tax which addresses the circum-

*An energy tax is not an entirely new idea. Severance taxes on particular fuels have been levied at the point of extraction in a number of states. The more general use of a severance tax as a natural-resource policy tool has been discussed in an important book on the economics of natural resources by Talbot Page.[16] Researchers at the University of Illinois have estimated how an energy tax shift might reduce energy demand and affect consumers.[17,18] Jay Forrester has proposed a tax shift involving a tax on oil and gas, to promote the conservation of these fuels.[19] And Congressman John Anderson proposed in 1979 a 50¢-per-gallon tax on gasoline to offset the social security tax.[20]

†Another component of the energy tax could be justified in a situation where price controls continue. In this case an energy tax shift could be imposed to raise the price of energy to the direct replacement-cost level. Rosenfeld and Fisher have proposed such a tax shift for electricity.[21]

stance that nonrenewable resources belong in part to future generations is needed because today's businessmen give insufficient weight to future values relative to present values.[22] A powerful illustration of this has been provided by Colin Clark in an analysis of the whaling industry.[23] One might think it would be desirable to maximize the sustained yield of whales. Clark shows, to the contrary, that under a wide range of conditions it is in the interest of the whaling industry to exterminate the whale! The well-known ruthlessness of the industry toward the stock of whales is explained by the need common to business for a quick payback on capital. With this constraint it pays to utilize capital as heavily as possible, giving up the whaling business when the resource is essentially exhausted.

In the case of oil, the requirement of a short payback period is reflected in the emphasis given to rapid exploitation of established fields. The prominent oil economist M. Adelman has written about Middle Eastern oil:[24]

> Depletion of reserves at the Persian Gulf is only about 1.5 percent a year. It is uneconomic to turn over an inventory so slowly.

The way the present economic system works, it pays business to sell off its oil resources as fast as possible.* As Harvard economist Marc Roberts[25] has said:

> . . . resource owners would have lost money by holding back on sales. It was far better from a narrow financial viewpoint to do what most of them did, develop their holdings and reinvest their profits. As long as a 7 percent return [in real terms—i.e., in addition to inflation] is available *elsewhere,* unless resource prices double [in real terms] every ten years, an owner loses money by not selling immediately. [Emphasis added.]

Recent advertising by the Mobil Oil Company is easy to understand in these terms. Mobil advocates rapid development and use of domestic oil resources.† (Not surprisingly, its statements are written as a defense of consumer interests.)

This very short-term view of resource values contrasts with that of the society as a whole. What the latter view is, or should be, is uncertain, but a very reasonable position has been stated by Robert Solow:[22]

*This is not true, of course, if the resource is controlled by a monopolist or by a cartel. Because the OPEC cartel is able to set and maintain high world oil prices, some OPEC nations can afford to hold back on oil production, because the oil revenues they receive are far in excess of what their economies can absorb.

†The firm does not need to stay in the oil business; it can adapt as oil is exhausted. Thus many oil companies have become energy companies. Mobil Oil has expanded beyond energy, by purchasing control of Montgomery Ward and the Container Corporation of America.

> In social decision-making there is no excuse for treating generations unequally and the time horizon is, or should be, very long. In solemn conclave assembled, so to speak, we ought to act as if the social rate of time preference were zero.

A social rate of time preference greater than zero would bias decisions about resource exploitation in favor of the present generation.

Because these arguments for an energy tax apply with varying force from one form of fuel to another, the energy tax need not be the same for all forms. Petroleum use should be discouraged with a high tax, both because of the high social cost of dependence on insecure sources of foreign oil and because it appears that petroleum resources are especially limited. Coal and nuclear fuels would also be taxed at a high rate, to reflect the serious side effects inherent in their use (see Chapter 5). The tax on natural gas could well be less than the tax on petroleum because gas is cleaner and probably more abundant. Domestically produced renewable resources should perhaps not be taxed. Thus the energy tax and other complementary conservation policies would discourage a shift to coal and nuclear energy, encourage the development of benign renewable resources, and, through reduced demand, extend substantially the life of oil and gas resources.

The exact level of the tax implied by the various side effects and resource-depletion considerations is not easily determined, because of fundamental problems involved in expressing these costs in monetary terms, as discussed above. Rather than use a pseudo-rigorous cost-benefit analysis to determine the energy-tax level, the level of the tax should be such that the estimated resulting pollution levels, rates of resource extraction and importation, and distribution of energy forms are in accord with the broad goals of an overall policy of natural-resource management—a policy which in the final analysis must be established in the political process.

Not Another Tax! Our proposal for an energy tax will certainly be viewed initially with alarm or skepticism. We are not proposing to increase tax revenues, however, but to introduce a tax shift: the energy tax would be accompanied by a coordinated rebate or reduction in other taxes, so that typical people and firms would stand still, so to speak. It is important to keep funds in the hands of energy users so that they will be in a position to invest in energy-efficiency improvements. While a tax shift would generate no new revenues, it would give the nation the opportunity to reduce those taxes which are economically inefficient in their impact, onerous to administer, or highly uneven in their application. Sales taxes, social security taxes, individual and corporate income taxes, and property taxes are candidates for reduction in favor of an energy tax.

A general energy tax levied on all fuels at the point of extraction would be slightly regressive. It would place a relatively higher burden on the poor—but only slightly so because total energy use—direct plus indirect—increases roughly in proportion to income (see Figure 8.11). A general energy tax would be no more regressive and might be less regressive than general sales taxes or the social security tax. Through design of the tax reductions or rebates accompanying the energy tax a net progressive effect can be achieved, as discussed below.

An energy tax used to offset income taxes would help stimulate demand for goods and services that are labor intensive, while inhibiting demand for energy-intensive products. An energy tax may also be more appealing than the property tax; being unrelated to current income and expenditures, property taxes create severe hardships for some, particularly in times of inflation.

The energy tax shift could be established initially by increasing the gasoline tax. The advantage of this approach is that the tax mechanism exists, and the viability of high taxes has been demonstrated in Europe, where the gasoline tax is roughly 10 times as large as the 12¢ a gallon U.S. tax. (In France and Italy the tax is especially large—$1.56 a gallon in 1979.)[26] By way of illustration, at the present rate of gasoline consumption, revenues generated by raising the gasoline tax by about $1.20 a gallon would be more than adequate to eliminate the social security tax.[27] Of course raising the price of gasoline by $1.20 a gallon suddenly could cause serious dislocations. Thus, the tax should either be introduced slowly over a period of a few years, so as to enable consumers to make the transition smoothly, or, if it is introduced rapidly, the tax revenues should be refunded in a manner that protects those consumers that would be most sensitive to a sudden price rise. After a few years the combined impact of the gasoline tax and automotive fuel-economy standards would be diminished demand for gasoline. As the gasoline tax base declined this way, the tax base could be extended to cover all nonrenewable fuels.

An important concern relating to the energy tax is its impact on different income groups. To illustrate this issue, let us suppose that an energy tax were applied equally to all fuels at a level that would be adequate to eliminate the social security tax. This across-the-board tax would be equivalent in energy terms to a gasoline tax of 22¢ a gallon.[28] We have estimated the impacts of this tax shift on three sample households with incomes in 1978 of $8000, $17,700, and $26,900. The $8000 income is at the 26th percentile level—i.e., 26% of all households have lower incomes. Similarly, the $17,700 and $26,900 incomes are at the 50th and 82nd percentile levels.[29] Because of the energy tax, the household with an income of $8000 would have paid in 1978 $398 more for energy and $106 more for nonenergy purchases. However, as shown in Table 15.3, these extra costs would have been approximately offset by the elimination of the social security tax. A similar result would have occurred at the highest income level considered here, while the middle income group would actually have come out ahead. This approximate balance at different income levels between the impacts of an energy tax and of a social-security-tax reduction arises from the fact that total energy consumption (direct plus indirect) is approximately proportional to income, as shown in Figure 8.11.

The approximate balance shown in Table 15.3 between the extra expenses and the savings associated with this tax shift would not continue to apply in practice. The principal reason is that many consumers would take advantage of opportunities to save energy and would come out well ahead at all income levels. The burden of the energy tax would fall on consumers who use energy inefficiently.

The energy tax might instead be used to offset individual and corporate income taxes. Jay Forrester has proposed that the revenues from a tax on oil and gas be returned to the economy in this manner. As

***Table 15.3** The effect of a hypothetical fuel-tax shift on three households in 1978. The fuel tax is nominally taken to be $1.75 per million Btu on nonrenewable energy sources, corresponding to a 57% increase in the average price of energy (1978$).*

HOUSEHOLD INCOME	$8000	$17,700	$26,900
Increase in Direct Energy Cost			
Residential	296	366	424
Automobile	102	255	401
Net Increased Cost for Nonenergy Purchases	106	334	268
Total Extra Expenses	504	955	1093
Employee Social Security Tax Reduction	484	1071	1071

SOURCE: See note 29.

Forrester states, a tax on energy and a corresponding reduction of "taxes on people" would establish a more favorable balance in the economy between use of energy and use of labor. Energy consumption in the U.S. is so high that if there were a uniform tax on all fuels at a level equivalent to 24¢ a gallon of gasoline, the tax at the 1978 level of energy use would have sufficed to offset half of personal income taxes plus all of corporate income taxes in that year.[30] With this proposal the residual personal income tax could be designed to retain the income-redistribution feature of the present tax structure.

An energy tax that would continue to offset income taxes in these proportions would of course have to be steadily increased in the energy-conserving future described in Chapter 13. The tax averaged over all forms of energy would have to increase to the equivalent of 65¢ a gallon of gasoline by 2010, if energy demand were to fall as projected in Chapter 13 and if revenue requirements were to increase in proportion to GNP.[32] With this tax the price paid by consumers for energy in 2010 would be roughly double what it would have been without the tax. However, *after imposing the tax, the fraction of GNP spent on energy (including the tax) would actually decline in the period to 2010*, because the increase in price would be more than offset by the reduction in energy demand.[33]

These dramatic results suggest that the commonly held view that an energy tax would be "politically unacceptable" does not take into account the tremendous political benefits that could be derived from this policy. We are inclined to agree with Jay Forrester's perception of the political acceptability of an energy tax shift:[19]

> I have been fascinated by reactions of people to this proposal for a high tax on petroleum products. On June 1, 1977 I testified along these lines before the Subcommittee on Energy and Power of the U.S. House of Representatives Commerce Committee. I had expected strong disagreement. To my surprise, most members expressed personal belief that such a move was necessary and in time must be taken. But, they said, how would they explain it to their constituents? Since that time, I have made the proposal to a wide cross section of those constituents. Almost all agree. But, they say, how would they explain it to their Congressmen?

These energy-tax-shift proposals imply a fundamental reform in taxation: replacement of taxation of labor, whether through social security or individual income taxes, by taxation of natural-resource exploitation. The opportunity is also created to modify corporate taxation to encourage capital investment. The nation's history suggests that such a shift is timely. As Robert Samuelson has noted, in an article endorsing

Congressman John Anderson's proposal for a gasoline tax to offset the social security tax:[34]

> . . . the question for the 1980's is simple: how do we tax ourselves? Do we tax labor via higher social security taxes (already scheduled to rise sharply) and higher marginal income tax rates? Do we tax capital and investment by keeping those rates high? Do we tax ourselves via inflation by running large deficits, which tend to spur higher money growth and more inflation? Or do we ease these tax burdens by taxing a critical commodity whose supply is scarce and, far more important, uncertain?

But even if the nation and average citizens would benefiit, what of the two main parties at risk, certain producers and the poor?

FUEL CONSERVATION AND THE PRODUCERS

The policies we propose would have a relatively minor impact on the daily lives of people as consumers. In an economic climate conducive to investment in conservation technologies and to technological innovation, people would be provided with the energy amenities they now seek, with much lower fuel consumption and at lower cost. The major economic impacts of the policies proposed here would fall on certain producers.

One impacted group would be energy-intensive industries like the steel, cement, paper, and chemical industries. Even though firms producing such products could pass cost increases on to their consumers, as taken into account in Table 15.3, many of these firms would not have access to capital to make desired efficiency improvements. Their financial position would be worse than ever. The detailed design of the tax shift should address this problem by following the principle that when the tax shift is implemented, the energy taxes are initially matched by tax reductions and/or rebates that leave every major category of firm in a neutral position.

Another impacted group would be the energy industry itself. Investments in energy supply would be sharply reduced from the level in the late 1970s, when energy supply investments accounted for more than 40% of all new plant and equipment expenditures. Such capital hogging by the energy industry, which has been assiduously encouraged by government, is damaging to the economy. In the economic climate we propose, the overall level of energy-related investments would be reduced; major investments would be made in less costly conservation technologies, while investments in energy supply would be primarily for replacement of retired capacity rather than for expansion.

The conventional energy-supply sector, which is dominated by large, growth-oriented corporations, would find this no-growth situation hard to accept at first. The resistance of these corporations would be the most formidable stumbling block for the changes proposed in this book. The industry's opposition to most elements of President Carter's National Energy Plan in 1977 was enough to reduce that plan to a motley collection of provisions that serve special interests. (For a fascinating glimpse of that lobbying activity, see the article by Elizabeth Drew, ref. 35.)

On the other hand, although resistance by the energy-supply sector is inevitable, firms in this sector could adapt handily once they could rely upon the new conservation policies. Building on their solid capital-raising position, many firms could become part of the fuel-conservation industry. For example, Exxon's plan to develop and market its electricity-saving alternating current synthesizer is a bold energy-conservation venture that will probably be extremely profitable.[36] More generally, the energy industries could be transformed into energy-service industries. Energy companies could become suppliers of services like industrial process heat, transportation, and space conditioning rather than of energy as such. We illustrate this concept with two important opportunities for energy utilities.

New Roles for Utilities: Most electric utilities have shown little interest in, and often have shown hostility toward, fuel conservation. The utilities launched a successful lobbying effort, for example, to eliminate a very modest fuel-conservation measure from the National Energy Plan submitted by President Carter to Congress in 1977—a measure that would have required merely that consumers' electricity bills reflect in a reasonable way true costs as these costs depend on time of day and season of the year. Even this simple reform aiming to make prices better reflect true costs was "too radical" a change for the powerful and often rigid electric utility industry. In many instances, however, utilities are in an advantageous position to participate in fuel conservation. We have stressed two important roles for utilities—industrial cogeneration and conservation investments for buildings.

Recall from Chapter 10 that a higher level of fuel savings and larger gains for the economy as a whole would result from utility ownership of on-site cogeneration facilities than from ownership by industrial firms which use process steam, because the most advantageous technologies need a strong connection to the utility grid. Major public policy changes are needed to encourage this role for utilities. "Institutional demonstration projects" are needed to explore the viability of utility ownership, or partial ownership, of cogeneration facilities and to test alternative schemes for pricing the produced steam and electricity.

The other important opportunity we have stressed for energy utilities is the financing of investments for fuel conservation in buildings. Utilities offer four special advantages here: (1) they have the ability to accumulate the large capital resources needed for conservation investments; (2) they already have a management structure for channeling this capital to consumers; (3) through their billing service, in which they can bill customers simultaneously for energy used and for the loan payment, they can enable customers in a very simple way to make judgments about conservation investments in terms of lifecycle costs; and (4) because utilities are regulated, they can be mobilized by government to finance and manage conservation programs.

Significant public policy changes would be required to promote utility involvement in conservation efforts—to motivate utilities and to ensure that the public interest would be served (e.g., local independent financial institutions should be able to have a role in the financing of conservation investments). It is particularly important to develop innovative schemes for utility financing of conservation investments. One approach would be to allow utilities to finance conservation investments and include these investments in the rate base (see Chapter 8). This arrangement could help make conservation investments attractive to utilities, and it would encourage conservation investments in those situations where the obstacles to lifecycle costing are especially great—as is the case with renter-occupied buildings and for low income groups (see Chapter 8).

Of course, utilities would undergo great changes as a result of the policies advocated here. Their planning would have to be done all over again—an especially repugnant prospect for them. But *electric utilities' planning will have to be fundamentally revised no matter what the public policies.* Present expansion plans of utilities, which continue to be based on electricity growth rates much higher than that of GNP, are unrealistic because of increasing real prices and the onset of saturation, discussed in Chapter 3. The salient point is that policy changes to promote fuel conservation would create major new opportunities for the energy utilities in spite of diminished opportunities for growth in energy sales. It is to be hoped that they will support such policy changes and seize the opportunities.

New Fuel-Conservation Businesses: The fuel-conservation industry would involve many parts of the economy. Vendors who today provide equipment for the energy-supply industry and other large manufacturers would create many new product lines. Efforts to save energy in space heating, for example, could create a huge demand for heat pumps, for heat-saving windows, and for air-exchange heat-recovery devices. The scale of these and most other conservation devices is well suited for technological improvement through competitive trials and through cost-cutting mass production.

In addition to providing opportunities for many large firms with mass-production, marketing, and financing capabilities, conservation offers the potential for creating many new businesses. In light of the diversity, the relatively small scale, and the regional variability of much of the technology, it is very likely that the fuel-conservation industry which evolves will include many small and medium-sized firms.

The variety of businesses which can participate in providing energy services reflects the fundamental technological fact that there are many ways to increase the efficiency of providing any energy service. The policies we propose would bring these technologies into more active competition with increased energy supply. The depth and diversity of these opportunities for competition would ensure that the market would function well—a complete reversal of present conditions. Consumers would have many choices in their purchases of energy services, and competition would lead to minimized costs for these services.

But this is a statement about an equilibrium condition, and we must examine how the poor can be adequately protected during the period of adjustment.

FUEL CONSERVATION AND THE POOR

The fuel-conservation program we propose, even if well designed, involves adjustments, including changed expectations, for many people. It would involve dislocations for poor people as well as for the energy industry. The country will certainly hear from the powerful energy industry; it may not hear so much from the poor.

Consider our tax-shift proposal. The energy tax must be complemented by other measures so as to avoid imposing a harsh burden on low income groups. The tax should be substituted for other taxes that are particularly regressive, or part of the tax revenues should be rebated directly to the poor. And the tax should be imposed in parallel with measures that provide for the improved performance of equipment used by the poor and of the buildings occupied by the poor. The creation of energy-service industries, as exemplified by the Oregon utility conservation plan, would hold down energy-related costs. This strategy should be supplemented by steps designed to effect energy-efficiency improvements for low income people—thermal performance standards for rental properties, energy-efficiency standards for energy-intensive appliances, and the strengthened automotive fuel-economy standards called for in our program.

But this "package" of proposals may not adequately protect the poor. One problem is that in the political process a package of proposals rarely survives in a coherent form. But even if the whole package were to make it through the political process, the program would likely fail to match the essential needs of certain affected groups, and there would likely be mismatches in timing between different parts of the program, resulting in temporary hardships. There are no easy answers here. Inevitably some dislocations will occur, but policies can be formulated to minimize the hardships. If careful analysis indicates that the package policy proposed here would not provide adequate protection for the poor, we would recommend that the energy tax shift be temporarily accompanied by direct steps to improve the housing or equipment at issue. *But under no circumstances should the energy-tax-shift concept, the cornerstone of an effective fuel-conservation strategy, be abandoned.*

While many individuals, as consumers, will encounter some difficulties in the initial years during which such a fuel conservation program works itself out, we believe that changes in the structure of employment associated with our proposals would be more important to lower-income people than changes experienced by them as consumers. Clark Bullard has shown that on balance, production associated with energy-conservation devices and low-energy-intensity products would be characterized by higher employment levels per dollar of economic activity than would the production that was replaced.[37] E. R. Berndt has pointed out that all of the econometric studies published to date show that energy and labor are substitutable inputs in the overall economy and that rising energy prices will lead to substitution of labor (employment) for energy.[38] Thus our proposed energy tax shift would tend to generate employment.

This new employment would also tend to be less specialized. For example, a case study by Bruce Hannon of reusable versus throwaway bottles shows that with reusable bottles, new jobs in retail stores, local delivery, and bottle cleaning would replace the smaller number of jobs lost in basic-materials processing, assembly-line operation, and long-distance shipping associated with the throwaway option.[39] Similarly, the demand for conservation-related services, such as repair work on durable products and construction and renovation of buildings to improve energy performance, would create new jobs requiring rather general skills. This increase in less specialized employment would, in our opinion, be a very favorable side effect of energy conservation in that it would tend to reduce structural unemployment. Since the more specialized employment is well established and more strongly unionized than the new employment that would be created, society might resist these employment-for-energy substitutions.[40] Nonetheless, we believe the policies proposed here—if sensibly enacted and administered—would help most of the poor.

While it is essential that programs to ease resource problems be designed to reduce the pains of adjustment felt by the poor, it would be quite another matter to fundamentally mis-shape natural resource policies in order to create special benefits for the poor. Effective policies to attack energy problems cannot also be expected to solve age-old income distribution problems.

CONCLUSIONS

In the past, society has force-fed economic growth with policies that encouraged the exploitation of natural resources. While these policies have led to a very high level of material affluence, they have also led to a rate of consumption of natural resources so high that the national security and environmental side effects of industrial and consumer activities are very damaging. The damage and threat of damage make necessary a new ecological approach to natural-resource management, in which consideration of the side effects becomes a major factor in charting the course for technical change. To date, the approach to ecological management has been merely to superimpose ad hoc control strategies upon an institutional structure that encourages resource exploitation. We have argued that this approach to ecological management is fundamentally flawed: even in favorable circumstances, this ad hoc control strategy will be only partially effective and will place severe constraints on industrial activity.

This book proposes a different approach to ecological management: pursuit of technical change that emphasizes resource conservation instead of resource exploitation. Through the example of energy, we have shown that the possibilities for innovation and economic growth through conservation are very great.

Whereas most discussions of conservation emphasize lifestyle changes and even sacrifice, our approach has focused entirely on technical change. Following our proposal, U.S. energy use would decline, but per capita energy use would still be high—in 2010 about 70% higher than in most of Western Europe and more than 10 times higher than in less developed contries (LDCs) today. Because it will be exceedingly difficult if not impossible to eliminate the consumption inequities between industrialized and less developed countries solely by raising consumption levels in LDCs, it is clear that in the long run it would be prudent for Americans to adopt lifestyle changes which would further reduce the impact of their activities on the world resource base and on the global environment. But we have not considered lifestyle changes in our discussion.

The challenge presented in this book is to *change the direction of technological development.* What are needed are new directions in education, research, and product development, and a more hospitable economic climate to facilitate the development and marketing of conservation technologies by the private sector. New policies are required to overcome the institutional barriers to the course we propose.

The proposed new direction for natural-resource policy provides the opportunity for producers, consumers, and environmentalists—groups now often locked in paralyzing struggles—to join in a mutual endeavor. For producers there would be new opportunities for providing energy services in an economic climate less constrained by regulations; for consumers the costs of energy services would be minimized; and for environmentalists there would be the prospect of a cleaner environment. This opportunity to reduce divisiveness is one of the major benefits of our proposal.

It is essential to begin the process of change now. Because of the way society is evolving, it will be much more difficult later to bring about a transformation of the U.S. economy. Dramatic change is much easier to bring about when economic growth is rapid than when it is slow.

Demographic trends are favorable for rapid economic growth during the next few years. As pointed out in Chapter 3, the labor force will continue to grow rapidly over the next decade. If rapid growth in employment were complemented by moderate growth in productivity, then the economy would grow rapidly over the next decade. Thereafter, the labor force will be growing extremely slowly and may even decline, so that economic growth will probably be slow as well. In short, the next decade, when the economy can potentially grow as fast as it did historically, may well be the last foreseeable opportunity for rapid growth of the U.S. economy.

Though demographic trends favor rapid economic growth in the near term, the prospects for growth in productivity are uncertain. The economy has been particularly sluggish over the last decade. A major cause of this slowdown appears to be the diminishing opportunities for significant innovations in the areas which contributed most to U.S. economic growth in the past—the internal-combustion engine, steel, man-made fibers, etc. *Initiating fuel-conservation policies now would markedly improve the prospects for economic growth* and thus would help create the favorable economic climate needed to sustain the effort. Fuel conservation offers major new theaters for innovation and opportunities for productivity gains. This potential benefit alone should persuade a thoughtful society to make the necessary changes in public policy.

America above all is its people. The people are at an unusual moment in their history. The large cohorts of young adults now moving into the labor force can, given forward-looking policies, be the architects of a strong economy which is compatible with the nation's resource base.

APPENDIX A.
ENERGY CARRIERS

An understanding of the roles of different forms of energy requires some notions about the transformation of primary energy (energy found in nature) into energy carriers (energy delivered to consumers in forms such as gasoline and electricity) and, especially, about the appropriateness of each carrier for various end uses. This sequence of energy forms is shown in Table A.1.

TYPES OF ENERGY CARRIERS

The physical forms in which energy is transported and stored before final conversion and consumption at the point of end use can be divided into four main categories: fluid fuels (gases or liquids), solid

Table A.1

PRIMARY ENERGY SOURCES (the forms of energy found in nature)

EXHAUSTIBLE	INEXHAUSTIBLE
Oil and Gas	Falling water, wind
Coal	Sunlight
Uranium*	Deuterium (for fusion)

ENERGY CARRIERS (the forms of energy transported and stored prior to conversion and final consumption at the point of end use)

Fluid chemical fuel
Electricity
Solid chemical fuel
Thermal energy

END USES FOR ENERGY

Heat
Cold
Motion
Electrolytic process
Feedstock

*Exhaustibility depends on technology.

fuels, electricity, and thermal energy.* We shall refer to these forms of energy as energy carriers. The carriers are usually different from primary energy forms (the forms in which energy resources are extracted from the ground), such as crude oil or uranium, but they need not be different. For example, natural gas is both a major primary form and a major carrier.

The properties of an energy carrier—especially its storability, its transportability, the ease of its conversion at the point of end use, and its environmental acceptability—establish its relative value for a particular end use. Table A.2 lists primary sources of supply, end uses, and selected characteristics for a number of energy carriers.

There is a wide range in the storage characteristics of different energy carriers. The data in Table A.3 show clearly that the storage characteristics of solid and liquid carriers (high energy density and low capital cost) are hard to beat.

As a general rule, the storability of a form of energy is closely related to its transportability. Fuels that are easily stored—coal, oil, uranium—can be economically shipped in batch mode—barge, railroad, truck, etc. Those that are expensive to store—electricity and natural gas—are most efficiently transported by line, or continuous, mode.

The systems required for distribution to, and conversion at, the site of end use are also important determinants of the appropriateness of a carrier. In small-scale applications (buildings, transportation, and small industry) problems involved with such systems inhibit the use of solid fuels. Coal and wood, however, can be used to heat individual houses in cases where (1) a distribution system for a more convenient carrier is not available, and/or (2) prices for energy in these forms are extremely low.

The storability of energy affects not only the physical structure of the delivery system but also the institutional structure. Since coal and oil can be brought to the retail market in batch mode, they can be sold competitively by multiple distributors. Since natural gas and electricity, on the other hand, entail the installation of an immobile distribution system, they are most efficiently distributed by a local monopoly. If methods were discovered to store these carriers in an inexpensive, compact form, they might also come to market in batch mode and be sold competitively.

* Thermal energy at moderate temperature is an important energy carrier. Although thermal energy is most often produced when and where it is used, it can be stored at moderate cost and, with appropriate design, for long periods. Transport of moderate-temperature thermal energy is practical over short distances, but is impractical over long distances. Since it is low-quality energy, its transport tends to involve large material flows. Steam and hot-water lines of 1 or 2 miles are common. Hot-water transport for tens of miles is being attempted.

Table A.2 Energy Carriers, Selected Characteristics

ENERGY CARRIER	PRIMARY ENERGY SOURCES	RELATIVE COST OF STORAGE	MODE OF TRANSPORTATION	RELATIVE COST OF DISTRIBUTION[a]	ENVIRONMENTAL IMPACT AT POINT OF END USE[b]	END USES
Fluid Fuels						
Liquid Hydrocarbons	Crude Oil, Well Gas, Coal, Biomass	Low	Batch, Line	Low	Moderate	Transportation, Heat, Feedstock
Methane	Well Gas, Crude Oil, Coal, Biomass	Moderate	Line	Low	Low	Heat, Feedstock
Hydrogen	Well Gas, Coal, Electricity, Biomass	Moderate	Line	Moderate (?)	Very Low	Heat, Feedstock, Transportation (?)
Solids						
Coal	Coal	Low	Batch	High[c]	High	Heat, Feedstock
Biomass	Biomass	Low	Batch	High[c]	Moderate	Heat, Feedstock
Electricity	Coal, Nuclear, Hydro, Biomass, Wind, Photovoltaic	High	Line	Moderate	Very Low	Electronic/Mechanical Activation, Lighting, Heat/cool
Thermal Energy	All	Moderate	Line	Moderate	Very Low	Heat/Cool

[a] Relative costs of distribution and handling at point of end use.
[b] Entries in this column can be interpreted as indicating relative costs of environmental control.
[c] The cost of handling solid fuel is much lower in very large installations.
SOURCE: Prepared by Joseph G. Asbury.

APPLICATIONS OF THE CARRIER CONCEPT

The properties and costs of the energy carriers determine their value in end-use markets. This section provides an overview of energy-use patterns in transportation, building heating and cooling, and industrial markets.

Table A.3 Energy-Storage Characteristics of Alternative Carriers

ENERGY FORM	DEVICE	SIZE	ENERGY DENSITY (kWh/kg)	CAPITAL COST ($/kWh)
Coal	Stockpile	10^5 ton	7	10^{-5}
	Bin	10 ton	7	.005
Fuel Oil	Tank	300 gal	11	.02
	Tank (auto)	15 gal	11	.06
Thermal	Water Tank	10^3 gal	.06	4
	Aquifer	10^7 gal	.05	.05
Electricity	Battery	1 kWh	0.5	50
	Pumped Hydro	10^7 kWh		10

SOURCE: Prepared by Joseph G. Asbury.

TRANSPORTATION

Liquid fuels have the advantages of high-energy density, low-cost storage, and the availability of relatively easily managed technologies for conversion at the point of final use. These advantages make liquid fuels ideal for transportation.

While considerable attention has been given to electricity as a fuel for transportation, electricity will be unable to compete with liquid fuels for most transportation applications without major breakthroughs in storage technology. (Table A.3 shows that the capital cost for storing energy in a battery is 1000 times as large as the cost of a gasoline tank.) Although batteries and mechanical storage are under development, there is little expectation that the advantages of liquid fuels will be overcome soon. Note, however, that low-performance vehicles and fixed-route vehicles have been economically powered by electricity.

The advantage of liquid fuels for transportation is apparent when one considers that automobiles with the best fuel economy now available, over 40 miles per gallon, could be supplied with a liquid fuel costing the equivalent of $3 per gallon of gasoline at the same energy cost per mile as characterized the average automobile on the road in 1979. Thus even with extremely high prices, at which liquid fuels would be available from many alternative sources, liquid-fuel-driven vehicles would still be "affordable."

BUILDING SPACE CONDITIONING

Space conditioning must typically be provided at low capital and maintenance costs. Although fluid fuels (oil and natural gas) are now the major energy carriers for this application, space conditioning is also provided by electricity and, in some dense areas, by thermal carriers (steam and hot water).

There is a significant problem with the use of electricity as the *major* carrier for building space conditioning: because of the high cost of storing electricity (see Table A.3) it is very costly to meet the temporal variation in demand. The demand for electricity on cold winter mornings or on hot summer afternoons greatly exceeds the average level of demand. The problem would become very serious in the northern region of the country if extensive use of resistive heating were to be adopted. There are "load management" strategies that can help. Solutions may also be sought with the help of thermal storage and fluid-fuel-driven peaking systems to provide energy at times of peak need (either onsite or tied into the electrical grid).

INDUSTRIAL LOW-TEMPERATURE HEAT

The overall inefficiency of electric resistive heating places this option at an extreme cost disadvantage for meeting industrial low-temperature-heating requirements. It seems likely that other alternatives to today's oil- or gas-fired boilers will become important. Some possibilities are coal-fired boilers (especially using fluidized-bed combustion), cogeneration of steam and electricity based on fossil-fuel sources, and heat pumps with industrial capabilities.

FEEDSTOCKS

Feedstocks are fuels used for nonenergy purposes, i.e., as the basic materials from which chemicals and construction materials are made. The principal feedstocks are now oil and natural gas. Feedstock prices could rise substantially without seriously affecting the economic viability of the product (with the exception of certain high-volume products, e.g., certain building materials, which would become less competitive). Thus oil and gas will continue to be major sources of feedstocks.

OVERVIEW OF THE DEMAND FOR CARRIERS

Because of the advantages of fluid fuels for most transportation, for feedstocks, and for certain heating applications, these energy carriers will continue to play a major role in the energy economy. We estimate that *at least* 40% of primary energy use will be associated with fluid fuels in the long run. We obtain this estimate by adding the present percentage of total fuel use accounted for by transportation and feedstocks to ¼ of all space-conditioning and process-steam requirements. (For low-temperature heating we project substantial but not complete displacement by other carriers as the eventual state of affairs.)

The remaining 60% of primary energy use will be associated with electricity and with solid chemical fuels and thermal energy. In an energy future where electricity is heavily promoted, much (but not all) of the nonfluid-fuel demand can be met with electricity. Because of substantial cost advantages in certain applications, it is likely that solid fuels and thermal energy will ultimately account for at least 10% of total primary-energy use. Thus in a "highly electrified" energy economy production of electricity would account for no more than about half of primary-energy use. (In 1978 electricity production accounted for 29% of primary-energy consumption.)

The critical figure here is the estimated 40% minimum share for fluid fuels. How dependable is this estimate? Possible changes in end-use technologies and in relative levels of demand for products and activities give rise to significant uncertainty. On the other hand, as stated above, most transportation, feedstock, and backup and peaking uses for fluid fuel would be justified in the long term even were the price of fuel to rise substantially. This general economic argument suggests that the 40% is a fairly "hard" number.

APPENDIX B.
ANNUAL CAPITAL-CHARGE RATES FOR CAPITAL INVESTMENTS

Suppose that an amount of money I_o is borrowed for N years to purchase capital equipment (or land) at an effective cost of money d (the discount rate) and that in the Nth year the equipment is sold for a salvage value I_N. If the loan is to be paid in N equal payments $\beta' I_o$, then

$$\beta' = \frac{1}{1-t}\left\{d + (1 - I_N/I_o)\left[\frac{(C(d,N)}{(1+d)^N} - \frac{t}{N}\right]\right\} \tag{B.1}$$

for a utility and

$$\beta' = C\,(r_d, N) - t\,\bar{r}_d - \frac{C(d,N)}{(1+d)^N}(I_N/I_o) \tag{B.2}$$

for a homeowner, where

$t \equiv$ income-tax rate (combined federal and state)
$r_d \equiv$ interest rate on a bank loan (debt)
$\bar{r}_d \equiv$ average (levelized) interest rate $= C\,(r_d,N)\left[1 - \frac{1}{(1+r_d)^{N+1}}\;\frac{C(d,N)}{C(d',N)}\right]$

$$d' \equiv \frac{d - r_d}{1 + r_d}$$

and

$$C(r,N) \equiv \text{capital-recovery factor} = \frac{r}{1-(1+r)^{-N}}$$

For a utility the effective cost of money is

$$d = \alpha_e r_e + (1-t)\alpha_d r_d \tag{B.3}$$

where

$\alpha_e\ (\alpha_d)$ = fraction of investment from equity (debt)
r_e = annual rate of return on equity

while for the homeowner the effective cost of money is

$$d = (1-t)\ r_d \tag{B.4}$$

In B.3 and B.4 the factor $(1 - t)$ occurs because debt interest is tax deductible.

There are other capital-related expenses besides the loan payment: The total annual capital-charge rate β is given by

$$\beta = \beta' + t_p + r_i + r_r \quad \text{(B.5)}$$

for utilities and

$$\beta = \beta' + (1 - t)\, t_p + r_i + r_r \quad \text{(B.6)}$$

for homeowners, where

t_p = property-tax rate
r_i = insurance rate
r_r = interim capital-replacement rate

In the case of the homeowner the property tax is multiplied by $(1 - t)$ because property taxes are deductible from income taxes.

In this book we have expressed all quantities in constant dollars, so that interest rates must be adjusted for inflation. Thus if r_d' and r_e' are the rates of return to debt and equity in current dollars, and if i is the inflation rate, then

$$r_d = \frac{1 + r_d'}{1 + i} - 1 \text{ and } r_e = \frac{1 + r_e'}{1 + i} - 1 \quad \text{(B.7)}$$

In this book we assume $i = 0.055$ and

$r_d' = 0.06$ for a municipal utility
$r_d' = 0.09$ for public utility debt and home mortgages
$r_d' = 0.12$ for home-improvement loans
$r_e' = 0.12$ for public utility equity

Thus

$r_d = 0.005$ for a municipal utility
$r_d = 0.033$ for public utility debt and home mortgages
$r_d = 0.062$ for home-improvement loans
$r_e = 0.062$ for public utility equity

We also assume

$t = 0.50$ for public utilities
$t = 0.0$ for municipal utilities
$t = 0.35$ for homeowners
$\alpha_e\ (\alpha_d) = 0.50\ (0.50)$ for public utilities
$\alpha_e\ (\alpha_d) = 0.0\ (1.0)$ for municipal utilities
$t_p = 0.02$ (all cases)
$r_i = 0.0025$ (all cases)
$r_r = 0.0035$ (all cases)

Using these parameters, the annual capital-charge rates used in this book are summarized in the following table for various cases:

	N	I_N/I_o	*d*	β
Land Investment*				
Public Utility	—	1	0.039	0.098
Municipal Utility	—	1	0.005	0.025
Homeowner	25	1	0.021	0.034
Capital Equipment				
Public Utility	10	0	0.039	0.171
	20	0	0.039	0.122
	25	0	0.039	0.113
	30	0	0.039	0.107
Municipal Utility	10	0	0.005	0.129
	20	0	0.005	0.079
	25	0	0.005	0.069
	30	0	0.005	0.062
Homeowner Improvement Loan	5	0	0.040	0.244
	10	0	0.040	0.142
Mortgage Loan	20	0	0.021	0.081
	25	0	0.021	0.071
	30	0	0.021	0.064

* For a land investment $r_r = r_i = 0$

ANNOTATED BIBLIOGRAPHY

The following is a selected reading list of books and reports that give emphasis to energy efficiency in energy policy for the U.S. economy as a whole.

1. *Alternative Energy Demand Futures to 2010,* the Report of the Demand and Conservation Panel to the Committee on Nuclear and Alternative Energy Systems (CONAES), National Academy of Sciences, Washington, D.C., 1979; this report was summarized in "U.S. Energy Demand: Some Low Energy Futures," *Science, 119,* April 14, 1978. This report develops relatively detailed alternative scenarios for future energy demand in the U.S., disaggregated by end uses. This analysis shows that while future energy demand is relatively sensitive to gradual energy price changes and to energy policy changes, future GNP growth is relatively insensitive to such changes. The scenario characterized by significant energy price increases and an aggressive energy conservation policy but only minor lifestyle changes involves an energy consumption level in 2010 approximately equal to that in 1975.

2. Roger W. Sant, "The Least-Cost Energy Strategy: Minimizing Consumer Costs Through Competition," published by The Energy Productivity Center of the Mellon Institute, Arlington, Virginia, 1979. This report is a preliminary across-the-board analysis of opportunities for technical changes that affect energy use. The concepts of energy service and cost of service are introduced. The goal set for energy policy is cost minimization rather than energy efficiency improvement as such.

3. *Energy Future,* the Report of the Energy Project at the Harvard Business School, edited by Robert Stobaugh and Daniel Yergin, Random House, New York, 1979. This timely book has been very effective in focussing public attention on the fact that energy efficiency improvements are far less costly and generally more promising than energy supply expansion.

4. *Energy Conservation and Public Policy,* edited by John C. Sawhill,

Prentice Hall, Englewood Cliffs, New Jersey, 1979. This book is a collection of papers by a number of experts in energy conservation policy and covers a range of important conservation issues.

5. Vince Taylor, "Energy: The Easy Path," January 1979, and "The Easy Path Energy Plan," September 1979, The Union of Concerned Scientists, Cambridge, Massachusetts. These two well written reports couple technical options with econometric analysis to project a low energy future which minimizes problems associated with expanding energy supply. In the projection, energy demand grows very slowly until 1985 and then declines.

6. Amory B. Lovins, *Soft Energy Paths: Toward a Durable Peace*, a Friends of the Earth book, Ballinger, Cambridge, 1977. Problems of present energy supply systems and their expansion and the advantages of solar based small scale supply systems are forcefully argued in this book. Energy efficiency improvement is an important aspect of Lovins' "soft energy path," but the primary emphasis is given to supply alternatives. The recommended path involves slow energy demand growth to the year 2000, followed by an absolute decline in the level of energy demand thereafter.

7. Denis Hayes, *Rays of Hope: the Transition to a Post-Petroleum World*, a Worldwatch Institute Book, W. W. Norton & Company, New York, 1977. This is a well written popular book calling for a transition from an energy economy based on fossil and nuclear fuels to one stressing energy conservation and renewable energy sources.

8. M.H. Ross and R.H. Williams, "The Potential for Fuel Conservation," *Technology Review*, p. 49, February 1977 and "Energy and Economic Growth," a study prepared for the Joint Economic Committee of the U.S. Congress, August 1977. In these two articles opportunities for technical change to improve energy efficiency are examined for all end-use sectors of the energy economy as it was in 1973. The principal emphasis is technology rather than economics. A program is described which would have reduced energy use 40% in 1973.

9. *A Time to Choose: America's Energy Future*, the Final Report of the Energy Policy Project (EPP) of the Ford Foundation, Ballinger, Cambridge, 1974. This was the first major study stressing energy conservation over energy supply expansion in the U.S. economy. One of the present authors (R.H.W.) was a coauthor of the EPP report. While this report was regarded as radical at the time it was published, many of the policies recommended have already been implemented. Experience and understanding obtained since the EPP report show that the conservation opportunities are much greater than was then thought

possible. For example, the provocative low-energy-growth scenario described there involves more rapid energy demand growth than the "Business-As-Usual" scenario described in Chapter 3 of the present book.

NOTES AND REFERENCES

CHAPTER 2

1. John Stuart Mill, *Principles of Political Economy.*

2. H.J. Barnett and C. Morse, *Scarcity and Growth: The Economics of Natural Resource Availability,* Johns Hopkins Univ. Press, 1963. V.K. Smith, ed., *Scarcity and Growth Reconsidered,* Resources for the Future, Johns Hopkins Univ. Press, 1979.

3. Foster Associates, Inc., *Energy Prices,* Ballinger, Cambridge, Mass., 1974. Table 9-d, p. 250.

4. D.A. Brobst and W.P. Pratt, eds., *U.S. Mineral Resources,* U.S. Geological Survey Professional Paper 820, 1973; and *Historical Statistics of the United States, Colonial Times to 1970,* Bureau of the Census, U.S. Department of Commerce, 1975.

5. These data, except for feedstocks, are obtained from *Historical Statistics of the U.S., Colonial Times to 1970,* and recent issues of *Statistical Abstract of the U.S.:*

Steel: Domestic raw steel production. *Hist. Stat.* series P265. Corrected beginning in 1965 for imports and exports as determined from data on steel mill products.

Lumber: Apparent consumption. *Hist. Stat.* series L77 plus L80. Converted to weight using the factor 16.7 tons/1000 cu. ft.

Cement: Hydraulic cement shipments by domestic producers. *Hist. Stat.* series M188. Converted to weight using the factor 376 lbs/bbl.

Paper and paperboard: Apparent consumption. *Hist. Stat.* series L174.

Three inorganics: Domestic production of sodium hydroxide, anhydrous ammonia and sulfuric acid. *Hist. Stat.* series P248, P249 plus P251.

Feedstocks refers to consumption of fuels by the industrial sector as raw materials (excluding lubes, waxes, and metallurgical coal). Data are obtained from *Energy Perspectives 2,* p. 71, U.S. Department of Interior, Bureau of Mines 1976; and Bureau of Mines press releases. Converted to weight using the factor 26.1 tons/billion Btu.

CHAPTER 3

1. Two government sponsored studies projected demand well below that shown in Table 3.1. The Institute for Energy Analysis (*U.S. Energy and Economic Growth, 1975-2010.* Oak Ridge, Tenn., Sept. 1976) projected low and high figures with a mid-point of 114 quads. The U.S. Dept. of Energy MOPPS study (Market Oriented Program Planning Study, Dec., 1977, Draft) also projected 114 quads for the year 2000.

2. The present fertility rate is 1.8 children per woman. If this rate rises to 1.9, if the labor force participation rate rises from the 1975 level of 0.62 to 0.65 and stabilizes (see note 3), and if the economy returns to "full employment" (4–5% unemployment) by 1985 and remains at the full employment level thereafter, then total population, the adult population (those aged 16 and over), the labor force, and employment (full time equivalent employees) are given by

DATE	TOTAL POPULATION (MILLIONS)	NON-INSTITUTIONALIZED ADULT POPULATION (MILLIONS)	PARTICIPATION RATE	LABOR FORCE (MILLIONS)	EMPLOYMENT (MILLIONS)
1975	213	153.4	0.62	94.8	74.4
1985	231	174	0.65	113	94
2000	254	194	0.65	126	104
2010	264	207	0.65	134	111

The population projection here is for a constant net immigration rate of 400,000 per year. The projection is roughly midway between that of the Census Series II (fertility returns to the replacement rate of 2.1) and Census Series III (fertility rate of 1.7). See Bureau of the Census, "Projections of the Population of the United States 1977 to 2050," *Current Population Reports: Population Estimates and Projections*, Series p. 25, No. 704, U.S. Department of Commerce, July 1977.

3. The labor force participation rate is the fraction of the noninstitutionalized working age population that is working or seeking work. The participation rate has remained remarkably constant over time (see Table below), despite the fact that it has steadily declined for men and increased rapidly for women. In this book we assume that both of these trends come to an end in about a decade.

Labor Force Participation Rate

	MEN	WOMEN	TOTAL
1890	0.843	0.182	0.522
1900	0.857	0.200	0.537
1920	0.846	0.227	0.543
1930	0.821	0.236	0.532
1940	0.825	0.279	0.552
1945	0.876	0.358	0.616
1950	0.868	0.339	0.599
1955	0.862	0.357	0.604
1960	0.840	0.378	0.602
1965	0.815	0.393	0.597
1970	0.806	0.434	0.613
1971	0.800	0.434	0.610
1972	0.797	0.439	0.610
1973	0.795	0.447	0.614
1974	0.794	0.457	0.618
1975	0.781	0.458	0.618

These data are the fraction of the total noninstitutional population in the labor force. The Armed Forces are included beginning in 1940. Data before 1947 include the institutional population and persons 14 years or older. Data from 1947 to the present are for persons 16 or older. Data for 1890 to 1970 are from Series D29–41 of the U.S. Department of Commerce, *Historical Statistics of the United States: Colonial Times to 1970*, Part I, U.S. Government Printing Office, Washington, D.C., 1975. Data for 1970 to 1975 are from the *Statistical Abstract of the United States*, various years.

4. Here the service sector is defined as services, government, communications, finance, insurance, and real estate, as these industries are defined by the Office of Business Economics, U.S. Department of Commerce. The goods sector is everything else. During the period 1950–1975 the productivity growth rates for these sectors were 2.4% per year for goods and 0.9% per year for services, resulting in an average productivity growth rate of 1.8% per year. During the same period

the fraction of the Gross Product Originating in the goods producing sectors declined at an average annual rate of slightly less than 0.2% (see National Income and Product Accounts in U.S. Department of Commerce, Survey of Current Business). If this gradual shift away from goods production continues then the goods production share of GNP (with all quantities measured in 1958$) would decline from 63.3% in 1975 to 60.5% in 2000 and 59.4% in 2010. Then if the productivity growth rates for goods and services production separately are the same in the future as the average values for the period 1950–1975 the average productivity growth rate in the future will slowly decline. Using the employment projections from note (2) we thus obtain

	GROWTH RATES (%/YEAR)		
	EMPLOYMENT	PRODUCTIVITY	GNP
1950–1975	1.7	1.8	3.5
1975–1985	2.4	1.7	4.1
1985–2000	1.0	1.5	2.5
2000–2010	0.65	1.45	2.1
1975–2000	1.35	1.6	2.95
1975–2010	1.15	1.55	2.7

While this GNP projection is based on a fertility rate of 1.9 children per woman, the projection through 2010 is fairly insensitive to the fertility rate. If the fertility rate rose to 2.1, the average employment rate (and hence the GNP growth rate) 1975–2010 would be only 0.1%/year higher.

5. Employees are full time equivalent employees as defined in the National Income and Product Accounts, U.S. Department of Commerce; see *Survey of Current Business*, July issues.

6. For two analyses of this decline see William Nordhaus, "The Recent Productivity Slowdown," *Brookings Papers on Economic Activities No.* 3, 1972, p. 493; and Edward F. Denison, *Accounting for Slower Economic Growth*, The Brookings Institution, 1979.

7. According to Table 6-79, p. 193 in Edison Electric Institute, *Economic Growth in the Future*, McGraw Hill Book Company, 1976, the gas price in the year 2000 is projected to be $1.92 per million Btu (in 1975 dollars).

8. In 1975 the average price of natural gas to consumers was $1.29 per million Btu and the average price at the wellhead was $0.44 per million Btu. (See Tables 90 and 91 in American Gas Association, *Gas Facts: 1976 Data*, 1977.) Thus the average cost of transmission and distribution was $0.85 per million Btu. Also new intrastate gas (which is deregulated) cost $1.79 per million Btu at the wellhead in 1977 (see Energy Information Administration, *Annual Report to Congress: Volume II*, 1977, April 1978, p. 162) or about $1.54 per million Btu in 1975 dollars. Because this gas is not regulated its price should be approximately the present replacement cost of gas. This can be converted into an average consumer replacement price of $1.54 + $0.85 = $2.39 per million Btu. Synthetic gas from coal would be twice as costly as this (see note 9).

9. In "America's First Coal Gas Plant Is Ready to Go," *Energy Daily*, vol. 6, No. 179, September 15, 1978 the "tailgate" price in 1983 for gas from the nation's first commercial scale high-Btu coal gasification plant in Mercer County, North Dakota was projected to be between $6.25 and $8.25 per million Btu. Taking the midpoint in this range and allowing for 5.5% annual inflation (the expected rate at the time of the announcement) this corresponds to $4.70 per million Btu in 1975 dollars or a consumer price of

$4.70 + $0.85 = $5.55 per million Btu.

10. According to Table 6-79 in ref. 7 the average price of electricity in 2000 would be 1.7¢/kwh. The actual average price in 1975 was 2.7¢/kwh (see "1978 Annual Statistical Report," *Electrical World*, March 15, 1978, p. 99).

11. In note 28 of Chapter 8 the replacement cost of electricity for an all electric house with a heat pump is estimated to be 8.22¢/kwh (1976 dollars), including the cost of the heat pump. Excluding the cost of the heat pump the cost is 6.64¢/kwh (1976 dollars) or 6.28¢/kwh (1975 dollars). For comparison the average price of residential electricity in 1975 was 3.21¢/kwh (see "1978 Annual Statistical Report," note 10).

12. According to *Long Term Economic Growth, 1860–1970,* prepared by the Bureau of Economic Analysis, U.S. Department of Commerce, June 1973, Series A175, manufacturing output per man-hour increased 9.9-fold between 1869 and 1969.

13. See Figure B.8, p. B-17 in ERDA, *A National Energy Plan for Energy, Research, Development and Demonstration* (ERDA-48, 1975).

14. According to Table 6.64, p. 182 in *Economic Growth in the Future* (see ref. 7) 47% of U.S. energy in the year 2000 would be committed to power generation. In the ERDA-48 projection (see ref. 13) the corresponding percentage is 50%.

15. This is in addition to the use of electricity for steel making in electric furnaces.

16. Today process steam is provided mainly by fossil fuel-fired boilers at industrial sites, where 75 to 85 percent of the fuel energy is delivered to process steam. If electricity were used instead to provide this process heat, about two-thirds of the fuel energy would be lost as waste heat in the generation of electricity. As a result, the use of coal for process heat via central station electricity generation, for example, would require more than twice as much coal and much more capital investment (for electricity generation and delivery to the industrial site) than would the direct burning of coal at industrial sites. The relative economics would not change significantly if the electricity were produced from nuclear fuel instead of coal, since most projections show the prices of coal and nuclear generated electricity to be comparable.

17. Dow Chemical Co., "Evaluation of New Energy Sources for Process Heat," prepared for the National Science Foundation, Sept., 1975.

18. Frank von Hippel and Robert H. Williams, "Energy Waste and Nuclear Power Growth," *Bulletin of the Atomic Scientists,* December 1976, p. 20.

19. In the ERDA/BNL scenarios (see ref. 13) automobiles are driven in 2000 a total of 2.05 $\times 10^{12}$ vehicle miles. In 1975 automobiles were driven a total of 1.03×10^{12} vehicle miles (see Table 9.1).

20. Vehicle miles driven (VM) by autos can be expressed as

$$VM = \frac{S \times t \times P}{1}$$

where

S = average speed of travel for all cars
t = average time spent in cars per capita per year
1 = load factor, the average number of people per vehicle
P = population

At present (1975)

S = 32.6 mph
t = 357 hours/year (59 minutes/day)
1 = 2.2
P = 213.5 million

There are no powerful reasons to expect S and 1 to change much in the future. Thus since the population will probably increase about 19% in the period 1975–2000 (see note 2) it follows that the ERDA projection that VM will double in this period implies that t must increase 67% to about 99 minutes/day.

21. Vehicle miles driven can also be expressed as

$$VM = P_A \times R_V \times (VM/V)$$

where

P_A = adult population (all people ages 16 and over)
R_V = vehicle ratio, the number of autos per adult
VM/V = number of vehicle miles traveled per vehicle per year

At present (1975)

P_A = 153.4 million
R_V = 0.75
VM/V = 9410 miles, a value which has remained essentially unchanged over time

The adult population will probably increase 26%, 1975-2000 (see note 2). Thus, the ERDA projection that VM will double in this period implies that R_V must increase 57% to 1.18.

22. See the "Annual Electrical Industry Forecast" in the September 15 issue each year of *Electrical World.*

23. From the "29th Annual Electrical Industry Forecast," *Electrical World,* September 15, 1978.

24. Scott Burns, *The Household Economy,* Beacon Press, 1977.

25. See e.g. E.L. Allen et al., *U.S. Energy and Economic Growth, 1975-2000,* (Oak Ridge, Tenn., Institute for Energy Analysis, ORAU/IEA-76-7, 1976); and Energy Research Group, University of Cambridge, "World Energy Demand, 1985-2020," (London, World Energy Conference, 1977).

26. The percentage change in energy demand resulting from a one percent increase in the price of energy is called the price elasticity of energy demand. Research on price elasticities has been reviewed recently in Robert S. Pindyck, "The Characteristics of the Demand for Energy," in ref. 27. Pindyck points out that most studies, which show relatively small elasticities, are developed from time series analyses for individual countries. In these analyses the time period has been too short for consumers to fully respond to price changes, so that what is measured is the "short run" elasticity, or the response of consumers to price changes before they have time to make changes in energy consuming equipment. Because energy prices vary so much from country to country, however, international comparisons between countries with very different price levels can be used to estimate "long run" elasticities. Pindyck suggests, on the basis of these international comparisons that good "working" estimates of these elasticities would be:

residential sector	−0.8 to −1.0
industrial sector	−0.8
gasoline demand	−0.9

These estimates are nearly twice as great as the value of −0.5 assumed here for the Business As Usual scenario.

27. John C. Sawhill, ed., *Energy Conservation and Public Policy,* Prentice-Hall, Inc., Englewood Cliffs, N.J., 1979.

28. William M. Bulkeley, "Baby Boom of 1947-57, A Key Economic Force, Now Spurs Home Sales," *The Wall Street Journal,* July 27, 1978.

29. We assume that there will be 2.4 persons per household in 2010 (compared to 3.0 in 1975). See ref. 33. Thus (see note 2) there will be 264/2.4 = 110 million housing units in 2010 corresponding to an average growth rate of 1.3%/year, 1975-2010. We assume that total personal income grows in proportion to GNP. (See note 4). Thus income per household increases in this period by a factor of 2.55/1.55 = 1.65, corresponding to an average annual growth rate of 1.4%/year. From a 1972 national survey of household energy use as a function of income (D.K. Newman and D. Day, *The American Energy Consumer,* a report to the Energy Policy Project of the Ford Foundation, Ballinger, Cambridge, 1975) we obtain the result that in the range of the high incomes, where the average will be shifting in this period, each 1% rise in income will give rise to

0.28% rise in household energy use. Thus the income effect corresponds to an average annual growth rate of 0.4%/year. See also Figure 8.7, which shows, using data from the early 1960's, that direct residential consumption of energy saturates as income rises.

30. E. Hirst and J. Carney, "The ORNL Engineering-Economic Model of Residential Energy Use," Oak Ridge National Laboratory Report ORNL/CON-24, July 1978. About 20% of all dwellings are centrally air conditioned, 10% have 2 or more room air conditioners, and 20% have one.

31. Food freezers and dishwashers are each present in about 35% of households. See "Residential and Commercial Energy Use Patterns, 1970–1990," prepared by the Interagency Task Force on Energy Conservation for the Federal Energy Administration Project Independence Blueprint Final Task Force Report, Nov., 1974.

32. Robert Herendeen and Anthony Sebald, "Energy, Employment, and Dollar Impacts of Certain Consumer Options," in *The Energy Conservation Papers*, Robert H. Williams, ed., Ballinger, Cambridge, 1975.

33. *Alternative Energy Demand Futures to 2010*, Report of the Demand and Conservation Panel to the Committee on Nuclear and Alternative Energy Systems, National Research Council, National Academy of Sciences, 1979.

34. This result is comparable to that projected by Hirst and Carney, ref. 30, with the additions by Eric Hirst, "Energy and Economic Benefits of Residential Energy Conservation RD & D," Oak Ridge National Laboratory report ORNL/CON-22, February 1978.

35. J.R. Jackson and W.S. Johnson, "Commercial Energy Use: A disaggregation by Fuel, Building Type, and End Use," Oak Ridge National Laboratory Report ORNL/CON-14, February 1978.

36. The following data show that the increase in commercial space has been proportional to the increase in the number of service sector employees over the last decade.

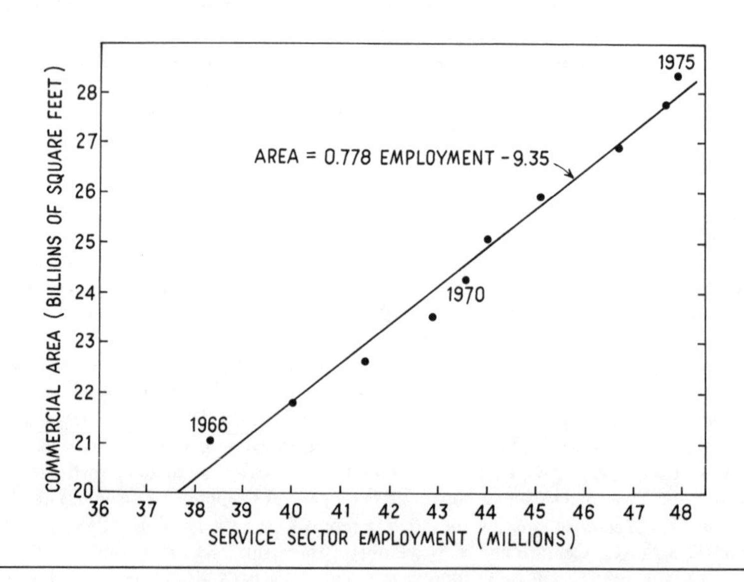

From this data we derive the following relationship:

Commercial area = 0.778 (Service Sector Employment) − 9.35, where commercial area is measured in billion square feet (data from ref. 35) and service sector employment (data from July issues of *Survey of Current Business*) is given in millions. In this application the definition of the service sector is slightly different from that given in note 4: wholesale and retail trade is included as well. Thus for the purpose of this analysis service sector employment was 48 million in 1975 (see *Survey of Current Business*, July, 1978). We shall assume that this expanded service sector grows at the same rate as the service sector defined in note 4, however. For the conditions outlined in note 4, this means an increase by a factor of 2.05, 1975–2010, or, according to the above formula, to an increase in commercial area by a factor of 2.35, corresponding to an average growth rate 2.5%/year.

37. We assume that automobiles with an average fuel economy of 26.3 mpg account for 80% of total VM and that light trucks with an average fuel economy of 21 mpg account for 20% of total VM. In 1975 light trucks accounted for 9% of total VM (see Table 9.1).

38. Between 1950 and 1970 GNP grew at 3.6% per year in the U.S., while freight haulage (measured in ton-miles) increased at 2.5% per year (see Eric Hirst, "Energy Intensiveness of Passenger and Freight Transport Modes, 1960–1970," Oak Ridge National Laboratory Report ORNL-NSF-EP-44, April 1973), and air travel (measured in passenger-miles) by scheduled air carriers increased at 13.6% per year (U.S. Federal Aviation Administration, *FAA Statistical Handbook of Aviation*, annual). From 1970–1977 the growth rate in air travel by scheduled carriers has been much slower: 5.6%/year compared to 3.1%/year for GNP.

39. Our measure of industrial energy intensity is industrial energy use per real dollar of Gross Product Originating, as reported by the Department of Commerce. Other indices of industrial value added, prepared by the Federal Reserve Board and Bureau of the Census, tend to grow more rapidly (making energy per dollar fall more rapidly) but these indices are less closely tied to GNP as a whole, so we have chosen to use the Commerce index. For an examination of the differences between these industrial indices, see Jack J. Gottsegen and Richard C. Ziemer, "Comparison of Federal Reserve and OBE Measure of Real Manufacturing Output, 1947–64," in J.W. Kendrick, ed., *The Industrial Composition of Income and Product*, National Bureau of Economic Research, Conference on Research in Income and Wealth, vol. 32, 1968.

40. Charles A. Berg, "Process Innovation and Changes in Industrial Energy Use," *Science 119*, February 10, 1978.

41. "Projections of the Population of the United States: 1977 to 2050," *Current Population Reports: Population Estimates and Projections*, Series P-25, No. 704, July 1977.

CHAPTER 4

1. The difference between demand projected for 2010 under Business-As-Usual conditions and the 1975 domestic production level is 42 quadrillion Btu. The Alaska pipeline has a carrying capacity of about 2 million barrels of oil per day or 4 quadrillion Btu per year. A 1 Gw(e) nuclear power plant operating on the average at 65% of capacity consumes about 0.057 quadrillion Btu of fuel per year.

2. To describe what resources are left in the ground it is useful to distinguish between *reserves* and *additional resources*. Established and inferred reserves are those discovered and measured resources which can be profitably extracted at present prices. Additional resources which are judged to be ultimately recoverable include both identified resources which could be profitably recovered at higher prices and estimated undiscovered resources which are expected to be profitably recoverable at some time in the future.

For oil and gas respectively (in quadrillion Btu) we have: production to date, 610 and 460; reserves, 420 and 440; and additional resources, 1270 and 540. For comparison the 50 year fluid fuel requirements in a Business-As-Usual future are estimated by us as 2600 quads.

The estimates of reserves and additional resources are taken from "Geological Estimates of Undiscovered Recoverable Oil and Gas Resources in the United States," U.S. Geological Survey Circular 725, 1975, and Commission on Natural Resources, "Mineral Resources and the Environment," National Academy of Sciences, 1975, respectively.

3. The National Academy of Sciences estimate represents the consensus judgment of a prestigious group of scientists assembled by the National Academy of Sciences. But a word of caution. There has been a long history of premature claims that the U.S. is running out of oil, as shown in the following table:

History of Domestic Oil Resources Predictions by Government Officials

1859: First oil well drilled in the United States.
1866: U.S. Internal Revenue Commission says synthetics available if oil production should end.
1883: Little or no chance for oil in California—U.S. Geological Survey (82 billion barrels of oil-in-place found to end of 1972)
1891: Little or no chance for oil in Kansas or Texas—U.S. Geological Survey. (15 billion barrels of oil-in-place in Kansas and 147 billion barrels in Texas to date)
1908: Maximum future supply of oil, 23 billion barrels—officials of the U.S. Geological Survey
1914: Total future production only 6 billion barrels—officials of the U.S. Bureau of Mines
1920: United States needs foreign oil and synthetics; peak domestic production almost reached—Director of U.S. Geological Survey (interest in shale oil was high during the 1920s—many oil companies purchase oil shale leases) (95 billion barrels have been produced since 1920)
1931: Must import as much foreign oil as possible to save domestic supply—Secretary of the U.S. Department of Interior (East Texas field discovered late in 1930 but potential not immediately recognized)
1939: U.S. oil supplies will last only 13 years—U.S. Department of the Interior
1947: Sufficient oil cannot be found in the United States—Chief of Petroleum Division, U.S. Department of State.
1949: End of U.S. oil supply almost in sight—Secretary of the U.S. Department of the Interior. (Interest in oil shale revived in the late 1940s.)

SOURCE: Resources for the Future, "Toward Self Sufficiency in Energy Supply," draft report to the Ford Foundation's Energy Policy Project, September, 1973 (unpublished).

4. Over the years there has been a wide range of estimates of undiscovered oil and gas resources in the U.S., with the United States Geological Survey giving relatively high estimates and veteran geologist M. King Hubbert giving relatively low estimates. In 1975, however, the U.S.G.S., in Circular 725 (see note 2), came out with resource estimates very close to those of Hubbert (see M.K. Hubbert, *U.S. Energy Resources: a Review As of 1972*, a background paper prepared for the Committee on Interior and Insular Affairs, United States Senate, 93rd Congress, 1974). One must not conclude, however, the experts are converging on the truth. Instead we are witnessing the need for institutions to eliminate disagreement and to converge on a "truth" of the moment.

5. One independent gas producer (Robert A. Hefner III, Managing Partner, The GHK Companies, "The Future for Conventional U.S. Natural Gas Supply," paper presented at the *Aspen Institute Workshop on R&D Priorities and the Natural Gas Option*, Aspen, Colorado, June 25-29, 1978) estimates that at prices up $4.30 (1977 dollars) per thousand cubic feet the amount of gas available in the U.S. is about 10 times that given in note 2. Heffner points out that

considerable quantities of gas can be expected in deep reservoirs. His company was a partner in drilling the deepest hole ever drilled (with a total depth of over 30,000 feet), and he claims that the technology is available to drill to 40,000-50,000 feet. The figure shown on p. 282 is Heffner's estimated supply curve for natural gas.

Of course Heffner's gas supply curve must be regarded as speculative, since relatively poor data are available at this time on unconventional gas sources. However, if he is even partly right the existing natural gas policy makes little sense. In particular the policy relating to industry should be changed from one of forcing industry away from natural gas use to one of encouraging industry to use gas more frugally. We discuss industrial natural gas policy in the context of a national fuels policy for an energy conserving economy in Chapter 13.

Geochemical considerations tend to support the notion that in deep wells gas should be favored over oil. The figure on p. 283 (from H.D. Klemme, "Geothermal Gradients, Heat Flow, and Hydrocarbon Recovery," in *Petroleum and Global Tectonics*, A.G. Fischer and S. Judson, eds., Princeton University Press, 1975) shows that the ratio of gas to oil in hydrocarbon deposits depends on the temperature of the formation. The higher the temperature the greater the ratio. Above about 300°F most of the hydrocarbon would be gas. Since the temperature increases with depth (at a global average rate of 1.4°F per 100 feet) it follows that deeper drilling should yield a greater ratio of gas to oil. Where the geothermal temperature gradient is about average (1.4°F/100 ft) mostly gas should be found at depths below about 16,000 feet. Where the geothermal gradient is twice the average, mostly gas should be found below 8000 feet.

6. Coal resource data are obtained from Paul Averitt, "Coal Resources of the United States, January 1, 1974, "U.S. Geological Survey Bulletin 1412, USGPO, Washington, 1975. The total remaining coal in the ground is estimated in this report to be 3.97×10^{12} tons or 82×10^{18} Btu. To obtain recoverable resources the following are subtracted from this total:

- all coal with more than 3000 feet overburden (0.39×10^{12} tons)
- coal with less than 3000 feet of overburden in thin seams: 14-28 inches thick for bituminous coal and anthracite and 2½-5 feet thick for subbituminous coal and lignite (0.78×10^{12} tons from identified resources)
- coal left after mining: 50% of coal in intermediate and thick seams with less than 3000 feet of overburden. (0.54×10^{12} tons from hypothetical resources and 0.50×10^{12} tons from identified resources)

Thus ultimately recoverable coal is estimated to total 1.03×10^{12} tons. The resources in quadrillion Btu are then: reserves 4,600; identified but difficult to mine, 5,600; and hypothetical, 10,200.

7. Thomas H. Maugh II, "Tar Sands: A New Fuels Industry Takes Shape," *Science 199*, p. 756, February 17, 1978. See also Eliot Marshall, "OPEC Prices Make Heavy Oil Look Profitable," *Science 204*, p. 1283, June 22, 1979.

8. Thomas H. Maugh II, "Oil Shale: Prospects on the Upswing . . . Again," *Science 198*, p. 1023, December 9, 1977.

9. John Harte and Mohamed El-Gassier, "Energy and Water," *Science 199*, p. 623 February 10, 1978.

10. Successful *in situ* extraction technology and technical improvements to reduce water requirements in converting the extracted product to useful energy carriers would relax these constraints somewhat. (*In situ* extraction means extracting the fuel in fluid form without direct excavation, e.g. by drilling holes, partial combustion of the hydrocarbons in place, and extraction of liquids or gases liberated by the heating.)

11. See Table III F-7 in U.S. ERDA, *Final Environmental Statement on the Liquid Metal Fast Breeder Reactor Program* (ERDA-1535, 1975).

Gas Supply Curve of R. A. Hefner

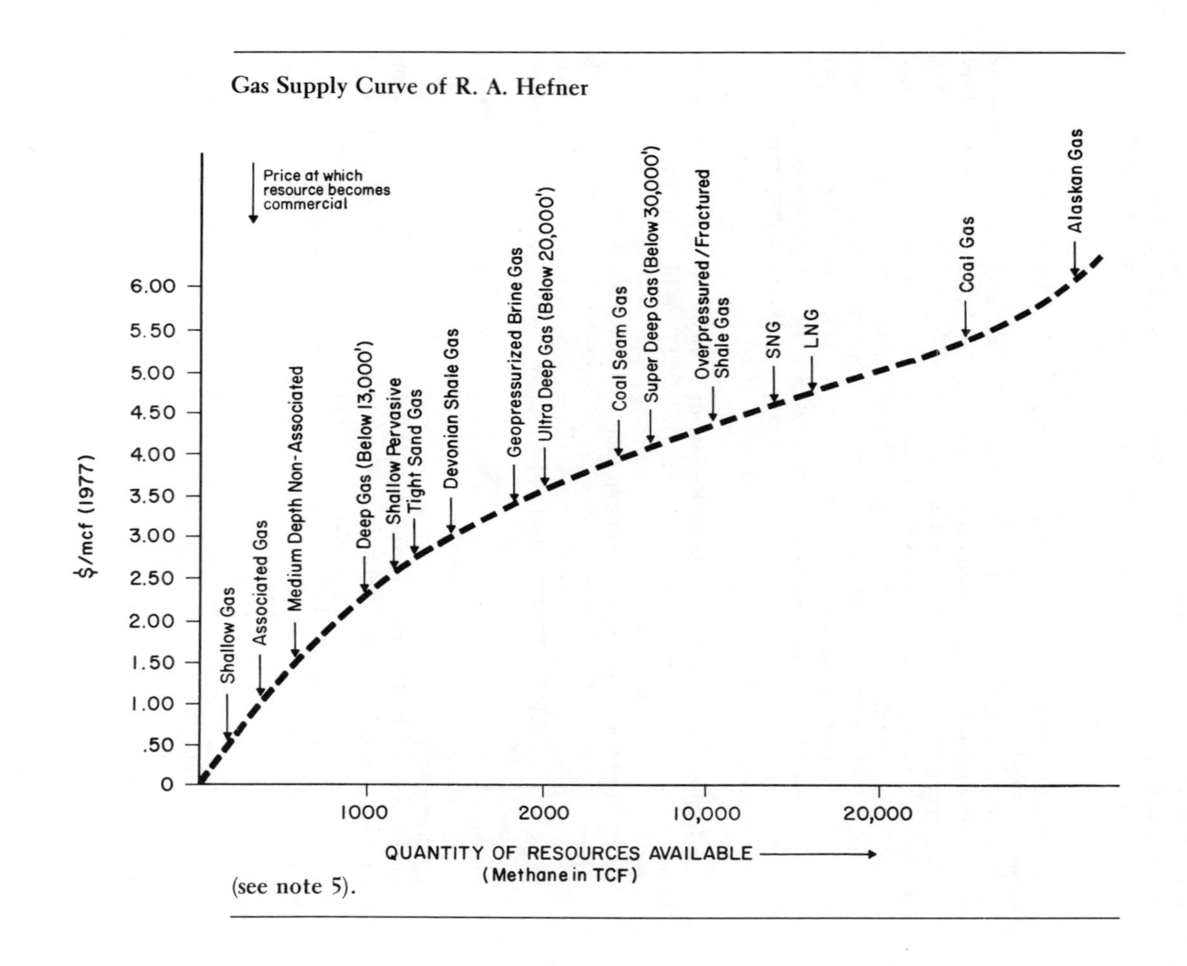

(see note 5).

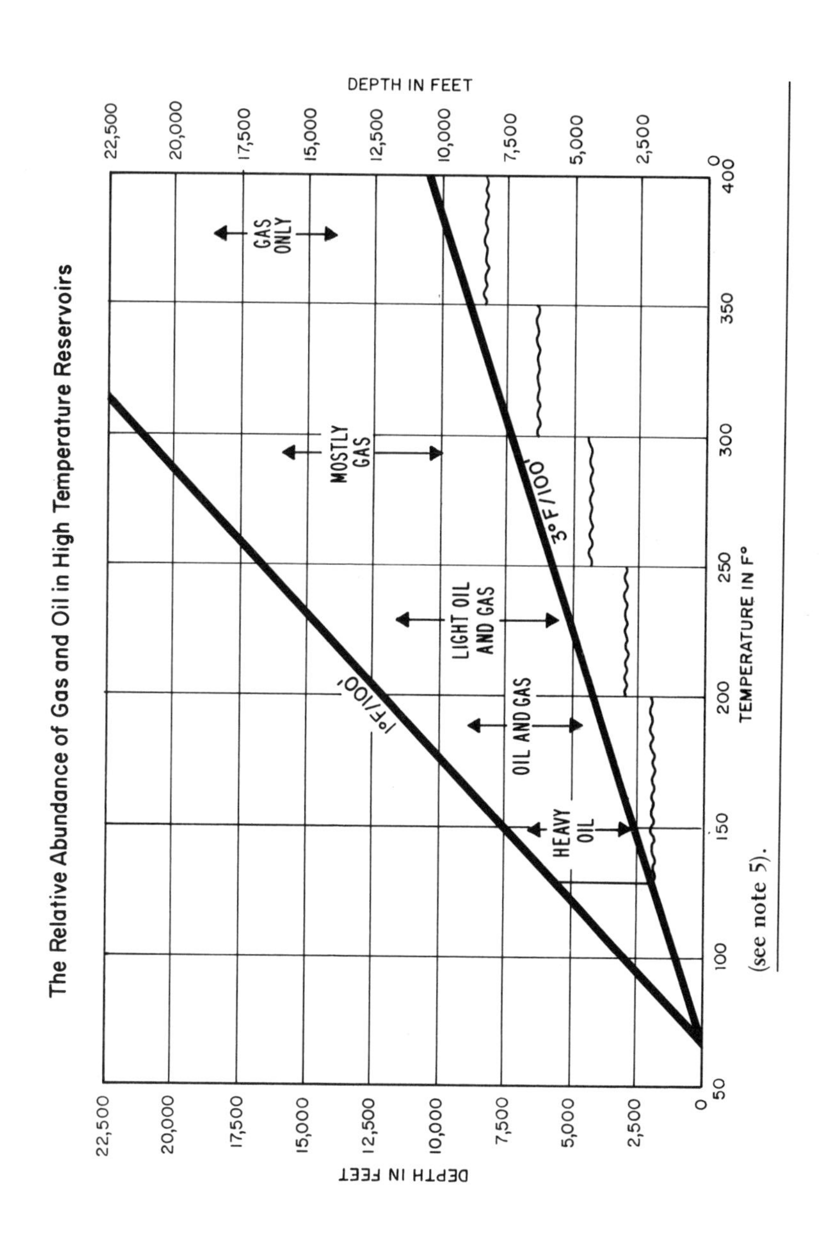
The Relative Abundance of Gas and Oil in High Temperature Reservoirs
DEPTH IN FEET
22,500
20,000
17,500
15,000
12,500
10,000
7,500
5,000
2,500
0
GAS ONLY
MOSTLY GAS
LIGHT OIL AND GAS
OIL AND GAS
HEAVY OIL
1°F/100'
3°F/100'
50
100
150
200
250
300
350
400
TEMPERATURE IN F°

(see note 5).

12. The DOE estimate of "low cost" uranium resources (forward costs of up to \$50/lb. of U_3O_8) was 4.3 million tons as of January 1, 1977. A 1 Gw(e) LWR operating at 65% of capacity requires 143 tons/year of U_3O_8 as make-up fuel.

13. Energy Information Administration, *Annual Report to Congress: Volume II, 1977*, DOE/EIA-0036/2, page 224, April 1978.

14. As pointed out in H.A. Feiveson, F. von Hippel, and R.H. Williams, "Fission Power: An Evolutionary Strategy," *Science*, January 26, 1979, advanced burner reactors (ABRs) based on already developed reactor types could be operated on a "once through" fuel cycle (i.e., without reprocessing spent fuel) with 40% less uranium than is required by present light water reactors. If all reactors built after 2000 are such ABRs and if nuclear power grows as projected in this *Science* paper, then cumulative uranium requirements over the next 100 years would be no more than the present estimate of low cost uranium resources.

15. Even the Department of Energy has been moving toward this realization. A report in *Inside D.O.E.* (McGraw Hill) for November 20, 1978 stated: "A DOE strategy paper on nuclear R&D reports that, based on current estimates of uranium supply and electric demand, there is little or no need to decide on the future of breeder reactors—pro or con—before 1990. . . . Moreover the mid-range uranium supply and growth projections place the need for such a push beyond the year 2020."

16. Capital expenditures in energy producing industries were obtained from Table 7.1, p. 128, in the Energy Information Administration's *Annual Report to Congress*, vol. 3, 1978. Excluded are expenditures for pipelines, tankers, chemical plants, marketing, and other "downstream" capital expenditures. Total new plant and equipment expenditures for the U.S. are obtained from "Economic Indicators," prepared for the Joint Economic Committee by the Council of Economic Advisors, various issues.

17. The electric utility industry is much more capital intensive than most other industries, as the following data for 1972 indicate:

INDUSTRY	GROSS PLANT PER DOLLAR OF SALES REVENUE	NET PLANT PER DOLLAR OF SALES REVENUE
All Manufacturing Corporations	\$0.60	\$0.31
Transportation Equipment	0.44	0.21
Primary Metals	1.19	0.59
Basic Chemicals	1.11	0.52
Petroleum Refining	1.36	0.75
Electric Utilities	4.51	3.58

The electric utility data shown here were obtained from Tables 51S and 52S of the Edison Electric Institute's *Statistical Yearbook of the Electric Utility Industry for 1977*, (October, 1978). The data for other industries were obtained from pages 34-49 of the Federal Trade Commission's *Quarterly Financial Report for Manufacturing Corporations* (Second Quarter, 1972).

18. W. David Montgomery and James P. Quirk, "Cost Escalation in Nuclear Power," Environmental Quality Laboratory, California Institute of Technology, EQL Memo 21, January 1978; Irvin Bupp and Claude Derian, *Light Water*, Basic Books, 1979.

19. The price of gas from the nation's first commercial scale high-Btu coal gasification plant has been projected to be about \$6.10 (1979\$). See note 9, Chapter 3. For comparison gas producers received an average of \$1.14 per thousand cubic feet in 1979.

20. If the demand for fluid fuel derived from coal (synfuel) were 50 quads/year, the present level of demand for fluid fuel, and if efficiency for converting coal to synfuel were 50% then 100 quads/year of coal would be required. At an average energy density of 20 million Btu's per ton for that date, 5000 million tons per year would be mined. In 1976, 670 million tons were produced.

CHAPTER 5

1. For an excellent overview of environmental problems posed by energy systems in the context of a discussion of a broad range of related problems facing modern civilization see P.R. Ehrlich, A.H. Ehrlich, and J.P. Holdren, *Ecoscience: Population, Resources, Environment*, W.H. Freeman and Company, San Francisco, 1977.

2. For an overview and extensive bibliography see Energy Policy Project of the Ford Foundation, *A Time to Choose: America's Energy Future*, Ballinger, Cambridge, 1974. Chapter 8.

3. William Ramsey, *Unpaid Costs of Electrical Energy*, Resources for the Future, Johns Hopkins University Press, 1979; Office of Technology Assessment, U.S. Congress, *The Direct Use of Coal; Prospects and Problems of Production and Combustion*, 1979.

4. Lester Lave and Eugene Seskin, *Air Pollution and Human Health*, published for Resources for the Future by the Johns Hopkins University Press, Baltimore, 1977.

5. G.E. Likens and F.H. Bormann, "Acid Rain: A Serious Regional Environmental Problem," *Science*, vol. 184, p. 1176, June 14, 1974.

6. The national average cost of generating electricity (in 1975$) from coal in a plant employing flue gas desulfurization technology is estimated as follows (in mills/kwh):

capital charge	7.9
inventory	0.1
fuel	8.7
operation and maintenancce	2.1
	18.8

SOURCE: Table 9.1, p. 275 in Sam H. Schurr et al., *Energy in America's Future: the Choices Before Us*, published for Resources for the Future by The Johns Hopkins University Press, Baltimore, 1979. In this estimate about 1.3 mills/kwh of the capital charge is associated with the costs of flue gas desulfurization (see Table 9.2, p. 276, op. cit.). Also about half the operation and maintenance expense is associated with flue gas desulfurization. Thus about 13% of the generation cost is associated with flue gas desulfurization. However, to the cost of generation one must add the cost of transmission and distribution, which in 1975$ is estimated to be about 17.3 mills/kwh. (The cost given for 1972 in M.L. Baughman and D.J. Bottaro, "Electric Power Transmission and Distribution Systems: Costs and Their Allocation," report of the Center for Energy Studies, University of Texas at Austin, July 1975, is converted to 1975 dollars using as a deflator the Handy-Whitman index of Public Utility Construction Costs.) Thus flue gas desulfurization is estimated to account for about 6½% of the cost of delivered electricity from a coal fired plant.

7. National Academy of Sciences, *Rehabilitation Potential of Western Coal Lands*, a report to the Energy Policy Project, Ballinger, Cambridge, 1974.

8. The miner who works 40 hours per week, 40 weeks a year for 45 years has an expectation of one tenth of a fatal accident at work during that time, if the fatal accident rate is 1.4 per million man hours. Roughly this rate applied in the 1960's.

9. "Chemicals and Health," report of the Panel on Chemicals and Health of the President's Science Advisory Committee, September, 1973.

10. E.A. Nephew, "The Challenge and Promise of Coal," *Technology Review*, December, 1973.

11. For an excellent overview of man's impact on global climate see S.H. Schneider, *The Genesis Strategy: Climate and Global Survival,* Plenum Press, New York, 1976.

12. G.M. Woodwell, R.H. Whittaker, W.A. Reiners, G.E. Likens, C.C. Delwiche, and D.B. Botkin, "The Biota and the World Carbon Budget," *Science 199,* p. 141, Jan. 13, 1978. Minze Stuiver, "Atmospheric Carbon Dioxide and Carbon Reservoir Changes," *Science 199,* p. 253, Jan. 20, 1978. U. Siegentheler and H. Oeschger, "Predicting Future Atmospheric Carbon Dioxide Levels," *Science 199,* p. 388, Jan. 27, 1978.

13. S. Manabe, "Estimates of Future Change of Climate Due to the Increase of Carbon Dioxide Concentration in the Air," in *Man's Impact on the Climate,* W.H. Matthews, W.W. Kellogg, and G.D. Robinson, eds., MIT Press, Cambridge, 1971. See also S. Manabe and R.T. Wetherald, "The Effect of Doubling the CO_2 Concentration on the Climate of a General Circulation Model," *Journal of Atmospheric Sciences,* Vol, 32. No. 1, pp. 3-15, January 1975.

14. *Inadvertent Climate Modification,* Report of the Study of Man's Impact on Climate (SMIC), MIT Press, Cambridge, 1971, p. 239.

15. Broecker has analyzed the ratio of the isotopes oxygen 18 and oxygen 16 from Greenland ice cores to obtain a speculative prediction of natural climatic (average temperature) cycles. The prediction is that in the absence of man's activities there would be a bottoming out in the early 1980's of the pronounced cooling of the global average temperature which has been observed since the 1940's. If this is correct and calculations of the impact of increased CO_2 in the atmosphere are reliable, then a warming trend should be observed after the mid '80's. Wallace S. Broecker, "Climatic Change: Are We on the Brink of a Pronounced Global Warming?" *Science 189,* p. 460, August 8, 1975.

16. W.W. Kellogg, "Is Mankind Warming the Earth?" *Bulletin of the Atomic Scientists,* p. 10, February 1978.

17. Cesare Marchetti, "On Geoengineering and the CO_2 Problem," *Climatic Change,* p. 59, 1977.

18. For an overview see *Nuclear Power Issues and Choices,* the report of the Nuclear Energy Policy Study Group, Spurgeon M. Keeny, Jr., Chairman, Sponsored by the Ford Foundation, administered by the MITRE Corp., Ballinger, Cambridge, 1977; Royal Commission on Environmental Pollution, Sir Brian Flowers, Chairman, *Nuclear Power and the Environment,* Her Majesty's Stationery Office, Sept. 1976; Christoph Hohenemser, Roger Kasperson and Robert Kates, "The Distrust of Nuclear Power," *Science 196,* p. 25, April 1, 1977; John P. Holdren, "Hazards of the Nuclear Fuel Cycle," *Bulletin of the Atomic Scientists,* p. 14, October 1974.

19. Jan Beyea, *Some Long Term Consequences of Hypothetical Major Releases of Radioactivity to the Atmosphere from Three Mile Island,* report to the President's Council on Environmental Quality, September 1979.

20. U.S. Nuclear Regulatory Commission, *Reactor Safety Study: An Assessment of Accident Risks in U.S. Commercial Nuclear Power Plants,* WASH-1400 (NUREG-75/014), October 1975; "Report to the American Physical Society of the Study Group on Light Water Reactor Safety," *Reviews of Modern Physics,* vol. 47 Supplement No. 1, Summer, 1975.

21. Frank von Hippel, "Looking Back on the Rasmussen Report," *Bulletin of the Atomic Scientists,* p. 42, February 1977; H.W. Lewis, R.H. Budnitz, H.J.C. Kouts, W.B. Loewenstein, W.D. Rowe, F. von Hippel, F. Zachariasen, *Risk Assessment Review Group Report to the Nuclear Regulatory Commission,* NUREG/CR-0400, September 1978.

22. See "Report to the American Physical Society of the Study Group on Light Water Reactor Safety," ref. 20.

23. J.O. Blomeke *et al.,* "Managing Radioactive Wastes," *Physics Today,* August, 1973; W.F. Bishop, I.R. Hoos, N. Hilberry, D.S. Matlay and R.A. Watson, "Essays on Issues Relevant to the Regulation of Radioactive Waste Management," U.S. Nuclear Regulatory Commission,

NUREG-0412, May, 1978; Office of Science and Technology Policy, Office of the President, "Isolation of Radioactive Wastes in Geologic Repositories: Status of Scientific and Technological Knowledge," a working paper for the Interagency Review Group on Nuclear Waste Management. July 1978, Draft.

24. Luther J. Carter, "Nuclear Wastes: The Science of Geologic Disposal Seen as Weak," *Science 200*, p. 1135, June 9, 1978, and references for Table 5.1.

25. *Report to the American Physical Society by the Study Group on Nuclear Fuel Cycles and Waste Management*, American Physical Society, July 1977.

26. Mason Willrich and T.B. Taylor, *Nuclear Theft: Risks and Safeguards*, a report to the Energy Policy Project, Ballinger, Cambridge, 1974.

27. John McPhee, *The Curve of Binding Energy*, Ballantine Books, New York, 1974.

28. "Security Agency Study, a report to the Congress on the Need for, and the Feasibility of, Establishing a Security Agency within the Office of Nuclear Material Safety and Safeguards," U.S. Nuclear Regulatory Commission, NUREG-0015, August 1976; Division of Safeguards, U.S. Nuclear Regulatory Commission, "Safeguarding a Domestic Mixed Oxide Industry Against a Hypothetical Subnational Threat," NUREG 0414, May 1978; M. Levinson and E. Zebroski, "A Fast Breeder System Concept, a Diversion Resistant Fuel Cycle," paper presented at the 5th Energy Technology Conference, Washington, D.C., February 27, 1978.

29. Harold A. Feiveson, Frank von Hippel, and Robert H. Williams, "Fission Power: An Evolutionary Strategy," *Science*, vol. 203, p. 330, January 26, 1979.

30. If even greater uranium savings are required than is possible with once through fuel cycles these advanced burner reactors could still operate with no weapons usable material in fresh reactor fuel if the fuel cycle were designed so that the recycled fuel were "isotopically denatured" uranium-233 and uranium-235 instead of plutonium. Both uranium-233 and uranium-235 can be isotopically denatured through dilution with uranium-238 so that the mixture is ineffective as nuclear weapons material. No such isotopic denaturant exists for plutonium. Thus when advanced burner reactors are operated in the "denatured recycle mode" the fuel would be a mixture of denatured uranium and thorium. Most of the fissile material (material fissionable by slow neutrons) produced in the reactor would be uranium-233, produced by neutron bombardment of thorium. Some plutonium would be produced from neutron bombardment of uranium-238 in the fuel. However, the plutonium would remain with the radioactive waste and would not be recycled.

31. Harold A. Feiveson, Frank von Hippel, and Robert H. Williams, *An Evolutionary Strategy for Nuclear Power (Alternatives to the Breeder)*, Center for Environmental Studies Report 67, Princeton University, September 1978.

32. Albert Wohlstetter et al, *Moving Toward Life in a Nuclear Armed Crowd*, prepared for the U.S. Arms Control and Disarmament Agency, 1976; Ted Greenwood, Harold A. Feiveson, and Theordore B. Taylor, *Nuclear Proliferation: Motivations, Capabilities, and Strategies for Control*, McGraw Hill, 1977.

33. David Burnham, "U.S. Documents Support Belief Israel Got Missing Uranium for Arms," *New York Times*, p. 3, November 6, 1977 and "Uranium Loss Brought No Security Changes in Decade, Report Says," *New York Times*, p. 14, February 25, 1979.

34. We assume a world of 8 billion people with per capita energy use half that of the U.S. today and with half the energy provided by nuclear power. In such a world the average thermal power in reactors would be 24,000 Gw(t), corresponding to 12,000 Gw(e) of installed electrical capacity, operating at ⅔ of maximum capacity on the average. The most proliferation resistant once through fuel cycle identified in ref. 29 is a high temperature gas cooled reactor using thorium and denatured uranium as fuel. For this technology a 1 Gw(e) power plant would discharge 22 kg of fissile plutonium (plutonium isotopes fissionable by slow neutrons) each year.

35. See, e.g., Joseph Nye, "Nonproliferation: A Long Term Strategy," *Foreign Affairs*, vol. 56, No. 3, page 601, April 1978.

36. Committee on Radioactive Waste Management, National Academy of Sciences, "Disposal of Solid Radioactive Wastes in Bedded Salt Deposits," U.S. Gov't. Printing Office, Washington, DC, November 1970.

37. *Science*, April 16, 1971, p. 249.

38. *Nuclear Industry*, August 1973, p. 20.

39. *Science*, April 25, 1975, p. 345.

40. *Science*, October 24, 1975, p. 361.

41. *The New York Times*, July 12, 1978, p. 13.

42. *Inside D.O.E.*, March 12, 1979.

CHAPTER 6

1. See, for example, Stanley W. Angrist and Loren G. Hepler, *Order and Chaos*, Basic Books, New York, 1967.

2. See, for example, Robert H. Williams, ed., *Toward a Solar Civilization*, MIT Press, 1979; William D. Metz and Allen L. Hammond, *Solar Energy in America*, American Association for the Advancement of Science, 1978; Denis Hayes, *Rays of Hope: The Transition to a Post-Petroleum World*, W.W. Norton, New York, 1977; B.J. Brinkworth, *Solar Energy for Man*, Halstead Press, 1972; Office of Technology Assessment, "Application of Solar Technology to Today's Energy Needs," June 1978, U.S. Government Printing Office; Frank Kreith and Jan Kreider, *Introduction to Solar Engineering*, McGraw-Hill Book Co., 1980.

3. W.A. Shurcliff, *Solar Heated Buildings of North America: 120 Outstanding Examples; Solar Heated Buildings: A Brief Survey*, 13th and Final Edition (19 Appleton Street, Cambridge, Massachusetts 02138, 1977); and "Active Type Solar Heating Systems for Houses: A Technology in Ferment," in *Toward a Solar Civilization*, ref. 2.

4. Battelle Columbus Laboratories, with Honeywell and Battelle Pacific Northwest Laboratories, Elton H. Hall program director, "Survey of the Applications of Solar Thermal Energy Systems to Industrial Process Heat," January 1977; Intertechnology Corporation, "Analysis of the Economic Potential of Solar Thermal Energy to Provide Industrial Process Heat," Warrenton, Va., February 1977.

5. See Richard Schoen *et al., New Energy Technologies for Buildings*, a report to the Energy Policy Project of the Ford Foundation (Cambridge, Mass., Ballinger, 1975). Between 1945 and 1960 some 50,000 solar hot water heaters were installed in southern Florida before the availability of cheap natural gas resulted in their being phased out. See also J.E. Scott, "The Solar Water Heater Industry in South Florida: History and Projections," *Solar Energy*, vol. 18, p. 387, 1976.

6. Energy Information Administration, U.S.D.O.E., a series of reports entitled "Solar Collector Manufacturing Activity," U.S. Government Printing Office. See, for example, the report for July 1977 through June 1978, published February 1979.

7. The comparative economics of solar heating is uncertain: collectors are still under development, installation and peripheral equipment costs are highly variable, and the costs of competing energy supplies are rising. We outline here a calculation in which we are optimistic about collector performance, we allow collector (but not installation and peripheral equipment) costs to be variable and we use today's replacement costs for other forms of energy. The general conclu-

sions reached in the text remain valid for substantial increases in these replacement costs and substantial decreases in present collector costs.

To compare solar energy with alternative fuels for water heating, we evaluate the capital cost per kw of delivered heat energy. In notes 8, 9, and 10 we show that for offshore oil, synthetic gas, and electricity these replacement costs are (in 1976$):

SOURCE	CAPITAL COST ($/KW DELIVERED)
Offshore oil	1400
Synthetic gas	2700
Electricity	2700

For the solar energy system we choose a high performance collector scheme: an evacuated tube collector (being developed by General Electric and Owens-Illinois). These collectors have much higher efficiency than ordinary flat plate collectors and are amenable to substantial cost reduction through mass production. We also choose an optimal arrangement for the collectors: a southfacing array tilted at the latitude angle plus 10°. Several of our assumptions are optimistic for the collectors: neglect of the effects of dirt, snow and possible shading, and the assumption that it would be economical to orient the collectors optimally. We assumed pessimistically, however, a somewhat higher temperature than necessary.

From the solar data given in note 11 for a New York City location we estimate the size of the solar collector for water heating by assuming the collector is large enough to provide 80% of the hot water demand at the average insolation rate in June (the month when the isolation rate is greatest). Thus the collector area is

$$\frac{0.8 \times (12 \text{ kwh/day})}{2.93 \text{ kwh/m}^2\text{/day}} = 3.28 \text{ m}^2$$

Following R.H. Bezdek et al., "Economic Feasibility of Solar Water and Space Heating," *Science*, p. 1214, March 23, 1979, we assume that the total capital cost is a fixed cost of $400 (storage and other costs independent of collector area) plus a cost proportional to collector area. Since the average delivered solar heat is (see note 11):

$$\frac{(2.38 \text{ kwh/m}^2\text{/day}) \times (3.28 \text{ m}^2) \times (365 \text{ days})}{8766 \text{ hours}} = 0.325 \text{ kw},$$

the breakeven solar collector cost C_o (in $\$/m^2$) in the case of electricity is given by

$$\frac{400 + 3.28\ C_0}{0.325 \text{ kw}} = 2700$$

or

$$C_o = \$150/\text{m}^2.$$

Similarly the total system breakeven capital cost per m² of collector is

$$C_T = C_o + (400/3.28) = \$270/\text{m}^2.$$

The breakeven capital costs for all the alternative systems considered here are

SOURCE	C_0 ($/M²)	C_T ($/M²)
Offshore oil	17	140
Synthetic gas	150	270
Electricity	150	270

Since installed evacuated tube collector costs are currently about $C_o = \$260/m^2$ in 1979 dollars or $\$210/m^2$ in 1976 dollars (communication between Howard Geller, Princeton University, and Pam Fay, General Electric Co., Valley Forge, Pa., August 1979), we see that solar water heating would not compete with alternative sources in the New York City area today. (Note that this reference to New York City means insolation and weather in that area, but costs on a national average basis. With its very high land, distribution, and maintenance costs, conventional utility energy is very costly in New York City, so that a rooftop collector could actually be highly competitive.) However, solar water heating would be competitive today with electricity in especially sunny areas such as Colorado and southern California (see R.H. Bezdek et al., op. cit.). In the future, however, the prospects are very good for making solar water heating competitive using evacuated tube collectors even in climates like that of New York City, since manufacturers believe collector costs can be reduced as much as 75% by increasing the production volume, by improving processes such as tube sealing, and by developing materials such as selective coatings.

8. New oil supplies from onshore sources in the lower 48 states (including refining, transportation, and bulk storage) are estimated to cost $44,000 per barrel of oil per day of capacity in 1977 dollars. (See "Resource Requirement, Impacts, and Potential Constraints Associated with Various Energy Futures," Bechtel Corp. for the U.S. Department of Energy, Aug. 1978). Deflating 10% to 1976 dollars, correcting for 12.6% conversion losses, and assuming a 90% capacity factor and a 65% oil water heater efficiency, we obtain a capital cost of

$$\frac{(\$44{,}000/\text{B/D})}{1.1 \times 0.874 \times 0.90 \times 0.65 \times (71\ \text{kw/B/D})} = \$1100/\text{kw}.$$

To this we must add the capital cost associated with the retail mark up, which was 9.2¢/gallon or $\$0.66/10^6$ Btu in 1976. Assuming an annual capital charge of 11.3% (see Appendix B) and a 65% efficient oil water heater this marketing capital cost becomes

$$\frac{(\$0.66/10^6\ \text{Btu}) \times (3413\ \text{Btu/kwh}) \times (8766\ \text{hours/year})}{0.113 \times 0.65} = \$270/\text{kw}.$$

Thus the total cost of offshore oil is $1370/kw.

9. Synthetic gas from coal is estimated to cost about $5.00 per million Btu (in 1976 dollars) at the coal gasification plant (see note 9, Chapter 3). To this must be added a cost for transmission and distribution (a national average of about $1.40 per million Btu in 1976—see note 5, Chapter 8). Assuming an annual capital charge rate of 11.3% (see Appendix B), 5% transmission and distribution losses; and a 65% efficient gas water heater, the capital cost for synthetic gas becomes

$$\frac{(\$6.40/10^6\ \text{Btu}) \times (3413\ \text{Btu/kwh}) \times (8766\ \text{hours/year})}{0.113 \times 0.95 \times 0.65} = \$2744/\text{kw}$$

10. There are three capital costs associated with an electrical system: the cost of the baseload power plant; the cost of providing the fuel; the cost of the transmission and distribution system.

The baseload power plant: A new nuclear power plant is estimated to cost $800/kw in 1976 dollars (see note 27d, Chapter 8). Such plants would be operated on the average at about 65% of rated capacity. Also losses in transmission and distribution amount to about 9%. Moreover electric water heaters are about 81% efficient (see R.H. Hoskins and E. Hirst, "Energy and Cost Analysis of Residential Water Heaters," Oak Ridge National Laboratory Report ORNL/CON-10, June 1977). Thus the cost associated with a baseload plant per average kilowatt delivered to the water in a domestic water heater is

$$\frac{\$800/\text{kw}}{.65 \times 0.91 \times 0.81} = \$1670/\text{kw}$$

The nuclear fuel system: According to note 27k, Chapter 8, the busbar cost for the nuclear fuel cycle is $0.00522 per kwh. Thus, assuming a capital charge rate of 11.3% (see Appendix B) the equivalent capital cost is

$$\frac{(\$.00522/\text{kwh}) \times (8766 \text{ hours/year})}{0.113 \times 0.91 \times 0.81} = \$549/\text{kw}.$$

Transmission and Distribution: Evaluation of transmission and especially distribution costs for electricity, for comparison with alternative energy technologies, is particularly difficult. The quantity of interest is the difference in distribution cost for a certain level of electrical service and a higher level of service. Substantial added costs could be incurred if the level of electric service were greatly increased, e.g., if a building were shifted from direct fossil fuel heating to electric heating. Unfortunately we do not have the detailed information needed to determine this difference.

The simple model we use here involves assigning a capital cost which is linear in the demand of $314/kw (see note 27i, Chapter 8) up to the maximum power required. Each kw of average hot water demand requires 1.18 kw of transmission and distribution capacity, however, since it is estimated that the average level of hot water demand is 18% higher in winter than on the average (45% higher in winter than in summer), according to R.W. Weatherwax and R.H. Williams, "Energy Conservation/Load Management Analysis for the Residential, Small Commercial, and Industrial Sectors of the Consolidated Edison Electric Franchise District," report to the New York Public Service Commission, March 1977. Thus the cost of transmission and distribution is

$$\frac{1.18 \times (\$314/\text{kw})}{0.91 \times 0.81} = \$503/\text{kw}.$$

[This result corresponds to a transmission and distribution capital charge for water heating (in ¢/kwh) which is about half the national average replacement cost associated with the capital component of transmission and distribution expenses for all residential and small light and power customers. See M.L. Baughman and D.J. Bottaro, "Electric Power Transmission and Distribution Systems: Costs and Their Allocation," Center for Energy Studies, University of Texas at Austin, July 1975.]

Thus the total capital cost for providing electricity is $2722/kw.

11. The following are the data on space and water heating demand per dwelling unit and the solar energy data used for the New York City application considered in the text:

	(1) WATER HEATING DEMAND[a] (KWH/DAY)	(2) SPACE HEATING DEMAND[b] (KWH/DAY)	(3) INSOLATION ON A TILTED COLLECTOR[c] (KWH/M²)	(4) COLLECTION AND DISTRIBUTION[d] EFFICIENCY	(5) COLLECTIBLE SOLAR ENERGY[e] (KWH/M²/DAY)	(6) USEFUL SOLAR ENERGY FROM 30 M² COLLECTOR[f] (KWH/DAY)
January	17.5	98.2	2.89	0.55	1.59	47.7
February	17.5	92.4	3.67	0.55	2.02	60.6
March	17.5	71.5	4.61	0.56	2.58	77.4
April	12.0	37.2	4.77	0.54	2.58	49.2
May	12.0	11.8	4.95	0.54	2.67	23.8
June	12.0	—	5.32	0.55	2.93	12.0
July	12.0	—	5.30	0.56	2.97	12.0
August	12.0	—	4.88	0.57	2.78	12.0
September	12.0	2.9	4.87	0.59	2.87	14.9
October	17.5	26.6	4.22	0.59	2.49	44.1
November	17.5	59.3	2.90	0.56	1.62	48.6
December	17.5	86.4	2.69	0.55	1.48	44.4
Average	14.8	40.6	4.26	0.56	2.38	37.1

NOTES

[a] From Chapter 8 we obtain the average daily demand for hot water to be

$$\frac{0.81 \times (6680 \text{ kwh/year})}{365 \text{ days/year}} = 14.8 \text{ kwh/day}.$$

Following the discussion in note 10, however, we assume that the daily winter demand is 45% higher than the daily summer demand.

[b] We assume that the total annual space heating demand is that for a single family house built to FHA standards (see Note 15, Chapter 8). The distribution of heating demand is assumed to follow the monthly distribution of heating degree days (with a base of 65°F).

[c] For a south facing collector tilted at the latitude angle plus 10°. Based on a computer model developed by Frank von Hippel, Princeton University. See Table G-4 in H.S. Geller, "Thermal Distribution Systems and Residential District Heating," Center for Energy and Environmental Studies Report, January 1980.

[d] For an evacuated tube collector in which the water is removed from the collector at an average temperature of 70°C (158°F). The collector efficiency ε_c is based on a computer model developed by Frank von Hippel (see reference in note c). The losses from the storage and the distribution system are assumed to be 10%. Thus the overall efficiency is assumed to be $0.9\ \varepsilon_c$.

[e] (5) = (3) × (4).

[f] (6) = the lesser of [30 × (5)] and [(1) + (2)].

12. Here we evaluate the economics of a solar system in New York City based on evaucated tube collectors and providing ⅔ of the total space and water heating demand.

The capital cost for space and water heating with offshore oil and synthetic gas would be about the same as for water heating alone (see notes 8 and 9), while the capital cost for electric space and water heating would be much higher than for water heating alone, since electric space heating poses a significant peak demand problem for electric utilities. The capital costs for these fuels are thus in 1976$ (see note 14):

SOURCE	CAPITAL COST ($/KW DELIVERED)
Offshore oil	1400
Synthetic gas	2700
Electricity	4700

The solar system would have a collector area of 30 m^2 (see note 11). Following R.H. Bezdek et al. (see note 7) we assume that the total system capital cost is $1300 plus a cost proportional to collector area. Since the average delivered solar heating rate (see note 11) is

$$\frac{(37.1 \text{ kwh/day}) \times 365 \text{ days}}{8766 \text{ hours}} = 1.55 \text{ kw},$$

the breakeven solar collector cost in comparison with synthetic gas is

$$\frac{1300 + 30\ C_o}{1.55 \text{ kw}} = 2700$$

or

$$C_o = \$100/m^2$$

and the total system cost is $40 higher:

$$C_T = \$140/m^2.$$

The breakeven capital costs for all the alternative systems considered here are

SOURCE	C_O ($/M²)	C_T ($/M²)
Offshore oil	30	70
Synthetic gas	100	140
Electricity	200	240

Thus we see that while combined solar space and water heating systems must be half as costly per m^2 as solar systems for water heating alone to compete with oil or synthetic gas, the total breakeven cost in the case of electricity is about the same for combined water/space heating systems as it is for water heating systems alone.

13. To be competitive with the cost of synthetic gas we estimate that collectors for solar water heating would have to cost just under $150/m²$ (note 7), while collectors for space and water heating would have to cost $100/m²$ or less (note 12), a reduction of ⅓. Alternatively collectors at $150/m²$ would become competitive with synfuel if the capital cost for the latter were about $3600/kw, or ⅓ higher than the $2700/kw cost assumed.

14. There are four capital costs associated with providing resistive space and water heat via electricity (for heat pump systems see note 15): The cost of the baseload power plant; the cost of the peaking power plant; the cost of providing the fuel; and the cost of the transmission and distribution system.

The baseload power plant: From note 11 the average annual demand at the power plant is 1.69/0.91 kw for space heat, and a total of

$$\frac{1}{0.91}[1.69 + 0.62/0.81] = 2.70 \text{ kw}$$

for space and water heat, so that the cost of baseload power per average kilowatt of heat delivered is

$$\frac{(\$800/\text{kw}) \times (2.70 \text{ kw})}{0.65\ [1.69 + 0.62]} = \$1439/\text{kw}$$

The peaking power plant: According to note 15, Chapter 8, the peak electrical demand for space heating is 10.51 kw at the house or 11.55 kw at the power plant. We assume that there is no water heating at the time of the peak. If 80% of all generating capacity is used at the time of the space heating peak then the peaking capacity needed is

$$\frac{1}{0.80}\left[11.55 - \frac{80}{65} \times 2.70\right] = 10.28 \text{ kw},$$

According to note 27h, Chapter 8, peaking capacity costs $282/kw. Thus the peaking capacity cost per average delivered kilowatt is

$$\frac{(10.28 \text{ kw}) \times (\$282/\text{kw})}{1.69 + 0.62} = \$1255/\text{kw}.$$

(We neglect capital costs associated with the fuel system for the peaking plant.)

The nuclear fuel system: As in note 10 the capital cost for the fuel system is given by

$$\frac{(\$0.00522/\text{kwh}) \times (8766 \text{ hours/year}) \times (2.70 \text{ kw})}{0.113 \times [1.69 + 0.62]} = \$473/\text{kw}$$

Transmission and distribution: The capital cost is determined by the peak demand:

$$\frac{(\$314/\text{kw}) \times (11.55 \text{ kw})}{1.69 + 0.62} = \$1570/\text{kw}.$$

[This result corresponds to a transmission and distribution capital charge for space and water heating together which is about 1½ times the national average replacement cost for the capital component of transmission and distribution expenses for all residential and small light and power customers. See note 10.]

Thus the total capital cost for providing electricity is $4737/kw in 1976$.

15. The total capital required per consumer would be about the same whether resistive heating, a conventional heat pump or an advanced heat pump were used, even though the electricity consumption would be very different in each case. See notes 27, 28, and 29 in Chapter 8.

16. In a system with 50% solar supply and electrical backup the average electricity demand for heating purposes would be half as large as in a 100% electrical system. However, the peak electrical demand for the 50% solar system would be about as large as the peak electrical demand in a 100% electrical system. Thus the ratio of the peak electrical demand to the average electrical demand would be about twice as large for the 50% solar system as for the 100% electrical system.

17. J.G. Asbury and R.O. Mueller ("Solar Energy and Electric Utilities: Should They Be Interfaced?" in *Energy II: Use, Conservation and Supply,* Philip H. Abelson and Allen L. Hammond, Eds., American Association for the Advancement of Science, 1978) have recognized this mismatch problem and have carried the argument one step further. In a solar system heat storage capacity is needed for overnight and for cloudy periods. Asbury and Mueller suggest that if this storage were instead used with the electric option without any solar heating, this storage capacity could be used for electrical load leveling, which could reduce the cost of the electric option. The economics of the electric option could indeed be significantly improved this way if the system peak were of relatively short duration. However, it is unclear whether a significant improvement could be achieved in areas where there are long winter "cold spells" (several weeks) in which the outdoor temperature rarely gets above, say 10°F. The first months of 1979 in the home-towns of the authors (Ann Arbor, Michigan and Princeton, New Jersey) had such cold spells.

18. The Energy Tax act of 1978 provides the individual with a credit against his income tax, associated with installation of residential solar energy equipment, of 30% of the cost up to $2000 and 20% of the costs which exceed $2000, but do not exceed $10,000.

19. The literature on owner-built low-cost house heating inventions is huge. One interesting article is: Sandra Oddo, "A Sunny Day in San Luis," *Solar Age,* p. 30, March 1979.

20. Theodore B. Taylor, "Low Cost Solar Energy," paper presented at the International Scientific Forum on an Acceptable Nuclear Energy Future of the World, Fort Lauderdale, Florida, November 11, 1977.

21. H. Tabor, "Solar Ponds," *Solar Energy,* vol. 7, p. 289, 1963; Ari Rabl and C.E. Nielsen, "Solar Ponds for Space Heating," *Solar Energy,* vol. 17, p. 1, 1975; A.B. Casamajor and R.E. Parsons, "Design Guide for Shallow Solar Ponds," Lawrence Livermore Laboratory, UCRL-5238, January 1978; M. Edesess and T.S. Jayadev, "Solar Ponds," Solar Energy Research Institute, March 1980; K.C. Brown, M. Edesess, and T.S. Jayadev, "Solar Ponds for Industrial Process Heat," presented at the Solar Industrial Process Heat Conference, Oakland, California, November 1979.

22. Whereas we assumed that the level of domestic hot water demand is nearly 50% higher in winter than in summer, the demand for industrial process heat would tend to be more uniform throughout the year in many capital intensive heat-using industries. Also substantial economies of scale could be captured in heat storage facilities for relatively low temperature process heat applications. These factors would both tend to make solar low temperature industrial process heat more economical than solar domestic water heat, in areas where land is available for collectors. Partially offsetting these advantages, are the facts that: (a) industry usually requires very short payback periods for its capital investments and (b) fuel costs are written off immediately as a tax deduction. See, for example, four articles on industrial process heat in the March 1979 issue of *Solar Age:*

"Industrial Process Heat/Creative Opportunities"; "Industrial Process Heat/Applications"; "Industrial Process Heat/GE at Riegel Textile"; "Industrial Process Heat/Concentrating Collectors."

23. The "median" industrial steam load is about 250,000 lb. of steam per hour per site, according to Dow Chemical Company, Environmental Research Institute of Michigan, Townsend-Greenspan and Company, Cravath, Swaine and Moore, *Energy Industrial Center Report,* report to the National Science Foundation, p. 449, 1975. This corresponds to a steam heating rate (@ 10^3 Btu/lb. of steam) of 73,000 kw. Let us suppose that a tracking, concentrating collector with an overall heat recovery efficiency of 55% is used. The collector might be sized so that during the summer period when insolation is the highest all the available solar heat is used. For the New York City area this would occur in June, when the average insolation is 233 watts/m^2 (see note 8, Chapter 11). Thus the collector area required is

$$\frac{73 \times 10^6 \text{ watts}}{0.55 \times (233 \text{ watts/m}^2)} = 0.57 \text{ km}^2 = 0.22 \text{ square miles}$$

Because of shading about 4 m^2 of land are needed for each m^2 of collector. Thus about 1 square mile is needed for the collector field. Since on the average the insolation is only 172 watts/m^2 in this area and since perhaps 10% of the area would be unused on the average due to maintenance, this system would provide only about ⅔ of the process heat requirements.

24. A firm that operates 10 hours a day and burns 100 barrels of oil per day in 80% efficient boilers produces steam at a rate of 46,000 lb. per hour, which is equivalent to 19,000 lb. of steam per hour on a 24 hour equivalent basis. Thus the area requirements for the collector field (including an allowance for shading) would be 0.17 km^2 or 43 acres.

25. This point was made in a study for D.O.E. entitled "Distributed Energy in California's Future," Paul Craig, private communication.

26. See for example, E.J. Francis and J. Seelinger, "Forecast Market and Shipbuilding Program for OTEC/Industrial Plant-Ships in Tropical Oceans," *Proceedings of the 1977 Annual Meeting of the American Section of the International Solar Energy Society,* pp. 24–28.

27. Electrolysis plants currently operate at an efficiency of about 60%. Advanced systems may achieve 80 percent efficiency. With this improvement the cost of hydrogen per million BTU would be

$$0.312 \times (\text{cost of electricity in mills/kwh}) + 0.227$$

Electricity from new nuclear baseload plants costs about 23.4 mills per kwh (see note 28). Thus hydrogen would cost $7.50 per million Btu. (Hydrogen from solar electric sources could be even more costly.) For details see ref. 29. For comparison synthetic methane from coal is estimated to cost about $5.00 per million Btu in 1976 dollars (see note 9, Chapter 3).

28. The busbar cost of baseload electricity from a new nuclear power plant is estimated to be (in mills/kwh):

capital[a]	15.9
fuel[b]	5.2
operation and maintenance[c]	2.3
	23.4

NOTES

[a] For a plant costing $800/kw, a 65% capacity factor, and a 11.3% annual capital charge rate (see notes 27a and 27d, Chapter 8).

[b] See note 27k, Chapter 8.

[c] See note 27l, Chapter 8.

29. R.H. Wentorf, Jr., and R.E. Hanneman, "Thermochemical Hydrogen Generation", *Science*, July 26, 1974, p. 311.

30. Mark S. Wrighton, "The Chemical Conversion of Sunlight", *Technology Review*, May 1977, p. 31.

31. From 1960 to 1972 fuelwood consumption declined at an average annual rate of 9% per year. But between 1972 and 1977 fuelwood consumption grew at 6%/year (compared to 1.2% for total energy use), until it reached a level of 635×10^6 cubic feet (0.23 quads) in 1977. See U.S. Dept. of Commerce, *Statistical Abstract of the United States*, 1978, p. 734. David Claridge of the Solar Energy Research Institute suggests (private communication) that total use of wood as a household fuel is much higher (although the above rates of change may not be misleading). Analysis of a survey by the Gallup organization by John O. Davies of Garden Way Inc., "Public Attitude and Actions Regarding Wood as a Home Heating Alternative," presented at the Wood Heating Seminar 5, St. Louis, Sept. 12, 1979 leads to an estimate of over 1 quad per year.

32. The estimate in Appendix A of the minimum need for fluid fuels as 40% of total energy consumption must be modified upward in the case of heavy dependence on solar energy for heating, because of increased requirements for backup energy based on stored fuel. We take 50 quads, of the total 102.5 quads, to be the minimum requirement for fluid fuels. (See note 33 for method of accounting.)

33. In this book solar energy is measured as the equivalent primary fossil fuel energy, obtained for a given energy carrier by dividing the amount of the delivered energy carrier by the present average efficiency for converting fossil fuel resources to the carrier in question. In particular, the fossil fuel equivalent energy associated with solar electricity is obtained by dividing the electrical energy delivered by 0.29.

34. In 1976 the U.S. farmers received, on the average, $2.15 per bushel of corn (see *Statistical Abstract*, 1978, p. 715). Assuming 56 lb. of corn per bushel, a 15% moisture content in the grain, and a fuel value of 7500 Btu/lb., this translates into a fuel equivalent price of $6.00 per million Btu. For comparison the average price of bituminous coal in 1976 was $0.84 per million Btu (see *Statistical Abstract*, p. 607).

35. In A. Poole and R.H. Williams, "Flower Power: Prospects for Photosynthetic Energy," *Toward a Solar Civilization* (see ref. 2) it is estimated that 100 to 150 million acres may be in this category. For comparison the U.S. has about 440 million acres of cropland.

36. Roderick Nash, "Problems in Paradise, Appropriate Technology and Environmental Values," presented at the Wingspread Symposium on Social Values and Technology Choice, June 1978. The main focus of this paper is the countercultural back-to-the-land movement, but it is nevertheless appropriate to the present discussion.

37. See, for example, H.A. Wilcox, "The Ocean Food and Energy Farm Project," paper presented at the 141st Annual Meeting of the American Association for the Advancement of Science, Jan. 29, 1975; I.T. Shaw Jr. *et al.*, "Comparative Assessment of Marine Biomass Materials," a report prepared for the Electric Power Research Institute by Science Applications, Inc./ Northwest, Bellevue, Washington, September 1979.

38. For 1972. See I.J. Bloodworth et al., *World Energy Demand to 2020*, report prepared for the Conservation Commission of the World Energy Conference by the Energy Research Group, Cavendish Laboratory, University of Cambridge, Cambridge, England, August 1977.

39. Most analyses which have been done envision electricity generation from high temperature solar heat sources. See, for example, Richard S. Caputo, "Solar Power Plants: Dark House in the Energy Stable," in *Toward a Solar Civilization*, ref. 2.

40. See Clarence Zener, "Solar Sea Power," in *Toward a Solar Civilization*, ref. 2.

41. David R. Inglis, *Wind Power and Other Energy Options,* University of Michigan Press, Ann Arbor, 1979; Percy H. Thomas, *Electric Power from the Wind,* Federal Power Commission, Washington, D.C., 1945; *Proceedings of the Second Workshop on Wind Energy Conversion Systems,* Washington, D.C., June 1975, prepared by the MITRE Corp., Frank R. Eldridge, Ed.

42. J.M. Leishman and G. Scobie, *The Development of Wave Power—A Techno-Economic Study,* prepared at the National Engineering Laboratory, East Kilbride, Scotland, February 1975.

43. See Bruce Chalmers, "The Photovoltaic Generation of Electricity," *Scientific American,* p. 34, October 1976; John Fan, "Solar Cells: Plugging into the Sun," *Technology Review,* p. 14, August/September 1978; Henry Kelly, "Photovoltaic Power Systems: a Tour Through the Alternatives," *Science,* p. 634, February 10, 1978; Henry Ehrenreich *et al., Principal Conclusions of the APS Study Group on Solar Photovoltaic Energy,* American Physical Society, New York, 1979.

44. In F. Von Hippel and R.H. Williams, "Solar Technologies," in *Toward a Solar Civilization,* ref. 2, it is pointed out that a global effort to put 10,000 Mw(e) of solar satellite power into orbit per year (equivalent to 10 large nuclear power stations) would require two space shuttle flights per day. Each shuttle flight would dump over 100 tons of chlorine in the form of hydrochloric acid into the stratosphere from the rocket exhaust. At two shuttle flights per day this activity would introduce chlorine throughout the stratosphere as fast as the maximum rate of stratospheric chlorine production in 1980 likely to be produced from the release of freon at ground level.

45. Peter E. Glaser, "Space Solar Power: An Option for Power Generation," paper presented at 100th Annual Meeting of the American Public Health Association, Atlantic City, New Jersey, November 14, 1972; Gerald K. O'Neill, "Space Colonies and Energy Supply to the Earth," *Science 190,* p. 943, December 5, 1975.

46. William G. Pollard, "The Long-Range Prospects for Solar Energy," *American Scientist 64,* p. 424, July-August 1976; Joseph St. Amand, "Satellite Solar Power Stations—Energy Source or Sink?" *Physics and Society,* The Newsletter of the Forum on Physics and Society, The American Physical Society, New York, Vol. 7, No. 3, December 1978; R.A. Herendeen, T. Kary and J. Rebitzer, "Energy Analysis of the Solar Power Satellite", *Science 205,* p. 458, Aug. 3, 1979; Janet Mitchell, "The Edsel of the Solar Age", *Environmental Action II,* July-Aug. 1979.

47. Assume a site with 400 watts/m^2 average wind energy at 150-200 ft. altitude (ref. 48) and net efficiency for conversion of wind to electrical energy of 20% (ref. 49). This implies average generation of 80 watts/m^2 electrical. A 200 ft. diameter machine then producces about 240 kw. The nominal rating of such a machine could be 1000 to 1500 kw.

48. Jack W. Reed, "Wind Climatology," *Proceedings of the Second Workshop on Wind Energy Conversion Systems,* Washington, D.C., June 1975, prepared by the MITRE Corporation, Frank R. Eldridge, ed.

49. C.G. Justus, "Annual Power Output Potential for 100 kw and 1 Mw Aerogenerators," *Proceedings of the Second Workshop,* ref. 48.

50. According to "The 29th Annual Electrical Industry Forecast." *Electrical World,* September 15, 1978, the (non-coincident) peak electrical demand in the U.S. in 1975 was 1.61 times the level of average electrical demand in the U.S. in that year. We now estimate the various components of the capital cost *per average delivered electrical kilowatt.*

The baseload power plant: For a nuclear system (see note 10) the cost of a baseload plant is

$$\frac{\$800/\text{kw}}{0.65 \times 0.91} = \$1352/\text{kw}.$$

The peaking power plant: If all capacity is operated at 80% of rated capacity at the time of this system peak then the peaking capacity cost is (see note 14):

$$\frac{1}{0.80 \times 0.91}\left[1.61 - \left(\frac{80}{65} \times 1.0\right)\right] \times (\$282/\text{kw}) = \$147/\text{kw}.$$

The nuclear fuel system: As in note 10, the capital cost for nuclear fuel is

$$\frac{(\$0.00522/\text{kwh} \times (8766 \text{ hours/year})}{0.113 \times 0.91} = \$445/\text{kw}.$$

Transmission and distribution: The capital cost for new transmission and distribution equipment was $0.0082 per kwh in 1972 (see M.L. Baughman and D.J. Bottaro, "Electric Power Transmission and Distribution Systems: Costs and Their Allocation," Center for Energy Studies Report, University of Texas at Austin, July 1975). Using the Handy-Whitman Index of Public Utility Construction Costs as a deflator, this becomes $0.013 per kwh in 1976 dollars. Since Baughman and Bottaro assume an annual capital charge rate of 13.5%, the equivalent capital cost is

$$\frac{(\$0.013/\text{kwh}) \times (8766 \text{ hours/year})}{0.135} = \$844/\text{kw}.$$

Thus the total replacement capital cost would be

baseload plant	1352
peaking plant	147
fuel	445
transmission and distribution	844
	$2788

51. In areas of typical sunniness in the U.S. the annual average insolation on a horizontal surface is 175 Watts/m². If the overall system efficiency for conversion of sunlight to delivered electricity were 12%, then about 50 square meters of solar collector would be required per kw(e) delivered.

52. If ½ of the primary energy use (see note 33 on method of accounting) in 2010 were for electricity production at central station plants where 10^4 Btu is required to produce a kwh, then the total average rate of electricity consumption in 2010, under Business As Usual conditions, would be (see Table 3.2):

$$\frac{0.91 \times 51.3 \times 10^{15} \text{ Btu}}{(10^4 \text{ Btu/kwh} \times (8766 \text{ hours})} = 532 \times 10^6 \text{ kw}.$$

Somewhat more than this much capacity would have to be purchased in the period 1975-2010, since electrical capacity lasts about 30 years. Thus the total capital cost required for electricity in this period would be slightly more than (see note 50)

$$(532 \times 10^6 \text{ kw}) \times (\$2800/\text{kw}) = \$1.5 \times 10^{12}$$

In this same period cumulative GNP would be about $\$93 \times 10^{12}$ (1976 dollars), assuming the 2.7% average annual GNP growth rate projected in Chapter 3. In the period 1971-1977 new plant and equipment expenditures averaged 7.5% of GNP. If this level of investment persists then cumulative expenditures for new plant and equipment would be $\$7 \times 10^{12}$ in this period. Thus the electric utilities share would be about 20%.

CHAPTER 7

1. Some recent studies of how energy is used are: Energy Policy Project of the Ford Foundation, *A Time to Choose*, Ballinger, Cambridge, Mass., 1974; E.L. Allen *et al., U.S. Energy and Economic Growth, 1975-2010*, vol. III of *Economic and Environmental Implications of a U.S. Nuclear Moratorium, 1985-2010*, Report ORAU/IEAO 76-7 prepared by the Institute for Energy Analysis for the Energy Research and Development Administration, September 1976; Demand and

Conservation Panel of the Committee on Nuclear and Alternative Energy Systems (CONAES), *Alternative Energy Demand Futures*, National Academy of Sciences, Washington, D.C., 1979.

2. The 1974 American Physical Society summer study on Technical Aspects of Efficient Energy Utilization applied the second-law efficiency concept to energy consuming activities throughout the economy. This study is available as W.H. Carnahan *et al.*, *Efficient Use of Energy, A Physics Perspective,* from NTIS (PB-242-773), or in *Efficient Energy Use*, Volume 25 of the American Institute of Physics Conference Proceedings. This work is briefly summarized by R.H. Socolow in *Physics Today*, August 1975. See also Marc H. Ross and Robert H. Williams, "The Potential for Fuel Conservation," *Technology Review* 79, p. 48, February, 1977.

3. The process of fuel combustion in itself results in a loss of about 30% of the capacity of the fuel energy to do work. Of the remaining potential work a little under ⅔ (i.e., about 40% of the fuel energy) can actually be transformed into electricity by a modern generating system. It is hoped to develop improved combined cycle (gas turbine–steam turbine) generators to achieve efficiencies up to about 55%, which would be about 0.8 of the maximum that could be achieved with a device based on combustion of a fuel. See the A.P.S. study for a discussion (ref. 2).

4. Making a universal judgment about the long term prospects is complicated by the fact that second-law efficiencies are sensitive to the complexity of the task and to the energy qualities in question. A transformation of one high-quality form of energy into another by a very simple step (as with an electric motor) may be characterized in the long run by an efficiency close to 100%. For systems which involve thermal energy at some stage, practical efficiencies tend to be lower. An example where the efficiency of thermal processes is expected to be relatively high was mentioned in note 3. The long run prospect for a process which proceeds by a complex series of steps, since it can be thought of as the product of subsystem efficiencies, is a second-law efficiency which may be much lower than the lower goal stated in the text. For a recent analysis of applications of second-law efficiency for evaluation of conservation opportunities, see B. Hamel, H. Brown, M. Ross, B. Hedman, J. Sweeney, S. Smith and M. Koluch, "Study of the Second Law of Thermodynamics as Related to Energy Conservation," a report to the Dept. of Energy by General Energy Associates, Cherry Hill, N.J., Nov. 1979.

5. A different way to look at the effect of modifying the task is to expand the boundary of the system whose efficiency is in question. Thus for example, instead of considering the efficiency of a particular heater in processing a food, one can in principle consider the efficiency of an entire series of processing steps, or that of processing, transport, warehousing, etc. Using this approach, Ayres and Narkus-Kramer (see ref. 6) have creatively confirmed that the technical potential for energy savings is, in principle, greater than indicated by the efficiencies of Table 7.1.

6. R.U. Ayers and M. Narkus-Kramer, "An Assessment of Methodologies for Estimating National Energy Efficiency," International Research and Technology, McLean, Va., June 1976.

CHAPTER 8

1. There were 71 million dwelling units in 1975, of which 52 million will still be in use in 2000 (see E. Hirst and J. Carney, "Residential Energy Use to the Year 2000: Conservation and Economics," Oak Ridge National Laboratory Report ORNL/CON-13, September 1977). We estimate that U.S. population will be 254 million (see note 2, Chapter 3) and that there will be 2.5 persons per household in 2000. Thus we estimate that there will be about 102 million households in the year 2000.

2. For example, in Table 9 of ref. (1) Hirst and Carney estimate that government programs to encourage "retrofits" could lead to reductions in building thermal integrities for space heating (relative to 1970 building performance) of 35% in single family housing units and 28% for multi-family housing units. As a second example, The Department of Energy estimate of the impact of

implementing the Residential Conservation Service (RCS) program established in Title II of the National Energy Conservation Policy Act (PL 95-619) is that 7% of households will request an audit each year, that 75% of audits will be followed up by conservation retrofits, and that the completed program will result in energy savings of 0.38 quads per year, which is less than 4 percent of residential space and water heating energy consumption in 1975. To estimate the expected savings for participating households, note that with a 5 year program about 26% of houses would be retrofitted, so that energy savings for space heating and water heating would average about 13% for the customers who would act on the advice of the household audit. See Department of Energy, "Residential Conservation Service program. Proposed Rulemaking and Public Hearing," p. 16577, *Federal Register*, Part II, March 19, 1979.

3. The energy characteristics of the house described here are based on a simple computer simulation of energy flows in the house. See L.W. Wall, T. Dey, A.J. Gadgil, A.B. Lilly, and A.H. Rosenfeld, "Conservation Options in Residential Energy Studies Using the Computer Program Twozone," Lawrence Berkeley Laboratory Report LBL-5271, August 1977. In actual practice fuel savings might not be as high as those predicted by the computer model.

4. In our estimate of fuel savings, unless explicitly mentioned, we are neglecting the corrections associated with the embodied energy of the changes made (e.g., the energy required to make insulation). In the cases we consider direct fuel use is so high that the embodied energy corrections do not alter any essential conclusions. See Bruce Hannon, Richard G. Stein, B.Z. Segal and Diane Serber, "Energy and Labor in the Construction Sector," *Science* 202, p. 837, November 24, 1978; and Robert Herendeen, "Notes and Comments on the Energy Cost of Building Construction," *Energy and Buildings I*, p. 95, May 1977.

5. The average price of gas to residential customers in California was $1.68 per million BTU in 1976. See Table 93 in American Gas Association, *Gas Facts: 1976 Data*, 1977. As the discussion of Chapter 4 indicated, the replacement cost of gas is not well established and will increase with time. For the purposes of illustration, however, we shall assume that the replacement cost of gas to residential customers, averaged over the life of the conservation investment, is equal to the cost of gas delivered to residential customers from new gas wells today.

We first note from *Gas Facts: 1976 Data* that for the U.S. as a whole the average wellhead price of gas in 1976 was $\$0.57/10^6$ BTU. But this low price was due to federal price controls. Uncontrolled intrastate gas sold for about $\$1.65/10^6$ BTU (See Energy Information Administration, *Annual Report to Congress:* Volume II, 1977, April, 1978, p. 162). Assuming a competitive market for new intrastate gas, this indicates that the replacement cost in 1976 was $\$1.08/10^6$ BTU higher than the average price. The average U.S. price to residential customers should rise this much from $\$1.98/10^6$ BTU in 1976 to $\$3.06/10^6$ BTU if the price were equal to the replacement cost. To get the replacement cost in California, one should subtract from the national replacement cost $\$0.19/10^6$ BTU, the difference (in 1976 dollars) between the prices of residential gas in the U.S. as a whole and in California in 1970. Thus the replacement cost in California in 1976 would be $\$2.87/10^6$ BTU.

(There are two problems with this calculation. First it neglects the fact that the cost of gas at the wellhead from new sources will increase over time—perhaps two- or three-fold by the turn of the century. Second it tends to overstate the present value of saved gas to the nation, because it includes the transmission and distribution cost; if the transmission and distribution system is already in place the incremental transmission and distribution costs are essentially zero. These two effects tend to cancel.)

6. If the price rises to the replacement cost (as estimated in note 5) over 10 years, the annual average rate of increase would be 5.4%/year, and the average future, or levelized lifecycle price of gas over this period is obtained by multiplying the 1976 price ($\$1.68/10^6$ Btu) by

$$\frac{C(d,\ 10)}{C(d - 0.054,\ 10)} = \frac{0.123}{.0925} = 1.33,$$

so that the average future price is $2.24/10^6$ Btu.

Here the capital recovery factor C (d, N) and the homeowners discount rate d are specified in Appendix B.

(This is certainly an underestimate of the average future price, since in January of 1979 the national average residential price of natural gas had already risen to $2.98/10^6$ Btu, corresponding to a California price, in 1976 dollars, of $2.29/10^6$ Btu.)

7. The following table gives a summary by region of the percentage of houses with various degrees of conservation equipment.

		US	NE	NC	S	W
Storm Doors	All doors covered	48	78	82	25	11
	Some doors covered	12	13	9	15	9
	No doors covered	39	8	8	59	79
Storm Windows	All windows covered	46	76	80	22	12
	Some windows covered	10	15	11	8	7
	No windows covered	43	8	8	69	80
Some Attic and Roof Insulation	Yes	74	79	84	67	67
	No	16	14	9	22	19
	Don't know	8	6	5	8	12

NOTE ON REGIONS: NE, Northeast; NC, Northcentral (Ohio through Great Plains); S, South (Delaware through Texas); W, West.

SOURCE: 1975 Annual Housing Survey, U.S. Bureau of the Census. (One to two percent of equipment totals were not reported.)

8. F.W. Sinden, "A Two Thirds Reduction in the Space Heat Requirement of a Twin Rivers Townhouse", *Energy and Buildings*, vol. 1, no. 3, April 1978, pp. 243-260.

9. S.D. Silverstein, "Efficient Energy Utilization in Buildings: The Architectural Window," *Proc. 10th Intersociety Energy Conversion Engineering Conf.*, IEEE, New York, 1975, p. 685; also William Shurcliff, *Thermal Shutters and Shades*, Brick House Publishing Co., Andover, Mass., 1980.

10. G.S. Dutt, J. Beyea and F.W. Sinden, "Attic Heat Loss and Conservation Policy," paper presented at the 1978 ASME Energy Technology Conference, Houston, Texas, November 1978.

11. F.W. Sinden, private communication.

12. Assuming a 70% efficient gas furnace the fuel savings rate for the test house would be 43 million Btu/year, according to reference (8). Thus the unit cost of saved fuel would be

$$\frac{.142 \times \$1245}{43 \times 10^6 \text{ Btu}} = \$4.11/10^6 \text{ Btu}.$$

13. The national average replacement cost of residential gas would be about $3.06/10^6$ Btu (see note 5). To this should be added $0.82/10^6$ Btu, the difference (in 1976 dollars) between the prices of residential gas in New Jersey and in the U.S. as a whole in 1970. Thus the replacement cost of gas in New Jersey in 1976 was about $3.88/10^6$ Btu, compared to a 1976 price of $3.16/10^6$ Btu. Following the procedure in note 6 one obtains an average future price of $3.50/10^6$ Btu.

14. M.H. Ross, "Residential Space Heating," a report prepared for Resources for the Future, March 16, 1978. See Sam H. Schurr *et al. Energy in America's Future, The Choice Before Us*, Resources for the Future, Johns Hopkins University Press, 1979, Part II, Chapter 5, Figure 5.2.

15. Following Sinden (ref. 8) we describe heating requirements as follows. The rate F (Btu/hour) at which a house must be provided heat from the heating system is given by

$$F = [L(T_i - T_o) - I - S]_+$$

where

L = thermal lossiness of the building shell (Btu/hour/°F)
I = internal heat loads from people and appliances, corrected for humidification of infiltrated air (Btu/hour)
S = solar gains through windows (Btu/hour)
T_i = indoor temperature (°F)
T_o = outdoor temperature (°F)

The subscripted plus sign on the bracket means "if positive, otherwise zero."

The total lossiness L is given by

$$L = L_A + L_C$$

where

L_C = conductive lossiness
L_A = air infiltration lossiness

The conductive lossiness is given by

$$L_C = \sum_S U_S A_S$$

where

U_S = conductance of surface S (Btu/ft²/hr/°F)
A_S = area of surface S (ft²)

The following table gives the components of the conductive lossiness for a house meeting the minimum construction requirements to obtain an FHA approved mortgage, for the model energy conserving house considered here (this model house is similar to that proposed in Raymond W. Bliss, "Why Not Just Build the House Right in the First Place," *Bulletin of the Atomic Scientists,* p. 32, March 1976), and for a house with the target design set forth in the recent American Physical Society study (see ref. 18) on the efficiency of energy use:

	U_S (BTU/FT²/HR/°F)				CONDUCTIVE LOSSINESS (BTU/HR/°F)		
SURFACE	FHA	MODEL	APS TARGET	A_S (FT²)	FHA	MODEL	APS TARGET
Roof	0.053	0.053	0.03	750	39.8	39.8	22.5
Walls	0.085	0.065	0.03	1520	129.2	98.8	45.6
Doors	0.65	0.30	0.20	60	39.0	18.0	12.0
Floors	0.084	0.025	0.025	750	63.0	18.8	18.8
Windows	0.65	0.54 (day) 0.085 (night)	0.30 (day) 0.085 (night)	180	117.0	63.1	37.9
				TOTAL:	388.0	238.5	136.8

Here the different day/night U-values for windows arise because we assume insulating shutters are slid over all windows at night (10 hours).

The air infiltration lossiness is given by

$$L_A = C\rho V\, E_A = 216\, E_A \text{ Btu/hour/°F}$$

where

$C = 0.24$ Btu/lb/°F is the specific heat of air
$\rho = 0.075$ lb/ft^3 is the air density
$V = 12{,}000$ ft^3 is the volume of our model house.
E_A = number of air exchanges per hour.

Thus the air infiltration numbers for the alternative houses considered are:

	FHA	MODEL	APS TARGET
E_A	1.0	0.5	0.2
L_A	216.0	108.0	43.2

The following are the principal internal heat sources in the house (based on consumption rates in note 24) corrected for humidification (which is assumed to be controlled by the availability of internal evaporative sources):

SOURCE	HEAT GAINS (BTU/HOUR) CONVENTIONAL HOUSE	MODEL HOUSE
Water Heater	520(a)	62(a)
Dishwasher	140(b)	421(b)
Refrigerator-Freezer	642	−89(c)
Cooking	311(d)	136(d)
Lighting	440	132
Clothes Dryer	57(e)	49(f)
Color TV	195	129
Other	195	195
People	514(g)	514(g)
Indoor Plants	−92(h)	−92(h)
TOTALS	2922	1457

NOTES

(a) We assume 20% of the heat leaks through the water heater walls in the conventional house. In the model house we assume 4% of the heat leaks through the walls and that the water heater COP = 2.9 (see note 23) so that the heat loss rate is

$$0.04 \times 2.9 \times (1380 \text{ kwh}) \times \frac{(3413 \text{ Btu/kwh})}{8766 \text{ hours}} = 62 \text{ Btu/hour.}$$

(b) Dissipated electricity plus 10% of water heat for the conventional house. In the model house dissipated electricity plus *all* water heat (the waste water is held up in a recovery tank until it cools to room temperature).

(c) When the refrigerator-freezer is integrated into the ACES, the heat lost to the refrigerator through the walls, door openings and gaskets amounts to 150 Btu/hour. But 61 Btu/hour are added to the room: 40% of compressor electricity input plus 60% of the heat from mullion and case heaters plus 80% of the heat from the evaporator fan. See the table in note 37.

(d) For a conventional stove we assume that ⅓ of the heat of cooking (see note 24) is used to evaporate water. We assume that the same amount of water is evaporated in the model energy conserving house.

(e) Assuming 10% of dryer heat leaks into the house.

(f) For waste heat recovery in winter, we assume that the hot damp air from the clothes dryer is fed directly into the heat distribution system. A good filter would be required to recover the lint from the dryer exhaust. Most of this heat goes into evaporating water from the clothes—about 4 lb. of water per day (see note 24f).

(g) For an average occupancy of 3. Assuming the average person engaged in moderate activity loses water through evaporation at a rate of 0.16 lb./hour, about half of the available body heat would be used to evaporate this moisture.

(h) For an indoor plant watering rate of 1 quart per day.

According to Table A-3 in Appendix A of *Saving Energy in the Home: Princeton's Experiments at Twin Rivers*, Robert H. Socolow, ed., the average solar gains in Btu/ft²/hour on vertical surfaces in the months November-March are

	DIRECT			TOTAL		
DIFFUSE	EAST	WEST	SOUTH	EAST	WEST	SOUTH
1.6	15.2	9.2	34.1	16.8	10.8	35.7

The solar gains through the windows are obtained from these insulation rates, the window distribution shown in Table 8.3, and an assumed window transmissivity of 0.8.

The following table summarizes the characteristic of the houses we have considered here:

	FHA HOUSE	MODEL HOUSE	APS TARGET HOUSE
Conductive Lossiness (Btu/hr/°F)	388	239	137
Air Infiltration Lossiness (Btu/hr/°F)	216	108	43
Humidification (Btu/hour)	892	1002	1002
Internal Heat Sources (Btu/hour)	3814	2459	2459
Solar Window Gains (Btu/hour)	2279	3739	3739
F (Btu/hour)	$[604(T_i - T_o) - 5201]_+$	$[347(T_i - T_o) - 5196]_+$	$[180(T_i - T_o) - 5196]_+$
$(T_o)_{max}$[a]	59°F	53°F	39°F
Length of Heating Season[b] (hours)	4760	3837	1780
Heating Degree Days[b,c]	5164	4702	2835
Heating Season Conductive Losses (10^6 Btu)	48.1	27.0	9.3
Heating Season Air Infiltration Losses (10^6 Btu)	26.8	12.2	2.9
Heating Season Humidification Requirements (10^6 Btu)	4.2	3.8	1.8
Heating Season Internal Heat Gains (10^6 Btu)	18.2	9.4	4.4
Heating Season Solar Window Gains (10^6 Btu)	10.8	14.3	6.7
Net Heat Required During Heating Season	50.1	19.3	2.9

	FHA HOUSE	MODEL HOUSE	APS TARGET HOUSE
Average Net Heating Rate During Heating Season [Btu/hour (kw)]	10,525(3.08)	5030(1.47)	1629(0.48)
Peak Net Heating Rate During Heating Season [Btu/hour (kw)]	35,870(10.51)	18,400(5.39)	7044(2.06)

NOTES

[a] $(T_o)_{max}$ is the outdoor temperature at which the heating system must be turned on. For an average indoor temperature of 68°F this is given by

$$(T_o)_{max} = 68 - \frac{(I + S)}{L}$$

[b] The annual distribution of temperature by hours for central New Jersey is given by the following figure, obtained from ref. 8:

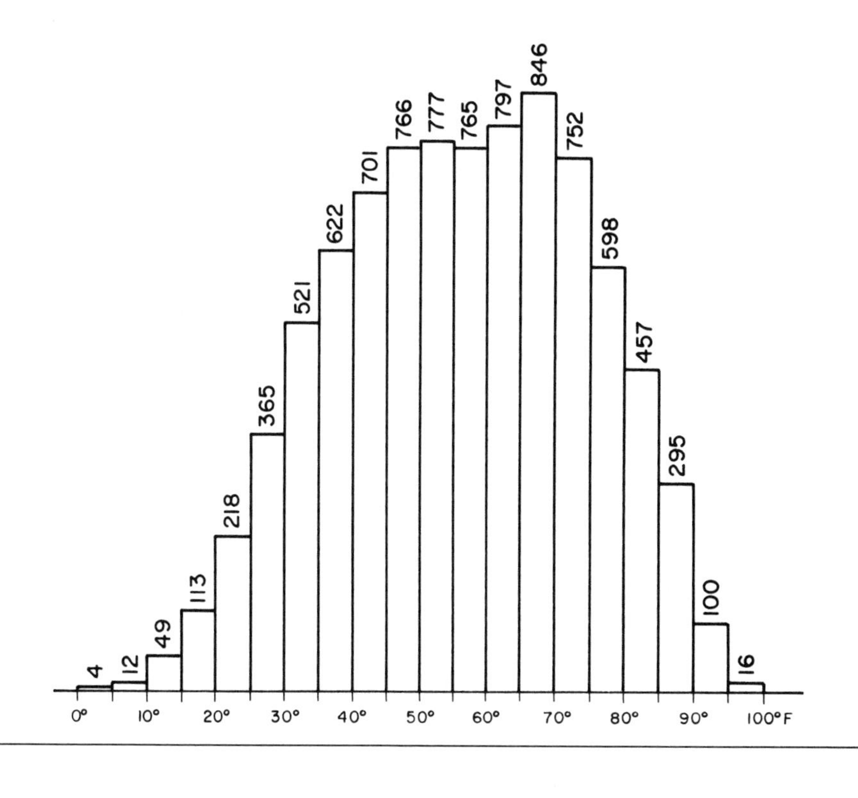

[c] Heating degree days are defined here (unconventionally) as

$$DD = \frac{1}{24} \sum_{T_o < (T_o)_{max}} t_o(68 - T_o)$$

where t_o is the number of hours in which the outdoor temperature is at T_o, obtained from the above figure.

16. Eugene H. Leger and Gautam S. Dutt, "An Affordable Solar House," paper presented at the 4th National Passive Solar Conference, Kansas City, 1979.

17. The Twin Rivers townhouse is similar in size (1420 ft^2) and in climate to the house we considered here. However, the heat required from the heating system in the Twin Rivers house was estimated to be 14 million Btu/year, compared to 19 million Btu/year for this model house. Of course one reason for the difference is that the Twin Rivers house is a townhouse whereas our model house is a single family dwelling with more outside walls. Also the Twin Rivers house has less air infiltration.

18. *Efficient Use of Energy, Part I—A Physics Perspective,* W.H. Carnahan et al., An American Physical Society study on the more efficient use of energy, American Institute of Physics Conference Proceedings No. 25, American Institute of Physics, New York, 1975. The target house suggested in this study, like the retrofitted Twin Rivers townhouse, would be very airtight: 0.2 compared to 0.5 air exchanges per hour for the model house we are considering here. Such a low air infiltration rate raises a number of health issues, which are discussed in Chapter 12.

19. See, for example, D.L. O'Neal, "Energy and Cost Analysis of Residential Heating Systems," Oak Ridge National Laboratory Report ORNL/CON-25, July 1978.

20. The pulse combustion furnace is expected to have an overall efficiency of 93%. See G.T. Hollowell, "Pulse Combustion—An Efficient, Forced Air, Space Heating System," *Proceedings of the Conference on Improving Efficiency and Performance on HVAC Equipment and Systems II,* Paper J.4, Purdue University, West Lafayette, Indiana, April 1976. The gas driven heat pump is expected to have a COP = 1.3 on the heating cycle, according to P.F. Swenson, "Competition Coming In Heat Pumps: Gas Fired May Be Best in Cold Climates," *Energy Research Reports,* p. 1, March 7, 1977. One goal of the American Gas Association heat pump research program is to have gasfired heat pumps available for the residential market by 1981, according to L.A. Sarkes, et al., "Gas Fired Heat Pumps: An Emerging Technology," *ASHRAE Journal,* pp. 36-41, March 1977.

21. Primary Space Conditioning Fuel Consumption[a] with Alternative Space Conditioning Systems in Model House

	(Relative Units)
Conventional Gas Furnace[b] + Electric Central A/C[c]	1.00[d]
High Efficiency Gas Furnace[b] + Electric Central A/C[c]	0.74
Gas Heat Pump System[e]	0.56
All Electric Systems	
Electric Resistance Heat + Central A/C[c]	1.73
Conventional Heat Pump[f]	1.04
ACES	0.62

NOTES:

[a] Primary fuel consumption includes generation, transmission, and distribution losses. Electrical generation losses are assumed to be 67% of fuel consumption for central station power plants. Transmission and distribution losses are assumed to be 10% for electrical systems and 5% for gas systems.

[b] The conventional gas furnace is assumed to be 61% efficient (see ref. 19). The high efficiency gas furnace is assumed to be a pulse combustion unit with an efficiency of 93%. For a description of such gas furnaces see G.T. Hollowell, ref. 20.

[c] The central A/C unit is assumed to have a COP = 2.5.

[d] For this "base case" annual fuel consumption would be 43.7 million Btu.

(e) The gas driven heat pump is assumed to have a COP = 1.3 for space heating and 0.9 for cooling. See P.F. Swenson, ref. 20.
(f) The conventional air-to-air heat pump is assumed to have a COP = 2.0 in both heating and cooling operations.

22. Air conditioning demand is much more difficult to estimate than space heating demand. For the house built to FHA specifications we assume the annual central A/C electricity requirements are the same as the 1970 average for the Mid-Atlantic region, 1.24 kwh/ft^2 of floor area, as given by S.H. Dole, *Energy Use and Conservation in the Residential Sector: A Regional Analysis,* RAND Corporation report R-1641-NSF, June 1975. Thus for a COP = 2, the annual cooling requirements are

$$2 \times (1.24 \text{ kwh/ft}^2) \times 1500 \text{ ft}^2 \times (3413 \text{ Btu/kwh}) = 12.7 \times 10^6 \text{ Btu}.$$

It has been estimated that in Twin Rivers, New Jersey, air conditioners operate about 700 hours during the year (see Appendix A in *Saving Energy in the Home: Princeton's Experiments at Twin Rivers,* Robert H. Socolow, ed.). Thus if the air conditioner is sized to meet the peak A/C load the peak A/C demand is

$$12.7 \times 10^6/700 = 18{,}000 \text{ Btu/hour}.$$

We estimate the seasonal and peak A/C demand for the model energy conserving house by taking into account the reduced air exchange rate and the reduced internal loads for this house. We do not attempt to estimate reduced conductive heat gains. We also assume that appropriate shading eliminates direct solar gains through the windows. Thus we obtain:

	SEASONAL COOLING LOAD (10^6 BTU)		PEAK COOLING LOAD (BTU/HOUR)	
	FHA HOUSE	MODEL HOUSE	FHA HOUSE	MODEL HOUSE
Air Exchange(a)	3.4	1.7	5500	2750
Internal Moisture Sources(b,c)	−1.5	−1.4	−510	−510
Internal Loads(b,d)	7.3	3.9	3020	1450
Other(e)	3.5	3.5	10,090	10,090
Total Cooling Load	12.7	7.7	18,100	13,780

NOTES:
(a) Air infiltration provides a heating load whenever the outdoor temperature is greater than the indoor temperature, which we assume is 72°F. According to the figure in note 15, this occurs about 1920 hours per year. The total air infiltration heat load is thus

$$H_{AI} = C\rho V\, E_A \sum_{T_o > 72°} t_o\,(T_o - 72)$$

where t_o is the number of hours the temperature is T_o, obtained from the figure in note 15.
(b) We assume the cooling season is the 1920 hours during which the temperature is higher than 72°F.
(c) The internal moisture sources are those given in note 15, except that there are no indoor plants in summer and in the model house the dryer is no longer exhausted into the house.
(d) The internal heat loads are the same as in note 15, except that in summer waste heat is not recovered from the dishwasher and dryer in the model house. Also we assume that during the peak demand period, the dishwasher, cooking, and the clothes dryer do not contribute.
(e) This is the residual, primarily conductive heat gains.

23. Because both the heat source (ice water) and the heat sink are maintained at constant temperatures, a much higher COP is possible than with a conventional heat pump which uses outdoor air as the heat source. With already developed heat pump technology the annual average coefficient of performance for the ACES can be

Function	COP
Space heating	2.65
Hot water	2.90
Air conditioning	10-13 (effective COP)

See E.C. Hise, "Performance Report for the ACES Demonstration House, August 1976 through August 1977", Oak Ridge National Laboratory Report ORNL/CON-19, March 1978. In our analysis we shall assume the effective COP for air conditioning is 10.

24. The total electricity requirements (kwh per year) are taken to be:

		MODEL HOUSE WITH ENERGY CONSERVING APPLIANCES		
	FHA ALL ELECTRIC HOUSE	RESISTIVE HEAT/ CENTRAL A/C	CONVENTIONAL HEAT PUMP	ACES
Space Heating	14,700	5645	2823[a]	2130
Air Conditioning	1,960[a]	900[b]	1125[a]	225
Hot Water	6,680	3765	3765	1380
Refrigerator-Freezer	1.650	740	740	420
Cooking	1,200[c]	750[d]	750[d]	750[d]
Lighting	1,130[c]	340	340	340
Dishwasher	270[e]	180[e]	180[e]	180[e]
Clothes Dryer	1,460[f]	580[f]	580[f]	580[f]
Color TV	500[g]	330[h]	330[h]	330[h]
Other	500	500	500	500
	30,000	13,700	11,100	6,800

NOTES:

[a] For a COP = 2.0.

[b] For a COP = 2.5.

[c] Average usage per household in 1970, according to S.H. Dole, "Energy Use and Conservation in the Residential Sector: A Regional Analysis," Rand Corporation Report R-1641-NSF, June 1975.

[d] Here it is assumed that a microwave oven is used for 75% of cooking and that for microwave cooking electricity requirements are only half of what they would be with a conventional electric range.

[e] Today's dishwashers (with input water @ 140°F) require about 0.75 kwh per wash (including 0.25 kwh for the drying cycle). We assume one wash per day. For the model energy conserving house we assume that the drying cycle is eliminated but that an extra 0.34 kwh per cycle is needed to heat the 7 gallons needed for the wash cycle from 120°F to 140°F. This extra 125 kwh/year is included in the electricity allocated to hot water.

[f] For a typical dryer today about 2 lb. of water are removed per kwh. We estimate today's consumption by assuming 8 lb. of wash per day and 1 lb. of water per lb. of wash. For the model house we assume that a dryer is used which is 75% efficient (2.5 lb. of water per kwh—about the best efficiency available in an electric dryer today) and that a high speed centrifugal extractor is used to remove half the water from clothes before they are put in the dryer.

[g] 1970 average.

[h] New color sets in 1974 required on the average 160 watts. (See Dick Furlong, "Technical Background Information for Appliance Efficiency Targets for Monochrome and Color Televisions", Appliance Efficiency Program, FEA, April 20, 1976). We adopt this value and the national average daily TV viewing time of 6 hours.

25. We estimate the cost of the ACES as follows:

ACES storage tank[a]	$1014
Insulation for tank[b]	376
Heat pump[c]	875
Ice bin coils[d]	254
Labor[e]	736
Overhead and profit[f]	814
TOTAL	$4069

while the costs for a conventional resistive heat/central A/C system for the same house would be

Resistance heating[g]		$ 362
Central A/C[g]		492
Overhead and profit[f]		214
	TOTAL	$1068

and the costs for a conventional air to air heat pump system for the same house would be

Heat pump[h]		$1242
Overhead and profit[f]		310
	TOTAL	$1552

NOTES:

[a] According to "Investigation of Alternative Types of Water Storage Components to be Utilized in Conjunction with the ACES System," report prepared by the office of Irwin G. Cantor for Oak Ridge National Laboratory (ORNL/Sub-78/14233/1), December 1977, the least costly tank was found to be one using preshaped polystyrene open cell blocks, between which are placed reinforcing steel and concrete, with the tank integrated into the foundation of the house. For this system the cost was estimated to be $135V^{-.45}$ per m^3. A 10,000 gallon tank, with a volume of 39 m^3, would thus cost $26/$m^3$.

[b] According to the *Application of Solar Energy to Today's Energy Needs*, prepared by the Office of Technology Assessment, June 1978 (vol. I, p. 496) polyurethane insulation can be blown in place for $102/$m^3$ of insulation. We assume 2″ of insulation on a 6′ high cylindrical tank.

[c] The peak heating demand is 5.4 kw(t) (see note 15). According to H.C. Fischer, "Cost of Early Production ACES," in *Summary of ACES Workshop I*, held October 29-30, 1975 at Oak Ridge, Tennessee, ORNL/TM-5243, July 1976, an ACES heat pump should cost $162/kw(t).

[d] H.C. Fischer estimates $414 for an 8.8kw(t) peak heat demand. This corresponds to (5.4/8.8) × $414 = $254 for our system.

[e] Estimate by H.C. Fischer, note (c).

[f] Taken to be 25% of direct material and labor costs.

[g] Baseboard resistive heaters are estimated to cost $67/kw and central air conditioners $123/kw(t), (see the OTA report, note (b), vol. II, p. 670). Thus the space heater costs

($67/kw) × 5.4 = $362.

Since the peak A/C demand is 4.0kw(t) (see note 22), the A/C cost is

($123/kw) × 4.0 = $492.

[h] According to the OTA report, note (b), vol. II, p. 670 an electric heat pump costs $800 per "ton" of capacity or $230/kw(t).

26. For new housing, the appropriate interest rate (in current dollars) would be about 9%, the mortgage interest rate. The appropriate capital charge rate in constant dollars with a 20 year loan would be 8.1% (see Appendix B).

The ACES costs $3000 more than the system with resistive heat and central A/C and leads to an annual savings of 6900 kwh. There is also an annual operation and maintenance cost of $50/year for the ACES, compared to $30/year for an A/C unit. Thus the cost of saved energy is

$$\frac{20 + (.081 \times 3000)}{6900} = \$0.038/\text{kwh}.$$

The ACES costs $2517 more than a conventional air to air heat pump and would lead to an annual savings of 4300 kwh relative to the heat pump system. We assume operation and maintenance costs are the same for both systems. Thus the cost of saved energy is

$$\frac{.081 \times 2517}{4300} = \$0.047/\text{kwh}.$$

27. The following table gives the replacement cost of electricity (in 1976$) for our model thermally tight all electric house with resistive heat and central air conditioning (see note 24):

	UNIT COST ($/KW DELIVERED)	CAPACITY (KW)	CAPITAL COST (1976$)	ANNUALIZED COST	
				$/YEAR[a]	MILLS/KWH[b]
Capital Equipment					
Baseload generation[c]	1350[d]	1.56[e]	2106	238.0	17.4
Peaking generation[c]	365[h]	3.75[f]	1369	154.7	11.3
Transmission and Dist.[c]	341[i]	5.79[e]	1974	223.1	16.3
Space Heating Unit	84[i]	5.4[e]	454	36.7	2.7
A/C unit	154[i]	4.0[g]	616	49.9	3.6
Subtotals			6519	702.4	51.3
Annual Expenses					
Fuel for power generation[k]				78.1	5.7
O & M for power generation[l]				34.3	2.5
O & M for Transmission and Dist.[m]				91.8	6.7
O & M for A/C unit[n]				30.0	2.2
Subtotals				234.2	17.1
TOTALS				936.6	68.4

NOTES:

[a] The annual capital charge rate is assumed to be 11.3% for utility owned capital and 8.1% for homeowner units covered by the mortgage (see Appendix B).

[b] Given by: (annual cost in $/year)/13,700 kwh.

[c] We assume 9% transmission and distribution losses. Thus 1 kw delivered power corresponds to 1.1 kw at the power plant.

[d] A new nuclear power plant is estimated to cost $800/kw in 1976 $ (see C. L. Rudisill, *Coal and Nuclear Generating Costs*, EPRI Special Report No. PS - 455 - SR, Electric Power Research Institute, Palo Alto, California, 1977). Thus for a 65% capacity factor the cost of delivered power is

$$\frac{800}{0.65 \times .91} = \$1350/\text{kw}$$

[e] The average demand is

$$\frac{13{,}700}{8766} = 1.56 \text{ kw(e)}$$

To determine the peak demand we assume that *all* water heating is done off peak and that the peak is a winter peak determined by the sum of:

- the average of all demand except space conditioning and water heating

$$\frac{3390 \text{ kwh}}{8766 \text{ hours}} = 0.39 \text{ kw(e)}$$

- the peak space heating demand of 5.4 kw(e) (see note 15).

[f] At the time of the system peak we assume baseload power plants are operated at 85% of capacity, so that the peaking capacity needed is

$$5.79 - \frac{85}{65} \times 1.56 = 3.75 \text{ kw(e)}$$

[g] From note 22 the peak A/C demand is 4.0 kw(t).

[h] According to ref. 25b, vol. I, p. 170, gas turbines for peaking purposes cost $282/kw. Thus if these units are used at 85% of capacity during peaking periods the cost per delivered kw is

$$\frac{282}{.85 \times .91} = \$365/\text{kw}$$

[i] According to the ref. in note 25b, vol. II, p. 721, distribution (exclusive of the substation) costs $157/kw(e) on the average. According to M. L. Baughman and D. J. Bottaro, "Electric Power Transmission and Distribution Systems: Costs and Their Allocation," Center for Energy Studies Report, University of Texas at Austin, July

1975, total T&D capital costs are twice those for distribution (exclusive of the substation). Thus, with 9% T&D losses, the T&D capital cost is

$$\frac{2 \times 157}{0.91} = \$341/\text{kw}$$

(For a discussion of this methodology for estimating the replacement cost for transmission and distribution, see note 10, Chapter 6. It should be noted that the estimate of the capital component of the replacement cost for T&D given here (16.3 mills/kwh) for an all electric house is comparable to the capital component of the *average* replacement cost for all residential and small light and power customers obtained by Baughman and Bottaro. Their estimate for 1972, 10.4 mills/kwh, when converted to 1976 dollars using the Handy-Whitman Index of Public Utility Construction Costs becomes 16.5 mills/kwh.)

(j) See note 25g (add 25% for overhead and profit).

(k) According to Study Group on Nuclear Fuel Cycles and Waste Management, *Nuclear Fuel Cycles and Waste Management*, American Physical Society, New York, 1977, the busbar cost for the nuclear fuel cycle is 5.22 mills/kwh—which becomes 5.74 mills per kwh when T&D losses are taken into account. The cost of fuel for peaking has not been included.

(l) In 1976 the operation and maintenance costs at nuclear plants were 2.3 mills/kwh at the busbar (2.5 mills/kwh delivered).

(m) The U.S. average O&M cost for T&D was 4.98 mills/kwh in 1972 (6.7 mills/kwh in 1976$), according to Baughman and Bottaro, (note i) above.

(n) According to the ref. in note 25b, vol. II, p. 670, operation and maintenance expenses for a central A/C unit amounts to about $30/year.

---

28. The following table gives the replacement cost of electricity (in 1976$) for our model thermally tight all electric house with a conventional (air to air) heat pump space conditioning system (see note 24):

	UNIT COST ($/KW DELIVERED)	CAPACITY (KW)	CAPITAL COST (1976$)	ANNUALIZED COST[a] $/YEAR	ANNUALIZED COST[a] MILLS/KWH
Capital Equipment					
Baseload Generation	1350	1.27	1715	194	17.5
Peaking Generation	365	3.75	1369	155	13.9
Transmission and Distribution	341	5.79	1974	223	20.1
Heat Pump	288[b]	5.4	1555	126	11.3
SUBTOTALS			6613	693	62.8
Annual Costs					
Fuel for Power Generation[c]				63	5.7
O&M for Power Generation				28	2.5
O&M for T&D				74	6.7
O&M for Heat Pump				50	4.5
			SUBTOTALS	215	19.4
			TOTALS	908	82.2

NOTES:

(a) The calculations here were carried out precisely as in note 27, except where indicated otherwise.

(b) Heat pumps are assumed to cost $230/kw(t) plus 25% for overhead and profit.

(c) The cost of peaking fuel has not been included.

---

29. The following table gives the replacement cost of electricity (in 1976$) for our model thermally tight all electric house equipped with an ACES (see note 24):

	UNIT COST ($/KW DELIVERED)	CAPACITY (KW)	CAPITAL COST (1976$)	ANNUALIZED COST ($/YEAR)	
Capital Equipment					
Baseload Generation	1350	0.78[a]	1053	119.0	
Peaking Generation	365	1.52	555	62.7	
Trans. and Dist.	341	2.54[b]	866	97.9	
ACES Unit			4069[c]	329.6	(496.4)[d]
SUBTOTALS			6543	609	(776)[d]
Annual Expenses					
Fuel for Power Generation[e]				39	
O&M for Power Generation				17	
O&M for T&D				46	
O&M for ACES				50	
TOTAL				761	(928)[d]

NOTES:

[a] This is the average rate of power consumption for an ACES house. See note 24.

[b] Since the space heating COP = 2.65 for an ACES unit (see note 23) the peak electricity demand for space heating in an energy conserving house with an ACES unit is 5.39/2.65 = 2.03 kw. To this must be added 0.51 kw(e), the average electrical demand for purposes other than space conditioning.

[c] See note 25.

[d] The first figure is for homeowner financing (8.1% annual capital charge rate). The figure in parentheses is for utility ownership and financing (12.2% annual capital charge rate). See Appendix B.

[e] The cost of fuel for peaking equipment has not been included.

30. The following table gives the replacement cost of electricity (in 1976$) for our model thermally tight house equipped with gas heat and electric central A/C.[a]

	UNIT COST	CAPACITY	CAPITAL COST	ANNUALIZED COST	
	($/KW DELIVERED)	(KW)	(1976$)	$/YEAR	MILLS/KWH
Capital Equipment					
Baseload Generation	1350	0.49[b]	662	74.8	17.3
Peaking Generation	365	1.37	500	56.5	13.1
Trans. & Dist.	341	2.01[c]	685	77.4	17.9
A/C Unit	154[d]	4.04[e]	622	52.9	12.2
		SUBTOTALS	2469	261.6	60.5
Annual Cost					
Fuel for Power Generation				24.6	5.7
O&M for Power Generation				10.8	2.5
O&M for T&D				28.9	6.7
O&M for A/C				30[f]	6.9
		SUBTOTALS		94.3	21.8
		TOTALS		355.9	82.3

NOTES:

[a] For the thermally tight house with energy conserving appliances described in notes 15 and 24. We assume both space heat and hot water are provided by gas but that the house has central electric air conditioning (A/C).

(b) The total annual electricity consumption is 4300 kwh (see notes 22 and 24). Thus the average delivered power is 4300/8766 = 0.49 kw.
(c) We assume air conditioning determines the peak demand. For a COP = 2.5 the peak A/C demand is 1.62 kw(e). To this we add an average consumption rate of 0.39 kw for other appliances. See notes 22 and 24.
(d) According to the ref. in note 25b, vol. II, p. 670, electric central air conditioners cost $430/ton ($123/kwt or $154/kw(t) including overhead and profit).
(e) From note 22, the peak thermal demand for the A/C unit is 4.04 kw(t).
(f) According to ref. in note 25b, vol. II, p. 670, an electric central air conditioner requires $30/year for maintenance.

---

31. The following table gives the total annual replacement cost of energy (in 1976$) for the house described in note 30:

Electrical System[a]		$356
Gas System		
Gas Furnace		
Capital[b]		28
Fuel[c]		80
Water Heater		
Fuel[c]		55
	TOTAL	$519

NOTES:
(a) See note 30.
(b) A gas furnace is estimated to cost $19 per 10^3 Btu/hour of capacity or $345. This estimate includes 25% overhead and profit. See the ref. in note 25b, vol. II, p. 670. If these items are included in the purchase price of a new home the appropriate annual capital charge rate is 8.1% for a 20 year mortgage. See Appendix B.
(c) The fuel required for space heating would be 26 million Btu with a 75% efficient gas furnace. Also about 18 million Btu of gas would be needed to provide hot water with a 70% efficient gas hot water heater. We assume that the replacement cost of gas in the U.S. is $3.06/$10^6$ Btu (see note 5).

---

32. J. G. Asbury, R. F. Giese and R. O. Mueller, "Residential Electric Heating and Cooling: Total Cost of Service," presented at Electric Power Research Institute Workshop on New Modes of HVAC: Economic Incentives and Barriers, Tampa, Florida, Jan., 1979. As an example of their analysis, in one storage scheme, a) baseload central station capacity would be heavily utilized, b) one-day thermal storage at the house would be required, and c) relatively little electrical peaking capacity would be needed. As with ACES a heat pump would be used, but the large ACES storage system would not be needed. Such storage schemes are very attractive with the present pattern of day and night electrical demand because the electricity for heating would be used at night, a period of low demand. This concept is employed to some extent in Europe. If, however, we consider electricity as the predominant or even a major source for space heat, demand patterns would be changed and the high cost of peaking power in cold weather would become important.

33. The electricity requirements for a typical electric water heater are estimated as follows. The base case corresponds to use of 71 gallons/day of water heated to 140°F in a 50 gallon heater; the annual electrical demand is 6680 kwh. See R. A. Hoskins and E. Hirst, "Energy and Cost Analysis of Residential Water Heaters," Oak Ridge National Laboratory Report ORNL/CON 10, June 1977.

Losses through the walls of the water heater and distribution pipe could be reduced by reducing the thermostat setting to 120°F. In so doing all uses can be accommodated except the wash cycle for the dishwasher, which requires 140°F water. For this purpose, however, an electric resistive heater can be installed in the dishwasher; 125 kwh per year would be required for this local heater to heat half the dishwasher water (7 gallons per day for one dishwasher load per day). With the thermostat setback plus jacket and distribution pipe insulation, heat losses are reduced to 4% of the electricity consumed, compared to 19% in the base case. (Data are from Hoskins and Hirst.)

The result of these steps is an annual electricity demand of 5760 kwh. When, in addition, the hot water tank is incorporated in the ACES the heat delivered to hot water would be 2.9 times the electrical energy consumed—that is, the COP for water heating would be 2.9. (ACES data are obtained from E. C. Hise, "Performance Report for the ACES Demonstration House August 1976 through August 1977," Oak Ridge National Laboratory Report ORNL/CON-19, March 1978). The net effect of all these measures would be to reduce electricity requirements 70% to 2070 kwh/year, with energy efficiency improvements only.

34. In R. H. Hoskins and E. Hirst (ref. 33), it is estimated that average daily hot water usage for a thermostat setting of 140°F would be 71 gallons. If the thermostat setting is reduced to 120°F and the demand for heat remains the same then the daily consumption would increase to 93 gallons. For a family of 4 we estimate that hot water usage could be reduced to 60 gallons per day, assuming that on the average water is available at the tap at 115°F. The components of this hot water budget might be roughly as follows:

Bathing: One 5-minute shower with 105°F water per day per capita with a shower head flow rate of 2 gallons per minute would require 33 gallons of hot water from the tap at 115°F. (Today flow rates vary from 3-10 gallons per minute. Thus a restricted flow shower head would have to be used.)

Laundry: Total water requirements for laundering an 8 lb. load of clothes range from 35-70 gallons, while the hot water requirements range from 5-25 gallons (see John G. Muller, "The Potential for Energy Savings Through Reductions in Hot Water Consumption, FEA, April, 1975). With a warm water wash and a cold water rinse it is feasible with washers available today to use only 5 gallons of hot water per 8 lb. load—which we adopt as a daily average. For some households hard water and/or heavily soiled clothes create problems which may be solved with more powerful chemicals. (The energy to manufacture the chemicals is relatively small, but there may be side effects.) In some cases more hot water will be required to obtain a satisfactory wash.

Dishwasher: A dishwasher consumes about 14 gallons of hot water per cycle, which we adopt as a daily average.

Kitchen, Bathroom, Utility Sinks: It has been estimated (see John G. Muller, cited above) that the average family consumes 25 gallons of water per day for sink uses. Perhaps 10 gallons a day would be water at 105°F (i.e., 8 gallons of tap water @ 115°F).

35. These changes are increased insulation thickness, improved insulation, removal of the fan from the cooled area, added anti-sweat heater switch, improved compressor, and increased surface areas for condenser and evaporator. See R. Hoskins and E. Hirst, "Energy and Cost Analysis of Residential Refrigerators," Oak Ridge National Laboratory Report, ORNL/CON-6, Jan., 1977.

36. For a purchase such as a refrigerator, we assume the consumer's discount rate is the same as the prime rate (9% in current dollars or 3.3% in constant dollars). Even though the mean lifetime of a refrigerator is 16 years, we assume here a lifecycle of 10 years (a 12% capital recovery factor) so that the cost of saving electricity is

$$\frac{.12 \times \$75}{910 \text{ kwh}} = 1¢/\text{kwh}.$$

37. The refrigerator-freezer energy use discussion is based on the following model data for a 16 cubic foot frost-free top-freezer refrigerator:

HEAT GAINS (BTU/HOUR)	CONVENTIONAL[a]	IMPROVED UNIT[a]		
		FREEZER	REFRIGERATOR	TOTAL
Through Walls	228.9	37.4	55.8	93.2
Door Openings	7.4	1.6	5.8	7.4
Gasket Area Infiltration	49.6	28.0	21.6	49.6
Food Load	23.4	19.7	3.7	23.4

HEAT GAINS (BTU/HOUR)	CONVENTIONAL[a]	IMPROVED UNIT[a]		
		FREEZER	REFRIGERATOR	TOTAL
Defrost Heater	42.7	42.7	—	42.7
Drain Heater	5.1	—	—	—
Mullion and Case Heaters	31.4	5.4	10.3	15.7
Evaporator Fan	19.9	1.4	1.4	2.8
	408.4	136.2	98.6	234.8

ELECTRICITY CONSUMPTION (KWH/DAY)	CONVENTIONAL UNIT[a]	IMPROVED UNIT[a]	ACES UNIT[b,c]
Compressor	3.16	1.34	.46
Refrigerator		(0.56)	(.07)
Freezer		(0.78)	(.39)
Condensor Fan	0.34	—	—
Case and Mullion Heaters	0.55	0.28	0.28
Defrost Heater	0.30	0.30	0.30
Evaporator Fan	0.14	0.10	0.10
Drain Heater	0.04	—	—
	4.53	2.02	1.14

NOTES:

(a) The numbers for the conventional unit are based on Hoskins and Hirst, ref. 35. The improved unit involves the proposed changes set forth in that report (all the changes except the elimination of the frost free feature and the evaporator fan). The heat gains shown here are slightly less than in that report because the indoor temperature is taken to be 68°F instead of 70°F.

(b) When the refrigerator freezer is incorporated into the ACES, circulated ice water from the ACES ice tank provides the cooling for the refrigerator compartment. As in the case of ACES air conditioning (see E. C. Hise, ref. 23) we assume this can be accomplished with an effective COP = 10. Since the ACES ice tank can replace room air as the heat sink for the freezing unit the temperature difference between the heat source and heat sink for the freezer heat pump is reduced in half. The COP for the freezer heat pump therefore should double.

(c) If the frost free feature were eliminated, the total electricity required would be reduced in half again, to 0.61 kwh/day.

38. One of these bulbs goes by the trade name of LITEK. (See "The Electrodeless Fluorescent Lamp," Energy Research and Development Administration Fact Sheet, April 14, 1976.) Suppose the LITEK bulb is introduced in 1980, when the price of residential electricity will average about 4¢/kwh. Over the period 1980–1990 the levelized cost of electricity should be about 4.5¢/kwh, assuming a 2% annual rate of real growth. The LITEK bulb is expected to last 20,000 cycles (about 20,000 hours over 25 years). However, we shall assume a life of only 10 years. We assume that the incandescent bulb lasts 800 hours (1 year).

If the LITEK bulb costs \$7.50 the annual cost (in \$1976) would be

$$.12 \times \$7.50 + (800 \text{ hours/year}) \times (.0285 \text{ kw/bulb}) \times \$.045/\text{kwh} = \$1.90/\text{year}$$

while the annual cost for the equivalent number of incandescent bulbs would be

$$0.50/\text{year} + (800 \text{ hours/year}) \times (0.10 \text{ kw/bulb}) \times \$0.045/\text{kwh} = \$4.10/\text{year}$$

Alternatively the unit cost of saved energy would be

$$\frac{.12 \times \$7.50 - \$0.50}{80 - 23} = \frac{\$0.40}{57 \text{ kwh}} = \$0.007/\text{kwh}$$

39. A. H. Rosenfeld, "Some Potentials for Energy and Peak Power Conservation in California," in *Energy Use Management, Proceedings of the International Conference,* R. A. Fazzolare and C. B. Smith, eds., Volume III/IV, p. 987, 1978.

40. The annual energy requirements shown in Figure 8.6 for the FHA and the model energy conserving houses are obtained from notes 15, 22, and 24.

The annual energy costs shown in Figure 8.6 for the thermally tight houses with energy conserving appliances consist of: (a) a loan payment of $160 on the investments needed to tighten the building shell and to improve the efficiency of appliances (see note 41) and (b) the annual costs for the energy supplies and the basic space conditioning equipment (see note 27 for the resistive heat case, note 28 for the heat pump case, note 29 for the ACES case, and notes 30 and 31 for the gas heat case).

The annual energy costs shown in Figure 8.6 for the FHA houses are obtained from the energy consumption data in notes 15, 22 and 24, the space conditioning appliance cost data given in notes 27-31, and the replacement electricity costs for alternative houses given in note 42. Again we assume that the replacement cost for gas is $3.06/10^6 Btu.

In the case of the Annual Cycle Energy System Figure 8.6 shows the annual cost for alternative types of financing: mortgage financing and private utility financing. While mortgage financing would be less costly to the homeowner, it may not be especially practical. As we have pointed out, ACES capital cost is so high (over $4,000 per house according to ref. 25) that a typical homeowner may not be willing to purchase it. But the utility should be interested, because investing in ACES would be less costly than investing in the alternative of building extra generation, transmission, and distribution systems. Thus utility ownership of annual cycle energy systems, though more costly to the homeowner, may be a more practical arrangement for these systems.

41. The incremental construction cost of the model conservation house is assumed to be $2000, with respect to a house built to FHA standards. (The FHA house is already much tighter than the average house).

If this investment were paid off as part of a 20 year mortgage the levelized annual cost would be $160 per year (see Appendix B).

The added investment includes improvements to the shell, especially improved window systems (tighter fitting windows, indoor shutters, external shades or awnings) and changes to reduce air infiltration. Much of the shell improvement is through redesign with little added cost. This investment also covers efficiency improvements in mass-produced appliances such as the refrigerator, the dishwasher, the dryer, the water heater, the air conditioner, and, for the gas heated house, the furnace.

In the case of the gas furnace, for example, we assume the overall efficiency is improved from 61% to 75%. This improvement could be achieved simply by introducing an automatic flue damper, by replacing the pilot light with an electronic ignition, and by increasing duct insulation to 2″. These measures are estimated to cost $195 (see ref. 19).

42. The national average electricity price paid by residential consumers in 1976 was 3.6¢/kwh. The replacement cost of electricity depends not only on the supply technology used but also on the nature of the demand. The replacement cost is higher the higher the ratio of the peak demand to average demand. In this book we have estimated the replacement cost to residential consumers from nuclear generated electricity for four different demand situations:

	RATIO OF PEAK DEMAND TO AVERAGE DEMAND	REPLACEMENT COST[a] (¢/KWH)
Gas heated house w/electric[b] air conditioning	4.1	6.3
All electric house w/resistive heat[c]	3.7	6.0

	RATIO OF PEAK DEMAND TO AVERAGE DEMAND	REPLACEMENT COST[(a)] (¢/KWH)
All electric house w/conventional heat pump[(d)]	4.6	6.6
All electric house w/ACES unit[(e)]	3.3	5.6

NOTES:
(a) The replacement costs given here include only the costs of generating, transmitting, and distributing electricity. The capital and maintenance costs of electricity consuming devices are *not* included.
(b) See note 30.
(c) See note 27.
(d) See note 28.
(e) See note 29.

43. The capital investment differences for the FHA house and model house (1976$) are:

OFF-SITE ENERGY SUPPLY CAPITAL (REPLACEMENT COST)	GAS HEAT, HOT WATER[(a)]		ELECTRIC RESISTANCE[(b)]	
	FHA	MODEL	FHA	MODEL
Baseload	1328	662	4632	2106
Peaking	852	500	2735	1369
T & D	1174	685	3887	1974
Fuel for elect	437	218	1523	691
Gas supply	3087	1193	—	—
Total supply capital	6878	3258	12,777	6140
Net incremental investment on-site[(c)]		1470		1370

NOTES:
(a) See notes 30 and 31.
(b) See note 27.
(c) The net on-site investment is $2000 for conservation improvements (see note 41), less the capital saved because the space conditioning equipment for the energy conserving house does not need to have as large a capacity as it does for the conventional house. This capital savings amounts to $530 for the gas heated house (see notes 30 and 31) and $630 for the resistively heated house (see note 27).

44. R. H. Socolow, "Energy Conservation in Existing Residences: Your Home Deserves a Housecall," paper presented at the conference *Energy Efficiency As a National Priority,* sponsored by Public Citizen, Inc., May 20, 1976, Washington, D.C.

45. Losses reported in ref. 10 were about 3 to 7 times the theoretically calculated loss rate.

46. D. T. Harrje, G. S. Dutt, J. Beyea, "Locating and Eliminating Obscure but Major Energy Losses in Residential Buildings," *ASHRAE Transactions,* vol. 85, part 2, 1979.

47. If inspections average 2 man-hours per house, one man could accomplish perhaps 700 inspections per year (3 per day for 230 days). To inspect all houses over 15 years would require

$$\frac{80 \times 10^6 \text{ inspections}}{15 \text{ years} \times 700 \text{ inspections/inspector/year}} = 8{,}000 \text{ inspectors}$$

An inspector could probably be supported (salary, benefits, equipment, administration) for $50,000 per year. Thus, a public house doctor program would cost $400 million per year. Residential fuel consumption amounted to 16.2 quadrillion Btu in 1975 and probably would average about 14 quadrillion Btu over the 15 year inspection program. Thus, the program could be supported by a residential energy tax of

$$\frac{\$400 \times 10^6}{14 \times 10^{15} \text{ Btu}} = \$0.03 \text{ per million Btu}$$

48. R. A. Grot and R. H. Socolow, "Energy Utilization in a Residential Community," in *Energy: Demand, Conservation and Institutional Problems,* Michael S. Mackrakis, ed., MIT Press, 1974.

49. F. W. Sinden, "Monitoring Household Energy Use," paper prepared for the Third Annual Conference and Exhibition on Technology for Energy Conservation, January 23-25, 1979, Tucson, Arizona.

50. Creative scientific study is yielding suggestive results about measuring building energy performance, particularly research which is improving our understanding of the thermal response of buildings to periodic variations in conditions. See Robert Sonderegger, "Dynamic Models of House Heating Based on Equivalent Thermal Parameters," Ph.D. thesis published as a report of the Center for Environmental Studies, Princeton University, Sept., 1977.

51. At the time of this writing the Department of Energy was in the process of promulgating energy performance standards for new buildings (Office of Solar Applications, DOE, "Energy Performance Standards for New Buildings," Advance Notice of Proposed Rulemaking and Notice of Public Meetings, November 1978). The proposed standards for residential buildings are based on a standardized computation of performance based on a highly simplified picture of the house. The standards are well motivated but, we fear, simulated performance may deviate widely from actual performance. (See, for example, note 45.) Moreover even this modest approach is beset by high costs. A program of R & D leading to more powerful methods of measurement and analysis, combined with replacement cost pricing of energy, appears to us much more promising.

52. Clive Seligman, John M. Darley and Lawrence J. Becker, "Behavioral Approaches to Encouraging Residential Energy Conservation," *Energy and Buildings,* Vol. 1, No. 3, 1978.

53. Richard Schoen, Alan S. Hirshberg and Jerome M. Weingart, *New Energy Technologies for Buildings,* Jane Stein, ed., Ballinger, 1975.

54. See "Oregon Utilities Can Put Home Insulation in Rate Base: Sound Economics," *Electrical Week,* July 17, 1978. It is estimated that the program will add 1.1 to 1.3¢/kwh to the price of residential electricity. This is considerably less than the difference between the estimated cost of residential electricity from new sources in Oregon (4.5¢/kwh) and the residential rate in 1978 (2.4¢/kwh).

55. One problem is that a regulation inhibiting new utility initiatives in financing of conservation improvements was written into the 1978 National Energy Act. In 1980 legislation was passed that relaxed this constraint.

56. Robert H. Williams and Marc H. Ross, "Drilling for Oil and Gas in Our Houses," *Technology Review,* March/April 1980.

57. A. H. Rosenfeld, D. B. Goldstein, A. J. Lichtenberg and P. P. Craig, "Saving Half of California's Energy and Peak Power in Buildings and Appliances Via Long-Range Standards and Other Legislation," paper submitted to California Policy Seminar, University of California at Berkeley, January 15, 1977.

58. Data on energy use and income and discussion of some issues can be found in R. Herendeen, "Affluence and Energy Demand," *American Society of Mechanical Engineers Technical*

Digest, 73-WA/Ener-8, Oct. 1974; Dorothy K. Newman and D. Day, *The American Energy Consumer*, a report to the Energy Policy Project of the Ford Foundation, Ballinger, 1975; and Denton E. Morrison, "Equity Impacts of Some Major Energy Alternatives, report to CONAES Risk/Impact Panel, May, 1977.

59. See several statements both in support of and opposing lifeline rates in "Panel on Lifeline Rate Proposals," pp. 333–523, in *Compilation of Statements of Witnesses Before the Subcommittee on Energy and Power, March/April 1976*, for the Committee Print *Electric Utility Rate Reform and Regulatory Improvement*, Hearings Before the Subcommittee on Energy and Power of the Committee on Interstate and Foreign Commerce, U.S. House of Representatives, April 1976.

60. A proposal for rationing, focussed on gasoline, is presented in Carter Henderson, "The Tragic Failure of Energy Planning," *The Bulletin of the Atomic Scientists*, p. 15, December 1978.

61. Willard Gaylin, Ira Glasser, Steven Marcus, and Daniel Rothman, *Doing Good: The Limits of Benevolence*, Pantheon, 1978.

CHAPTER 9

1. The energy budget for the round trip, Princeton, New Jersey to Airlie House, Virginia, was:

		ENERGY CONSUMED (GALLONS OF GASOLINE)
9:30 AM	Home to Princeton Airport (car)	0.2
10:00 AM	Princeton to Newark (air)[(a)]	10.0
11:10 AM	Newark to National (air)[(b)]	13.3
1:00 PM	Taxi, National to Airlie House[(c)]	8.0
5:00 PM	Chauffered Ride, Airlie House to National[(d)]	8.0
7:20 PM	National to Newark (air)[(b)]	13.3
9:00 PM	Newark to Princeton (air)[(a)]	2.5
9:20 PM	Princeton Airport to Home (car)	0.2
	TOTAL	55.5

NOTES:

(a) According to the pilot of the plane (which has an 8 passenger capacity) 10 gallons of fuel are consumed per flight between Princeton and Newark. In the morning, the author was the only passenger. Returning at night, there were 4 passengers.

(b) For a Boeing 727 traveling at 60% capacity, the energy intensity is 7800 Btu per passenger mile. The trip is 210 miles one way. At 125,000 Btu/gallon fuel consumption was

$$\frac{7800 \times 210}{125{,}000} = 13.3 \text{ gallons}$$

(c) The trip was about 60 miles with only one passenger in a behemoth consuming about 15 mpg. The taxi returned empty. Thus the energy consumption was

$$\frac{2 \times 60 \text{ miles}}{15 \text{ mpg}} = 8 \text{ gallons}$$

(d) Same conditions as (c).

2. *Alternative Energy Demand Futures to 2010*, Report of the Demand and Conservation Panel to the Committee on Nuclear and Alternative Energy Systems, National Research Council, National Academy of Sciences, 1979.

3. D. B. Shonka, A. S. Loebel, and P. D. Patterson, *Transportation Energy Conservation Data Book: Edition 2*, Oak Ridge National Laboratory Report, ORNL-5320, Oct., 1977.

4. Harry E. Strate, "Nationwide Personal Transportation Survey, Automobile Occupancy, Report No. 1," U.S. Dept. of Transportation, Apr., 1972. This survey conducted in 1969–70 found that 78% of all trips involved 1 or 2 occupants and 6% of all trips involved more than 4 occupants. Another survey showed a lower level for occupancy. William Hamilton, "Prospects for Electric Cars," a report by General Research Corp., Santa Barbara, Cal., Nov. 1978, reports that 1967/68 surveys in Los Angeles and Washington, D.C. showed that 91 and 93%, respectively, of all trips involved 1 or 2 occupants, and that 4 and 3%, respectively, involved 4 or more occupants.

5. We adopt a nominal 10,000 miles per year for the average automobile (see Table 9.1). According to Table 9.2 a typical cost of operating an automobile is 17.3¢/mile. Thus the annual budget for a car in the mid-1970s was about $1700.

6. American Physical Society Summer Study on Technical Aspects of Efficient Energy Utilization, 1974. Available as: W. H. Carnahan et al., *Efficient Use of Energy, A Physics Perspective,* from NTIS (PB-242-773), or in *Efficient Energy Use,* Volume 25 of the American Institute of Physics Conference Proceedings.

7. "Program Plan Report for Applied Research in Road Vehicles," prepared by the Energy Systems Group, Energy Systems Planning Division, TRW, McLean, Virginia, July 29, 1977; and "Data Base on Automobile Energy Conservation Technology" by the same group, Sept. 25, 1979.

8. Jim Dunne, "VW's 60-mpg Turbo-diesel Safety Car," *Popular Science,* Oct. 1977, p. 104. See also Ullrich Seiffert, Peter Walzer, and Hermann Oetting, "Improvements in Automotive Fuel Economy," presentation by Volkswagenwerk AG at the First International Automotive Fuel Economy Research Conference, Arlington, Va., Oct./Nov. 1979.

9. The efficiency used here is an "effective" efficiency, defined as the ratio of the heat content of the fuel produced to the heat content of the raw biomass. This effective efficiency in some cases can be greater than 100% because some of the applied solar heat gets converted to stored chemical energy in the fuel produced. See M. J. Antal, Jr., *Biomass Energy Enhancement,* a report to the President's Council on Environmental Quality, prepared at Princeton University's Department of Mechanical and Aerospace Engineering, July 1978.

10. With conventional technology the needed heat is usually provided by burning some of the raw biomass—a very wasteful use of the high quality chemical energy stored in biomass. Thus, for example, the efficiency of converting wood to methanol by conventional means has been estimated to be only 37%. See T. B. Reed, "Efficiencies of Methanol Production from Gas, Coal, Waste, or Wood," paper presented at the 171st Meeting of the Division of Fuel Chemistry of the American Chemical Society Symposium on Net Energetics of Integrated Synfuel Systems, April 5–9, 1976, New York, NY.

11. According to Poole and Williams (see note 13) recoverable crop residues amount to 4.2 quadrillion Btu/year. At 70% conversion efficiency this corresponds to 2.9 quadrillion Btu of useful liquid fuel. Since the average car in 1975 was driven about 10,000 miles this liquid fuel would support

$$\frac{50 \text{ mpg} \times (2.9 \times 10^{15} \text{ Btu/year})}{(1.39 \times 10^{5} \text{ Btu/gallon}) \times 10{,}000 \text{ miles/year}} = 104 \text{ million cars.}$$

There were 98 million cars in use in 1975. The biomass requirements per car are

$$\frac{(1.39 \times 10^{5} \text{ Btu/gallon}) \times (10{,}000 \text{ miles/year})}{0.7 \times 50 \text{ mpg} \times (15 \times 10^{6} \text{ Btu/ton})} = 2.7 \text{ tons/year,}$$

for biomass with a heating value of 15×10^{6} Btu/ton.

12. Alan D. Poole and Robert H. Williams, "Flower Power: the Prospects for Photosynthetic Energy," *Bulletin of the Atomic Scientists,* p. 49, May 1976.

13. In May 1980 Gulf and Western Industries announced the development of a new zinc-chloride battery system which they claim would permit the manufacture of a car able to go 150

miles at 55 mpg between recharges. (See A. J. Parisi, "Electric Car System Is Introduced," *The New York Times*, June 7, 1980.) The range would be considerably less in urban driving. Because of their high costs and range limitations relative to efficient fuel-driven cars, electric vehicles will remain special-purpose vehicles in the foreseeable future.

14. Frank W. Davis, Jr. and L. F. Cunningham, "Will the Reaction of the Auto Transportation System to the Energy Crisis be Technological or Institutional?" *Transportation Engineering*, p. 26, Feb. 1978.

15. The relative energy performance of alternative passenger transport modes shown in Figure 9.5 is based on the following data (in 10^3 BTU/PM)[a]:

	URBAN	INTERCITY
Auto (1974 average)[b]	8.1	3.6
VW Rabbit Diesel[c]	2.8	1.25
Urban Bus[d]	3.6	—
Urban Rapid Rail[e]	3.2–5.2	—
Airplane (1974 average)[f]	—	8.5
Train (1974 average)[g]	—	3.1

NOTES:

[a] This includes refinery losses in producing gasoline (10% of oil input) and diesel fuel (3% of oil input). See J. P. Longwell, "Synthetic Fuels and Combustion," in *Sixteenth Symposium (International) on Combustion*, the Combustion Institute, 1976. Also for electrically driven systems the fuel consumed is assumed to be 11,000 Btu/kwh.

[b] The average fuel consumption for autos in 1974 was 13.7 mpg (see Figure 9.1) or about 12.3 mpg in urban areas and 16.3 mpg for intercity transportation. According to "National Personal Transportation Study Report No. 1: Auto Occupancy," U.S. Department of Transportation, April 1972, the urban and intercity load factors average about 1.4 and 2.4 PM/VM respectively. Thus (since the energy content of gasoline, including refinery losses is 139,000 Btu/gallon) in 1974 the average specific energy required for urban auto transportation was

$$\frac{139{,}000 \text{ Btu/gallon}}{12.3 \text{ mpg} \times 1.4} = 8100 \text{ Btu/PM}$$

while that for intercity travel was

$$\frac{139{,}000 \text{ Btu/gallon}}{16.3 \text{ mpg} \times 2.4} = 3600 \text{ Btu/PM}$$

[c] The VW Diesel has an EPA rated fuel economy of 40 mpg (urban) and 53 mpg (intercity). We derate these by 10% for actual driving. Thus (since the energy content of diesel fuel, including refinery losses, is 143,000 Btu/gallon) the specific energy required by the VW diesel is

$$\frac{143{,}000 \text{ Btu/gallon}}{40 \text{ mpg} \times 1.4 \times 0.9} = 2840 \text{ Btu/PM}$$

for urban transportation and

$$\frac{143{,}000 \text{ Btu/gallon}}{53 \text{ mpg} \times 2.4 \times 0.9} = 1250 \text{ Btu/PM}$$

for intercity transportation.

[d] According to M. F. Fels, "Energy Thrift in Urban Transportation: Options for the Future," in *The Energy Conservation Papers*, R. H. Williams, ed., Ballinger, Cambridge, 1975, a typical diesel fueled 40 seat urban bus has a fuel economy of 5 mpg and a 20% load factor. Thus the specific energy is

$$\frac{143{,}000 \text{ Btu/gallon}}{5 \text{ mpg} \times (8 \text{ passengers/bus trip})} = 3600 \text{ Btu/PM}$$

(e) In a comparison of four systems by M. F. Fels, "Breakdown of Energy Costs for Rapid Rail Systems," Center for Environmental Studies Report No. 44, Princeton University, January 1977, the lowest energy requirements were for the New York City subway (0.29 kwh/PM) and the highest were for BART (0.47 kwh/PM).

(f) According to the Aerospace Corporation, *Characterization of the U.S. Transportation Systems, vol. 1—Domestic Air Transportation*, Los Angeles, March 1977, the average fuel consumption was 7623 Btu/PM (@ 57% load factor) in 1974. The number given in the table above also includes refinery losses.

(g) For an average loading of 22 passengers per rail car. From the Aerospace Corporation, *Characteristics of the U.S. Transportation System, vol. 4—Railroads (Freight and Passenger)*, Los Angeles, March 1977.

16. Harbridge House, "Corporate Strategies of the Automotive Manufacturers" for the National Highway Traffic Safety Administration, 1979. This is in 1978 dollars.

17. The sum of $45 billion is the value of gasoline, at $1.00/gal., saved by a fleet of autos averaging 27.5 m.p.g. instead of 14 m.p.g., with 125 million vehicles at the present level of miles/vehicle year. These numbers are roughly appropriate for the 1990's.

18. Assuming same number of miles driven as the present average and a cost of $1.00/gallon.

19. Frank von Hippel, "Forty Miles a Gallon by 1995 at the Very Least! Why the US Needs a New Automotive Fuel Economy Goal," Princeton University/Center for Energy and Environmental Studies Report #104, April 17, 1980.

20. Robert H. Williams, "A $2 a Gallon Political Opportunity," Princeton University/Center for Energy and Environmental Studies Report #102, April 14, 1980.

CHAPTER 10

1. See note 15 in Chapter 8.

2. Available heat driven air conditioners appear to be impractical for automobiles. See J. R. Akerman, "Automotive Air Conditioning Systems with Absorption Refrigeration," *SAE Transactions*, vol. 80, p. 132 (1971). Better, more compact, more efficient designs are needed.

3. See EPA, *Fuel Economy and Emission Control*, monograph, U.S. Environmental Protection Agency, 1972.

4. In 1976 74% of new domestically produced automobiles were equipped with air conditioning. See Motor Vehicle Manufacturers Association, *Motor Vehicle Facts and Figures '77*, Detroit, Michigan, 1976, p. 13.

5. In 1974 total fuel consumption by large industrial boilers (those with fuel consumption rates greater than 100 million Btu/hour) was about 4×10^{15} Btu (M. H. Farmer et al., *Application of Fluidized Bed Technology to Industrial Boilers*, report to the Federal Energy Administration by the Exxon Research and Engineering Co., Linden, N.J., 1976). The byproduct power production rate for cogeneration ranges from about 60 kwh for steam turbines to 400 kwh for Diesels for each 10^6 Btu of process steam (see Table 2 of R. H. Williams, "Industrial Cogeneration," *Annual Review of Energy*, vol. 3, p. 313, 1978). Thus assuming boilers are 80% efficient on the average the annual electricity generation potential would be 190×10^9 kwh for steam turbines to 1300×10^9 kwh for diesels. These correspond respectively to the outputs of 34 Gw(e) and 225 Gw(e) of baseload nuclear generating capacity operating at 65% of rated capacity on the average.

6. Note also in Table 10.1 that while low E/S ratio steam turbine cogeneration technology produces more valuable products from a given amount of fuel than does conventional process steam production, steam turbine cogeneration generates products which in most cases are less valuable than the product of central station power generation. This reflects the fact mentioned above that steam turbine systems have been optimized to produce power alone. In conventional central station power generation, based on the use of steam turbine systems, 30–40 percent of the fuel energy is converted into electricity and the waste heat is rejected to cooling water at a temper-

ature of about 100°F. In cogeneration applications the steam is exhausted from the turbine at a higher temperature (400°F, say). By increasing the exhaust temperature so that the "waste heat" becomes useful, however, the percentage of the fuel energy converted to electricity is reduced to 10-15 percent. No such reduction occurs with gas turbines or diesels because the "waste heat" is hot in ordinary power generation applications (600 to 1100°F). Despite the high temperature of this waste heat, a large fraction of the fuel consumed by these devices is converted to electricity because the input temperature (of gases) to the gas turbine and diesel is also much higher than it is to the steam turbine.

7. See Table 6.32 of S. E. Nydick et al., *A Study of Implant Electric Power Generation in the Chemical, Petroleum Refining, and Paper and Pulp Industries*, report to the Federal Energy Administration by Thermo Electron Corporation, Waltham, Mass., 1976.

8. For a brief discussion of advanced cogeneration technologies see pp. 340-347 in R. H. Williams, "Industrial Cogeneration," note 5.

9. See Chapter 4, note 5.

10. Federal Energy Regulatory Commission, "Final Rules on Small Power Production and Cogeneration Facilities; Regulations Implementing Section 210 of the Public Utility Regulatory Policies Act of 1978," *Federal Register*, vol. 45, No. 38, February 25, 1980, pp. 12214-12237.

11. Albert Smiley, "A Regulatory Policy for Industrial Cogeneration," Center for Energy and Environmental Studies Report, Princeton University, November, 1979.

12. For detailed discussion on the deregulation of power generation see *Promoting Competition in Regulated Markets*, Almarin Phillips, ed., The Brookings Institution, 1975; *Electric Power Reform: the Alternatives for Michigan*, William H. Shaker, ed., Institute of Science and Technology, University of Michigan, Ann Arbor, 1976.

CHAPTER 11

1. William G. Pollard, "The Long-Range Prospects for Solar Energy," *American Scientist*, p. 424-429, July-August 1976.

2. "Hirsch Casts Doubt on Many Aspects of ERDA's Solar Electric Program," *The Energy Daily*, October 21, 1976.

3. If a nuclear plant costing $800 per kilowatt of installed capacity (1976$) operates on the average at 65% of rated capacity, the cost per average kilowatt at the busbar is (see note 27d, Chapter 8):

$$\frac{\$800/\text{kw}}{.65} = \$1250/\text{kw}$$

4. The appropriate capital charge rate for utility electricity is 11.3% per year (see Appendix B). To the capital charge of 1.6¢/kwh for nuclear electricity one should add 0.8¢/kwh for fuel and operation and maintenance (see notes 27k and 27l of Chapter 8).

5. Henry Kelly, "Photovoltaic Power Systems: A Tour Through the Alternatives," *Science*, p. 634, February 10, 1978.

6. See, for example, Office of Technology Assessment, *Application of Solar Technology to Today's Energy Needs*, volume I, June 1978, p. 406.

7. The average energy demands per household for the community served by the photovoltaic cogeneration system are assumed to be those for the model energy conserving new houses

described in Chapter 8. The average house is assumed to have 4 inhabitants with the following heat and electrical demands:

	WINTER ($5\frac{1}{4}$ MO.)		SUMMER ($6\frac{3}{4}$ MO.)		ANNUAL
	PEAK (KW)	TOTAL (KWH)	PEAK (KW)	TOTAL (KWH)	TOTAL (KWH)[a]
Electricity	0.81[b]	1600[b]	0.52[b] (0.66[c])	1760[b] (1860[c])	3420 (3520[c])
Space Heat	5.39[d]	5660	—	—	5660
Hot Water	—	2120[e]	—	1880 (1310)[e]	4000
A/C	—	—	4.04[f]	2260	2260
Total Electrical Equivalent Demand[g]	5.82[h]	6610	4.39[i]	4770	11,380

NOTES:

[a] See note 24, Chapter 8.

[b] The ratio of peak to average demand and the ratio of summer to winter demand are based on R. W. Weatherwax and R. H. Williams, "Energy Conservation/Load Management Analysis for the Residential, Small Commercial, and Industrial Sectors of the Consolidated Edison Electric Franchise District" report to the New York Public Service Commission, March 1977.

[c] Includes 0.14 kw(e) for the auxiliary equipment to operate the heat driven air conditioner.

[d] See note 15, Chapter 8.

[e] The average demand for water heating is 0.65 kw. However, according to R. W. Weatherwax and R. H. Williams (note b) the hot water consumption rate is 45% greater in winter than in summer (0.55 kw vs. 0.38 kw). In the summer however, we assume that for the 1500 hour air conditioning season hot water is provided by burning fuel in the home. For an 80% efficient water heater the fuel consumed for this purpose is 2.4×10^6 Btu per annum to provide 570 kwh of hot water, in addition to the 1310 kwh provided by district heating.

[f] See note 22, Chapter 8.

[g] The total effective electrical demand is the demand for an all electric house with the same amenities in which space conditioning is provided by an air to air heat pump having a COP = 2 on the average.

[h] We assume that the peak occurs on a cold winter day and is equal to the peak demand for space heating (for which COP $\rightarrow$ 1 on a cold day) plus the *average* electrical demand for purposes other than water heating.

[i] We assume that the summer peak is equal to the peak demand for air conditioning (COP = 2) plus the *average* electrical demand for purposes other than water heating.

8. Direct normal solar insolation in New York City is as follows:

MONTH	INSOLATION (WATTS/M^2)
January	117
February	137
March	192
April	192
May	204
June	233
July	196
August	179
September	221
October	167
November	108
December	117

The average winter insolation (November 1 through first week of April) is 136 watts/m^2. The average for the rest of the year is 200 watts/m^2. The annual average direct normal insolation is 172 watts/m^2. Thus 35% of the direct sum occurs in winter and 65% in summer.

Source: E. C. Boes *et al.*, "Distribution of Direct and Total Solar Radiation Availabilities for the U.S.A.," *Sharing the Sun: Solar Technology in the Seventies*, vol. 1, pp. 238-263, Proceedings of the Joint Conference of the American Section, International Solar Energy Society and the Solar Energy Society of Canada, Inc., Winnipeg, Canada, August 15-20, 1976.

9. Using the insolation data from note 8, we obtain the annual solar electricity and heat production per square meter of collector as:

$$170.9 \text{ kwh/m}^2 = \varepsilon \times \alpha_2 \times \alpha_1 \times (0.172 \text{ kw/m}^2) \times 8766$$

for electricity and

$$780.0 \text{ kwh/m}^2 = \alpha_3 \times (1 - \varepsilon) \times \alpha_2 \times \alpha_1 \times (0.172 \text{ kw/m}^2) \times 8766$$

for heat, where

α_1 = efficiency of concentrating collector (0.76)
α_2 = absorptivity of receptor (0.95)
α_3 = heat recovery efficiency (0.85)
ε = efficiency of photocell (0.157 at 100°C)

From note 7 we obtain the following heat and electricity requirements in kwh (assuming that A/C is provided with a heat driven chiller having a COP=0.7):

	WINTER	SUMMER
heat	7780	4540
electricity	1660	1860

We assume losses for the system are as follows:

- Seasonal storage losses amount to 20% of storage output
- Thermal distribution losses are 10% of the system output
- Battery losses are 25% of battery output, which we assume accounts for ½ of electricity use.
- Electrical distribution losses are 4.5% of electrical output
- About 42% of the waste heat from the diesel is lost.

We assume that the diesel engine is 35% efficient, so that the ratio of heat (H) to electricity (E) recovered from the diesel unit is

$$\frac{H}{E} = \frac{.58\,(1-0.35)}{0.35} = 1.07$$

The following energy balance equations determine the area of the collector (A, in m²), the seasonal storage required (S, in kwh/m²), and the electricity produced by the diesel, exclusive of what is needed for long periods of cloudy weather (E_w in winter and E_s in summer, in kwh/m²):

$.9\,[273.0 + 1.07\,E_w + S] = 7780/A$ (heat in winter)
$.955\,[59.8 + E_w] = 1.125 \times (1660/A)$ (electricity in winter)
$.9\,[507.0 + 1.07\,E_2 - 1.2S] = 4540/A$ (heat in summer)
$.955\,[111.1 + E_s] = 1.125 \times (1860/A)$ (electricity in summer)

Solving these equations we obtain A = 16.54 m² and the following energy balances in kwh/m² of collector area:

	WINTER (5¼ MONTHS)		SUMMER (6¾ MONTHS)	
	HEAT	ELECTRICITY	HEAT	ELECTRICITY
Collector Output	273.0	59.8	507.0	111.1
Diesel Output	62.5	58.4	22.9	21.4
Seasonal Storage	187.3	—	−224.8	—
Heat Losses in Delivery	−52.3	—	−30.5	—
Battery Storage Losses	—	−13.1	—	−14.7
Electrical Distribution Losses	—	−4.7	—	−5.3
Demand	470.5	100.4	274.6	112.5

The above energy balances do not adequately take into account electricity requirements for long periods of overcast skies in summer. The diesel electricity production given above is adequate to provide electricity for only one day in six on the average during the summer. We assume that the diesel engine must provide in summer 50% more electricity without waste heat recovery (10.7 kwh/m²) for long cloudy periods. (Excess electricity produced in this amount in summer would be sold.)

Total fuel consumption for the diesel thus amounts to

$$\frac{1}{0.35} \times [(58.4+32.1)\ \text{kwh/m}^2] \times (16.54\ \text{m}^2) \times 3413\ \text{Btu/kwh} = 14.6 \times 10^6\ \text{Btu}.$$

Extra fuel is also needed for winter space heating in severe winters. The standard deviation in the winter heating demand is about 10% of the average, and the average excess demand would not be more than 5%. If this extra heat is provided via supplementary firing of the waste heat boiler (at 58% efficiency) then the extra fuel required per year is

$$\frac{1}{0.90} \times \frac{0.05}{0.58} \times (5660\ \text{kwh}) \times (3413\ \text{Btu/kwh}) = 1.9 \times 10^6\ \text{Btu}.$$

Thus the total fossil fuel needed annually (including that needed for domestic hot water during the air conditioning season) is 18.9×10^6 Btu.

10. The table on p. 327 gives the replacement costs of energy (in 1976$) per household for energy conserving houses supplied with energy by a community scale photovoltaic cogeneration system.

	UNIT COST ($/M² OF COLLECTOR UNLESS OTHERWISE INDICATED)	QUANTITY (M² OF COLLECTOR UNLESS OTHERWISE INDICATED)	CAPITAL COST (1976$)	ANNUALIZED COSTS			
				MUNICIPAL UTILITY		PRIVATE UTILITY	
				$/YEAR	MILLS/KWH[a]	$/YEAR	MILLS/KWH[a]
Land for Collector Field[b]	$2.42/m² of land ($10,000/acre)[c]	73m² of land[d]	177	4.4	0.4	17.3	1.5
Capital Equipment[e]							
Collector (except cell)	80–135[f]	18.20[g]	1456–2457	100.5–169.5	8.8–14.9	164.5–277.6	14.5–24.4
Photocell	33.3[h]	18.20[g]	606	41.8	3.7	68.5	6.0
Collector Installation and Miscellaneous	47[i]	18.20[g]	855	59.0	5.2	96.6	8.5
Battery System	30.3[j]	16.54	501	64.6	5.7	85.7	7.5
Diesel Backup	12.8[k]	16.54	212	14.6	1.3	24.0	2.1
Seasonal Thermal Storage	72.7[l]	16.54	1203	83.0	7.3	135.9	11.9
Thermal Distribution	61.4[m,n]	16.54	1015	70.0	6.1	114.7	10.1
Electrical Distribution	8.05[o]	16.54	132	9.1	0.8	14.9	1.3
Air Conditioning[p]	16.7	16.54	276	19.0	1.7	31.2	2.7
Overhead and Profit[q]	—	—	1565–1815	115.7–132.8	10.1–11.6	184.2–212.4	16.2–18.7
Subtotals			7821–9072	580–664	50.8–58.4	920–1062	80.9–93.3
Annual Expenses							
Backup Fossil Fuel[r]				94.5	8.3	94.5	8.3
O & M[s]				64.2	5.6	64.2	5.6
Sale of Excess Electricity[t]				−7.1	−0.6	−7.1	−0.6
Totals				736–820	64–72	1089–1231	96–108

NOTES:

[a] This "effective electricity price" is the annual cost divided by the electricity consumption (11,380 kwh per year-see note 7) of an equivalent all electric house (one with the same conservation features and amenities) provided with space conditioning using a conventional air to air heat pump having a COP = 2.0.

[b] The annual capital charge rate for land investments is 2.5% for a municipal utility and 9.8% for a public utility [See Appendix B].

[c] We have selected this as a nominal land price. Land could be much less or much more costly in different circumstances. Where land is more expensive the collectors could be installed on land that is simultaneously used for other purposes. For example the collectors could be mounted on the roof of a shopping mall and the storage system could be maintained beneath the shopping mall.

[d] Because of shading effects the required land area amounts to 4 m^2 per m^2 of collector.

[e] For all equipment except batteries we assume a 25 year equipment life. The appropriate annual capital charge rate is 6.9% for a municipal utility investment and 11.3% for a public utility investment. For batteries we assume a 10 year equipment life, so that the appropriate annual capital charge rates are 12.9% and 17.1% for municipal and public utilities respectively. [See Appendix B.]

[f] According to ref. 6, vol. II, p. 676, a "high cost" 2-axis concentrating collector with a geometric concentration ratio of 100 for a silicon cell array would cost $135/$m^2$ and a "medium cost" unit $80/$m^2$.

[g] We assume that the collector has 10% excess capacity to allow for maintenance.

[h] We assume that solar cells with 18% efficiency at 28°C in a 1 kw/m^2 solar flux cost $15/peak watt or $2700/$m^2$. We assume only 90% of the diameter of a cell (81% of its area) would be used in a concentrator. Thus the cost per m^2 of concentrating collector with 100-fold concentration ratio would be

$$\frac{(\$2700/m^2) \times (.01\ m^2/m^2 \text{ of collector})}{.81} = \$33.3/m^2$$

[i] This includes (see ref. 6, vol. II, pp. 688–697):

Installation and Plumbing of Collector	$20/$m^2$
Foundations	5
Site Preparation	1
Collector Pipefield	18
Shipping Costs	3
	$47/$m^2$

[j] The battery system output must be adequate to meet half the daily electrical consumption in winter. Thus, allowing for losses of 25% the storage capacity of the battery must be

$$\frac{1.25 \times (5.2\ \text{kwh/day})}{16.54\ m^2} = 0.39\ \text{kwh}/m^2$$

To determine the required delivery rate we assume that the winter peak is a coincident peak, that the reserve margin must be 30%, and that the batteries must be able to meet half the peak (the diesel engine must meet the other half). Thus the battery delivery rate must be

$$\frac{0.5 \times 1.3 \times 0.81\ \text{kw}}{16.54\ m^2} = 0.032\ \text{kw}/m^2$$

According to ref. 6, vol. II, p. 696 the present market price for a community scale battery system (including installation and other costs) is $70/kwh. In addition there would be a power conditioning cost of $95/kw peak. Thus the total battery capital cost is

$$0.39 \times (\$70/\text{kwh}) + 0.032 \times (\$95/\text{kw}) = \$30.3/m^2$$

[k] According to S. E. Nydick et al., *A Study of Inplant Electric Power Generation in the Chemical, Petroleum Refining, and Paper and Pulp Industries*, report to the Federal Energy Administration by Thermo Electron Corporation, Waltham, Mass., 1976, the installed cost of a diesel set with a waste heat boiler is about $500/kw, or $400/kw without overhead & profit. We assume the diesel capacity equals that of the batteries (see note 10j). Thus the cost is

$$(\$400/\text{kw(e)}) \times (0.032\ \text{kw}/m^2) = \$12.8/m^2$$

[l] For storage we consider a large steel tank. For this tank the capacity is determined by the interseasonal demand. The temperature of the storage reservoir is 180°F. The water is returned from consumers at 90°F. Thus the storage capacity required is

$$\frac{(187.3 \text{ kwh/m2})}{(180\text{–}90)} \times (16.54 \text{ m}^2/\text{house}) \times (2500 \text{ houses}) \times (3413 \text{ Btu/kwh})$$

$$= 2.94 \times 10^8 \text{ lb.} = 1.33 \times 10^5 \text{ tonnes } (1.33 \times 10^5 \text{ m}^3)$$

According to ref. 6 (vol. I, p. 437) a buried steel tank of this capacity would cost $20/m³ or $2.65 × 10⁶ ($1062/house). This cost is site dependent. If there were wet soil conditions, the tank would have to be insulated. According to ref. 6 (vol. I, p. 496) polyurethane foam with conductivity k=0.13 Btu-inch/hour ft²/°F can be foamed in place for $102/m³. To determine the maximum amount of insulation required we note first that the allowed annual heat loss from storage per household is (see note 9):

$$[(224.8 - 187.3) \text{ kwh/m}^2] \times (16.54 \text{ m}^2) \times 3413 \text{ Btu/kwh} = 2.12 \times 10^6 \text{ Btu.}$$

We assume the tank is cylindrical with a height of 10 meters (and thus has a surface area of 30.6 × 10³ m² or 330 × 10³ ft²). The insulation thickness d is determined (assuming that the reservoir is in contact with running ground water at 55°F and that the mean time the hot water is in storage is 6 months) by:

$$(2.12 \times 10^6 \text{ Btu/house}) \times (2500 \text{ houses}) = \frac{0.13 \times (8766/2) \times (330 \times 10^3 \text{ft}^2) \times (180 - 55)°\text{F}}{d}$$

Thus we obtain d = 4.43 inches of polyurethane foam, corresponding to 3447 m³ of insulation, at a cost of $141/house.

(m) The thermal distribution system is a design developed in Howard Geller, "Thermal Distribution Systems and Residential District Heating," Center for Energy and Environmental Studies Report No. 97, Princeton University, January 1980. The distribution system consists of one delivery pipe and one return pipe. The system delivers hot water (@ 180°F) for space heating and water heating in the winter and cold water (@ 46°F) for air conditioning in the summer cooling season. During the cooling season (1500 hours) hot water is provided by burning fuel at the household level. The return pipe carries warm water (@ 90°F) in the heating cycle and cool water (@ 64°F) in the cooling cycle.

The pipes are sized based on a constant pressure drop of 5 × 10⁻³ psi/ft at peak water flow. The peak cooling load determines the pipe size. Small pipes (0.75-2.5" diameter) are made of CPVC while larger pipes are lined asbestos-cement pipes.

The pipes are insulated with the optimum amount of polyurethane foam assuming $6/10⁶ Btu as the value of lost heat and municipal utility ownership. The insulated pipes are wrapped in polyethelene for waterproofing.

(n) According to Howard Geller (see note m) the installed cost breakdown per house for the thermal distribution system for a new development is (in 1979 $):

Pipes	$393
Insulation	178
Trenches	188
Valves and Fittings	11
Instrumentation and Control	8
Pumps	5
Btu Meter	81
Heat Exchanger for Space Conditioning	304
Heat Exchanger for Hot Water	88
	$1256

This becomes $1015 in 1976 $. Here the heat exchanger for space conditioning is a water to air unit sized to meet the summer cooling load.

(o) The peak electrical demand (see note 7) is 0.81 kw in the winter. According to ref. 6, vol. II, p. 721, the cost of distribution exclusive of the cost of the substation is $157/kw. If distribution losses are 4.5% then the cost of electrical distribution is

$$\frac{\$157/\text{kw(e)} \times 0.81 \text{ kw(e)}}{0.955 \times 16.54 \text{ m}^2} = \$8.05 \text{ m}^2$$

(p) For a single effect absorption air conditioner using 85°C water as a heat source, the cost of an air conditioning unit (with COP=0.7) in sizes greater than about 170 tons is about $200/ton or $57/kw(t) (see ref. 6, vol. I, p. 520). Since the air conditioning peak demand is 4.04 kw (t) (see note 7) the air conditioner cost is (with a 20% installment cost)

$$\frac{1.2 \times (\$57/\text{kw}) \times (4.04 \text{ kw})}{16.54 \text{ m}^2} = \$16.7/\text{m}^2$$

(q) Assumed to be 25% of direct capital cost.

(r) For fuel @ $\$5/10^6$ Btu the annual cost is (see note 9):

$$18.9 \times 5 = \$94.5/\text{year}$$

(s) Assumed to be $\$4/m^2$ of collector.

(t) From note 9 excess electricity amounting to 10.7 kwh/m^2 of collector could be sold. At 4¢/kwh the annual revenues would be

$$.04 \times (16.54\ m^2) \times (10.7\ kwh/m^2) = \$7.1/\text{year}$$

---

11. We obtain the following breakdown of the *total* collector costs (exclusive of the cost of solar cells) for medium cost and high cost collectors (in 1976 $\$/m^2$):

Direct Cost (see note 10f) for Concentrating, Tracking Collector	80	135
Installation (see note 10i)	47	47
Profit and Overhead (@ 25%)	32	47
	$159	$229

12. The figure on p. 331 shows the potential price reduction for one type of concentrating collector (a Fresnel lens) as a function of production volume.

13. Ref. 6, vol. II, p. 676.

14. See Ref. 6, vol. I, p. 289. Also the 1979 American Physical Society report on photovoltaic technology suggests a goal of $\$50/m^2$ may be achievable.

15. According to E. Hirst and J. Carney, "Residential Energy Use to the Year 2000: Conservation and Economics," Oak Ridge National Laboratory Report ORNL/CON-13, September, 1977, average energy use per household in the U.S. in 1975 was 228 million Btu. From note 9 fossil fuel consumption associated with the solar cogeneration system amounts to 19 million Btu per year.

16. According to note 10l the storage capacity required per house is 53 m^3 (14,000 gallons). Buried tanks of this size cost about $\$80/m^3$ (see ref. 6, vol. I, p. 437). Thus the cost of the tank at the household level would be $4200. This cost would be site dependent. If these were wet soil conditions the tank would have to be insulated. To determine the cost of insulation we note that the allowable annual heat loss is 2.12×10^6 Btu (see note 10l). For a cubical tank (with a surface area of 911 ft^2) the thickness d is given by

$$2.12 \times 10^6 \text{ Btu} = \frac{0.13 \times (8766/2 \text{ hours}) \times (911\ ft^2) \times (180 - 55)°F}{d}$$

Thus we obtain 30.6 inches, corresponding to 66 m^3 of insulation. The cost of this much insulation would be $6730.

17. The table on p. 332 gives the replacement cost of electricity (in 1976$) per household for energy conserving houses supplied with electricity only by a community scale photovoltaic electricity generating system.[a]

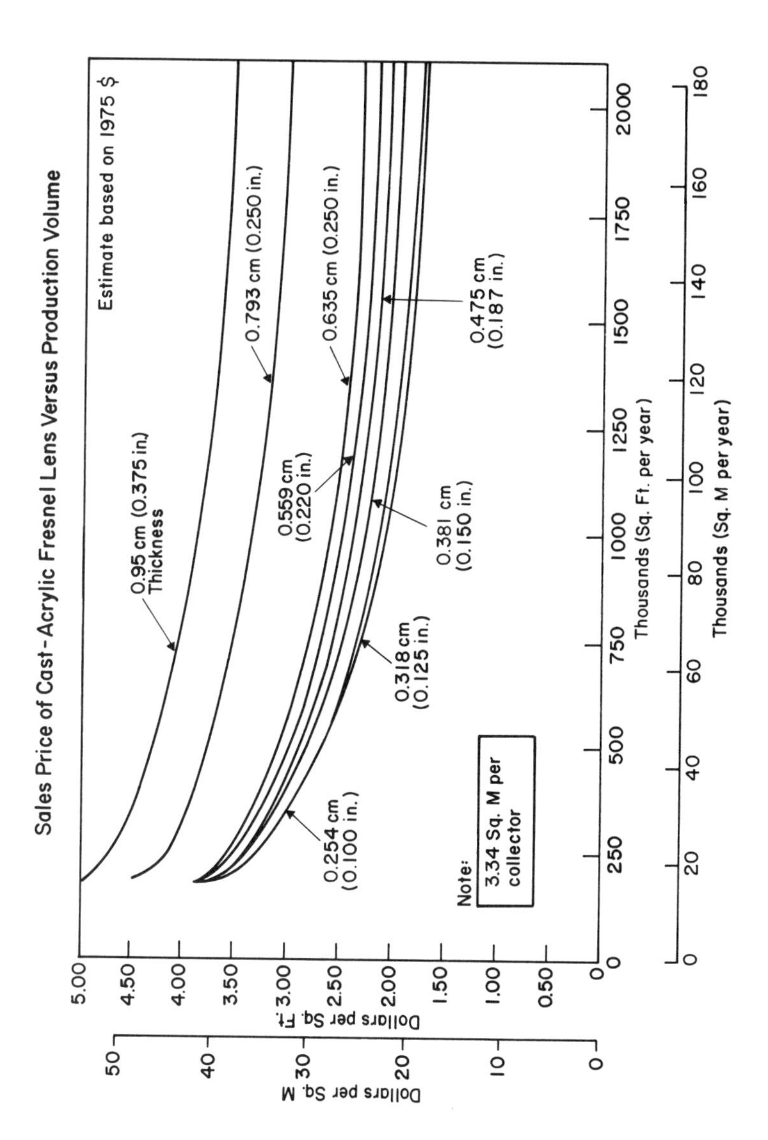

SOURCE: Unsolicited Technical Proposal for Cast-Acrylic Lenses for Solar-Cell Energy Generation, Swedlow Report No. 873, September 30, 1976, Swedlow, Inc., Garden Grove, California.

	UNIT COST ($/M² OF COLLECTOR UNLESS OTHERWISE INDICATED)	QUANTITY (M² OF COLLECTOR UNLESS OTHERWISE INDICATED)	CAPITAL COST (1976$)	ANNUALIZED COSTS			
				MUNICIPAL UTILITY		PRIVATE UTILITY	
				$/YEAR	MILLS/KWH[b]	$/YEAR	MILLS/KWH[b]
Land for Collector Field	$2.42/m² of land	73 m² of land	177	4.4	1.3	17.3	5.1
Capital Equipment							
Collector (except cell)	80–135	18.20	1456–2457	100.5–169.5	29.4–49.6	164.5–277.6	48.1–81.2
Photocell	33.3	18.20	606	41.8	12.2	68.5	20.0
Collector Installation and Miscellaneous	19[c]	18.20	346	23.9	7.0	39.1	11.4
Battery System	30.3	16.54	501	64.6	18.9	85.7	25.1
Diesel Backup	9.6[d]	16.54	159	11.0	3.2	18.0	5.3
Electrical Distribution	8.05	16.54	132	9.1	2.7	14.9	4.4
Overhead and Profit	—		800–1050	62.7–80.0	18.3–23.4	90.4–118.7	26.4–34.7
			4000–5250	314–400	92–117	481–623	141–183
Annual Expenses							
Fuel				73	21.3	73	21.3
O & M				32.1[e]	9.4	32.1[e]	9.4
Sale of Excess Electricity				−7.1	−2.1	−7.1	−2.1
Totals				416–502	122–147	596–738	175–227

NOTES:

(a) All quantities in this table except those indicated are calculated as in the table of note 10.

(b) This system provides the same electrical needs as the system described in note 7. Space conditioning and hot water are provided by some other energy source. Thus to calculate the cost of electricity the annualized cost ($/year) must be divided by the total electricity requirements (3420 kwh—see note 7).

(c) The installation and miscellaneous expenses are estimated to be $28/m^2 less than with a cogeneration system. According to ref. 6, vol. II, pp. 688-697, the installation cost could be reduced $10/m^2 for photovoltaic systems without plumbing. Also $18/m^3 is not needed for the collector pipefield (see note 10).

(d) Because a waste heat boiler is not needed the diesel costs only $300/kw.

(e) We assume that the O&M expenses are half as much per m^2 of collector as with the cogeneration system.

---

18. See note 41, Chap. 8.

19. See note 5, Chap. 8.

20. See note 42, Chap. 8.

21. See notes 30 and 31 in Chap. 8.

22. See note 27, Chap. 8.

23. See note 28, Chap. 8.

24. See note 29, Chapter 8.

25. The number of electrically heated houses increased from 1.0 million (1.9% of all houses) in 1960 to 4.9 million (7.7% of all houses) in 1970, and to 9.1 million (12.6% of all houses) in 1975. See D. L. O'Neal, "Energy and Cost Analysis of Residential Heating Systems," Oak Ridge National Laboratory Report ORNL/CON-25, July 1978.

26. *Electrical World* projected in 1978 (see "29th Annual Electrical Industry Forecast," *Electrical World*, p. 61, September 15, 1978) that the number of electrically heated homes would be 35 million by 1995, or 26 million more than in 1975. *Electrical World* also projected that the number of households will increase by 26 million in the period 1975-1995.

CHAPTER 12

1. C. D. Hollowell, J. V. Beck, C. Lin, W. W. Nazaroff, and G. W. Traynor, "Impact of Energy Conservation in Buildings on Health," *Changing Energy Use Futures*, Rocco A. Fazzolare and Craig B. Smith, eds., vol. II, pp. 638-647, Pergamon Press, New York, 1979.

2. G. D. Roseme, C. D. Hollowell, A. Meier, A. Rosenfeld, and I. Turiel, "Air-to-Air Heat Exchangers: Saving Energy and Improving Indoor Air Quality," *Changing Energy Use Futures*, vol. III, pp. 1229-1235.

3. B. O'Neill, M. Ginsburg and L. Robertson, "The Effects of Vehicle Size on Passenger Car Occupant Death Rates," Society of Automotive Engineers, 1977.

4. D. F. Fennelly, H. Kleman, and R. R. Hall, "Coal Burns Cleaner in a Fluid Bed", *Environmental Science and Technology*, No. 3, p. 244, 1977

CHAPTER 13

1. Demand and Conservation Panel of the Committee on Nuclear and Alternative Energy Systems of the National Academy of Sciences, "U.S. Energy Demand: Some Low Energy Futures," *Science*, p. 142 (April 14, 1978). See also *Alternative Energy Demand Futures to 2010*, Report of the Demand and Conservation Panel to the Committee on Nuclear and Alternative Energy Systems, National Research Council, National Academy of Sciences, 1979.

2. If as estimated by Robert Pindyck (see note 26, Chapter 3), the long run price elasticity of residential energy use were at least −0.8, then a ⅔ reduction in energy intensity as a pure price response would arise if real energy prices (including the energy tax) were to increase four-fold during the period 1975–2010. Since we are not assuming such a large energy price increase, our analysis depends in part on removal of non-price barriers.

3. In 1975 fuel use for space heating in commercial buildings averaged 140,000 Btu/ft^2 (see J. R. Jackson and W. S. Johnson, "Commercial Energy Use: A Disaggregation by Fuel, Building Type, and End Use," Oakridge National Laboratory Report ORNL/CON-14, February 1978). In contrast residential fuel consumption for space heating in 1975 averaged 100,000 Btu/ft^2 (see E. Hirst and J. Carney, "Residential Energy Use to the Year 2000: Conservation and Economics," Oak Ridge National Laboratory Report ORNL/CON-13, September 1978).

4. General Service Administration, Public Buildings Service, *Energy Conservation Design Guidelines for New Office Buildings,* July 1975.

5. The analysis of the office buildings performed for the BEPS program is based on a random sample of 11 large new office buildings distributed throughout the country. The buildings were redesigned by teams of professional architects and engineers so that the redesigned structures would represent (in the judgment of the redesign teams) "the maximum practicable degree of energy conservation." The redesign was not intended to minimize lifecycle costs. Rather the average estimated incremental cost of redesign ($1.07/ft^2) corresponds to a cost of saved primary energy equal to 24¢ per gallon of oil equivalent energy, with a 10% real discount rate and a lifecycle of 15 years. If lifecycle costs were minimized the energy savings via redesign would have been greater. In 1978 the average price of energy to commercial customers was $2.98 per million Btu or 41¢ per gallon oil equivalent energy (see note 31, Chapter 15). The replacement cost is much higher: Consider first natural gas. With deregulation of gas prices the price of natural gas in 1978 would have been about $1.00 per million Btu more (see note 5, Chapter 8). In the case of electricity the replacement cost would be somewhat less than that calculated for residential customers—perhaps 6¢/kwh in 1978 dollars (see note 42, Chapter 8). In the case of oil we take the price of heating oil in late 1979 as an estimate of the replacement cost, about 70¢ per gallon in 1978 dollars. The average replacement cost for 1978 thus becomes about 66¢ per gallon of oil equivalent.

6. The investment for retrofitting McCampbell Hall amounted to 50¢/ft^2. This corresponds to an average cost of saved energy of 3¢ per gallon of oil equivalent energy, with a 10% real discount rate and a lifecycle of 10 years.

7. Most of the improvement would be in new buildings. In 2010 the total amount of building space would be 2.35 times the amount in 1975 (see note 36, Chapter 3). Buildings last only about 45 years on the average. Thus in 2010 only about 46% of pre-1975 vintage buildings would still be standing and only about 20% of all commercial space would be pre 1975 vintage.

8. In reference 1 the total transportation budget for 2010 for Scenario II (which involves aggressive conservation, "aimed at maximum efficiency plus minor lifestyle changes") is 15.7 quads. To get the Scenario II estimate of the energy demand for "other transportation" we must subtract off the auto and light truck contribution. Since in Scenario II people spend 30% more time in cars in 2010 than in 1975 and the average automotive fuel economy is about 35 mpg, compared to our 45 mpg, the auto and light truck energy budget in Scenario II is about

$$1.30 \times \frac{45}{35} \times 4.2 = 7.0 \text{ quads.}$$

Thus fuel consumption for "other transportation" amounts to 8.7 quads in Scenario II. Scenario II involves only a doubling of GNP, 1975–2010, whereas we assume a 2.55 fold increase, however. Correcting the Scenario II estimate for this difference in the GNP projections we obtain

$$\frac{2.55}{2.0} \times 8.7 = 11.1 \text{ quads}$$

as our projection for the energy demand in 2010 for "other transportation".

9. Industrial energy consumption including utility losses with respect to real gross product originating in industry as reported by the Dept. of Commerce declined 8% 1973–1978. A somewhat greater decline, of 14% for 1972–1978 for the ten most energy intensive industries, is reported by the Industrial Energy Efficiency Program, U.S. Department of Energy, Dec. 1979. The even larger declines frequently reported are based on the Federal Reserve Board index of production. See Chapter 3, note 39.

10. We quote from "Research Project: Long Life Automobile," *Porsche Panorama,* p. 15, Nov. 1973:

Today the Porsche Development Center seeks ways and means of adjusting automobile development to changing environmental conditions.

The result of these studies is the longlife automobile research project. This starts from the premise that an automobile with doubled life span would approximately halve the problem. Building cars to last twenty years rather than ten (the statistical average life span now) seems solvable. Thus this study concentrates on two points: design measures which reduce wear and consequent selection of materials.

Design-wise, a long life can be achieved, for instance, by large volume engines of low specific output, by hydraulic clutch and electric ignition systems. The prototype shown at Frankfurt displayed larger than normal components such as gears, bearings and springs; absolutely no trim in order to eliminate the rust-prone holes used to attach the trim; foam filling for all body cavities, a rather sophisticated oil-feed system and a dust filter of the type now used only on race cars.

The choice of materials parameter includes corrosion-resistant bodywork (or aluminum, stainless steel or synthetics), ozone-resistant silent bushings, special alloy brake discs, and even scratch-proof window panes. Taken together all these would result in automobiles of new character—in utilitarian, robust, everyday cars built to quality standards so far unknown.

A secondary goal in choosing materials is that of intensifying recycling. The object here is to increase the proportion of re-usable materials and to lower that of the throwaways. The double gain comes in the form of raw materials saved and reduced scrap heaps.

The longlife-automobile also gives the motorist hope of a favorable economic position. According to cost projections by Porsche technicians, an automobile lasting twice as long as heretofore would cost only some 30% more when purchased. Economic calculations for the operation of the longlife autos suggest a saving of at least 15% based entirely on current costs. Assuming shortages and price increases for raw materials this figure would automatically improve.

Such longlife-autos would naturally effect many changes in the motor industry as well as its sales and service branches. With a bow to ship wharves and aviation plants car firms would have to change from pure production to a large percentage of overhaul and partial-modernization work. The result would be a shift from material usage to manpower use.

11. See, for example, Robert T. Lund, "Making Products Last Longer," *Technology Review,* p. 49, Jan. 1977.

12. Denis Hayes, "Repairs, Reuse and Recycling—First Steps Toward A Sustainable Society," Worldwatch Paper 23, Sept. 1978.

13. See, for example, Bruce Hannon, "Energy, Labor, and the Consumer Society," *Technology Review* 79, March/April 1977.

14. John G. Meyers et al., *Energy Consumption in Manufacturing,* a report prepared for the Energy Policy Project by the Conference Board, Ballinger, Cambridge, Mass., 1974.

15. A sample of suggestions can be found in the series of studies for the Federal Energy Administration on Energy Efficiency Improvement Targets. See Vol. 1 in each of the reports for the fabricated metals, petroleum, paper, primary metals, food, chemicals, stone, clay and glass, machinery, transportation equipment, and textile industries.

16. H. L. Brown (Drexel University) and B. Hamel, *Industrial Applications Study* (5 Vols.), A study for U.S. Department of Energy, March 1978.

17. Charles A. Berg, private communication.

18. J. Chase, N. J. D. Lucas, and W. Murgatroyd, "Industrial Energy Use-I: Power Losses in Electrically Driven Machinery," p. 179-196, and N. Ladomatos, N. J. D. Lucas, and W. Murgatroyd, "Industrial Energy Use-II: Energy Use in a Light Engineering Factory," p. 375-388, *Energy Research*, vol. 2, 1978.

19. N. Ladomatos, N. J. D. Lucas, and W. Murgatroyd, "Industrial Energy Use-III: The Prospects for Providing Motive Power in a Machine Tool Shop from a Centralized Hydraulic System," pp. 19-28, *Energy Research*, vol. 3, 1979.

20. D. J. BenDaniel and E. Z. David, Jr., "Semiconductor Alternating Current Motor Drives and Energy Conservation," *Science 206*, November 16, 1979, p. 773.

21. Gordian Associates, Inc., "Industrial International Data Base Pilot Study: The Cement Industry, "Committee on the Challenges of Modern Society of the North Atlantic Council, 1976.

22. Julian Szekely, "Toward Radical Changes in Steelmaking," *Technology Review 81*, p. 29, Feb. 1979; Y. Austin Chiang, W. M. Danver, and J. M. Cigan, Eds., *Energy Use and Conservation in the Metals Industry*, The Metallurgical Society of AIME, New York, N.Y., 1975.

23. Charles A. Berg, "Process Innovation and Changes in Industrial Energy Use," *Science 199*, Feb. 10, 1978.

24. Charles Berg, "Energy Conservation in Industry: the Present Approach, The Future Opportunities," report prepared for the Council on Environmental Quality (May 1979).

25. If, as estimated by Robert Pindyck (see note 26, Chapter 3) the long-run price elasticity of industrial energy demand were −0.8, then a 50% reduction in energy intensity as a pure price response would arise if real energy prices were to increase 2.4 fold during the period 1975–2010.

26. J. Darmstadter et al., *How Industrial Societies Use Energy: A Comparative Analysis*, published by Resources for the Future by the Johns Hopkins University Press, Baltimore. See also Lee Schipper, "Energy Use and Conservation in Industrialized Countries," in John C. Sawhill, ed., *Energy Conservation and Public Policy*, Prentice-Hall, Inc., 1979.

27. Alan D. Poole and Robert H. Williams, "Flower Power: Prospects for Photosynthetic Energy," *Bulletin of the Atomic Scientists*, May 1978, p. 49.

28. The following is the distribution of organic waste resources estimated for 2010, based on an average energy content of 15 million Btu per ton:

	BIOMASS RESOURCES (QUADS)	USEFUL FUEL (QUADS)
Crop Residues[a,b]	7.5	4.2
Manure[c,b]	1.9	1.0
Waste Wood in Forests[d]	4.5	2.0
Pulp Waste[e]	1.4	1.4
Urban Refuse[f]	1.3	0.9
	16.6	9.5

[a] The quantity of crop residues is estimated to be 500 million tons of dry biomass.

(b) We assume that 80% of the biomass is collectable and that this is converted to useful fuel at 70% efficiency. In converting raw biomass to high quality liquid fuels conversion efficiencies are usually 40% or lower (see note 10, Chapter 9). However, by using solar energy as a heat source to drive the chemical reactions for making such fuels, the overall effective efficiency (defined as the heating value of the liquid fuel divided by the heating value of the raw biomass) can be 70% or higher. See note 9, Chapter 9.

(c) Manure production from animals in confinement is estimated to be 125 million tons/year.

(d) Forest waste resources are estimated to be 300 million tons/year. We assume this is used directly as boiler fuel.

(e) The level of pulp waste is estimated to be 95 million tons in 2010. We assume this is used directly as boiler fuel.

(f) The level of collectable urban refuse in 1975 was about 1.3 quadrillion Btu. We assume this doesn't increase in the future. We assume it is converted to useful fuel @ 70% efficiency.

29. J. Huetter, Jr., "Status, Potential and Problems of Small Hydroelectric Power Development in the United States," prepared by Energy Research and Applications for The Council on Environmental Quality, March 1978.

30. Stephen Rosen, *Future Facts*, Simon and Schuster, New York, 1976.

CHAPTER 14

1. Richard D. Lyons, "Experts Upset by Drop in Innovative Research," *New York Times*, p. D1, May 31, 1978.

2. National Science Foundation, "National Patterns of R&D Resources, Funds and Manpower in the United States, 1953-1977," NSF 77-310, 1977.

3. Freeman J. Dyson, "The Hidden Cost of Saying No!" *Bulletin of the Atomic Scientists*, p. 23, June 1975.

4. See Table 3 in Carl Djerassi, "Birth Control After 1984," *Science*, p. 941, September 4, 1970.

5. Editorial by Philip H. Abelson, "The Federal Government and Innovation," *Science*, p. 487, August 11, 1978.

6. E. J. Hobsbawm, *Industry and Empire*, Penguin Books Limited, London, 1969.

7. Charles Berg, private communication.

8. Robert Gilpin, "*Technology, Economic Growth, and International Competitiveness*," a report prepared for the Subcommittee on Economic Growth of the Joint Economic Committee of the United States Congress, July 9, 1975.

9. M. H. Farmer et al., *Application of Fluidized Bed Technology to Industrial Boilers*, Report to the Federal Energy Administration by Government Research Laboratories, Exxon Research and Engineering Company, Linden, N.J., 1976.

10. Robert W. Hagemeyer, (ed.), *Future Technical Needs and Trends in the Paper Industry—III*, presented at the TAPPI meeting of March 12-14, 1979, Technical Association of the Pulp and Paper Industry, Atlanta, Georgia, 1979.

11. Harold B. Finger, "Energy Development—A Need for New Arrangements in Public and Private Collaboration," pp. 157-166, in the *Proceedings of the Symposium on Implications of Energy Conservation and Supply Alternatives*, sponsored by the Division of Industrial and Engineering Chemistry of the American Chemical Society, Colorado Springs, January 30-February 2, 1978.

12. "President Seeks Big Energy Drive: Backs $100 Billion Program for U.S. Self Sufficiency in a Decade or Less," *New York Times*, page 1, September 23, 1975.

13. *Energy Security Act, Conference Report*, Senate Report No. 96-824, June 19, 1980.

14. H. W. Lewis et al., "Report to the American Physical Society by the Study Group on Light Water Reactor Safety," *Reviews of Modern Physics*, vol. 47 supplement no. 1, Summer 1975.

15. Charles E. Lindblom, *Politics and Markets*, Basic Books, p. 298, New York, 1977.

16. See, for example, ref. eight. Also a number of other recent studies have discussed the difficulties of large government supported demonstration projects. *In Analysis of Federally Funded Demonstration Projects*, a report prepared by the RAND Corporation for the National Bureau of Standards, vol. 1, April 1976, NTIS No. PB-2 53108, it is concluded that:

> Demonstration projects appear to be weak tools for tackling institutional and organizational barriers to diffusion [of technology] . page iv.
>
> Large demonstration projects with heavy federal funding are particularly prone to difficulty . page v.

See also *The Role of Demonstrations in Federal R and D Policy*, report prepared by the RAND Corporation for the Office of Technology Assessment, July 1978, USGPO Stock No. 052-003-00557-3 and "Applications of R and D in the Civil Sector," Office of Technology Assessment Report, June 1978.

17. James M. Utterback *et al.*, "Strategic Alternatives for Energy Conservation," Center for Policy Alternatives, Mass. Inst. of Technology, February, 1977.

18. William J. Abernathy and James M. Utterback, "Patterns of Industrial Innovation," *Technology Review*, p. 41, June/July 1978.

19. J. Herbert Hollomon, "Technical Change and American Enterprise," Report No. 139 of the National Planning Association, October 1974.

20. John Jewkes et al., *The Sources of Invention*, second edition, W. W. Norton and Company, New York, 1969.

21. *Technological Innovation: Its Environment and Management*, U.S. Department of Commerce, Washington, D.C., January 1967.

22. Frederick M. Scherer, *Industrial Market Structure and Economic Performance*, pp. 361-362, Chicago, Rand McNally, 1970.

23. John H. Gibbons, "Long-term Research Opportunities," in *Energy Conservation and Public Policy*, John C. Sawhill, ed., published for The American Assembly, Columbia University by Prentice-Hall, Inc., Englewood Cliffs, New Jersey, 1979; J. Herbert Hollomon, "Government and the Innovation Process," *Technology Review*, May 1979; Energy Advisory Committee of the Association of American Universities, Jack M. Hollander, chmn., "University Contributions to the National Energy Program: New Knowledge, New Talent, New Integration," 1979; "Direct Federal Support of Research and Development Report," in the *Advisory Committee on Industrial Innovation*, final report to the Dept. of Commerce, USGPO, Sept. 1979.

24. An illustrative list of research topics relating to fuel conservation organized according to type of application and type of activity is given in Marc H. Ross and Robert H. Williams, "The Potential for Fuel Conservation," *Technology Review*, Feb. 1977, p. 48. This list is based on ref. 25.

25. An extensive list of applied research topics is given in American Physical Society summer study on Technical Aspects of Efficient Energy Utilization, 1974. Available as W. H. Carnahan et al., *Efficient Use of Energy, A Physics Perspective*, from NTIS (PB-242-773), or in *Efficient*

Energy Use, Volume 25 of the American Institute of Physics Conference Proceedings. This work is briefly summarized by R. H. Socolow in *Physics Today*, August 1975.

26. Jerry Grey, George W. Sutton and Martin Zlotnick, "Fuel Conservation and Applied Research," *Science 200*, p. 135, Apr. 14, 1978.

27. Martin Zlotnick, private communication.

28. W. C. Reynolds, *ERDA Fossil Energy Review, June 28-29*, Department of Energy, Washington, D.C., 1977.

29. See the Introduction of ref. 25.

30. Laura Fermi, *Atoms in the Family*, University of Chicago Press, 1954, p. 185.

31. See comments on The Office of Naval Research Program, *Advisory Committee on Industrial Innovation*, p. 209, ref. 23.

32. For a description of the rationale for this program see "Federal Support for Public Interest Science," testimony by Frank von Hippel, presented on April 21, 1975 before the Subcommittee on the National Science Foundation of the Senate Committee on Labor and Public Welfare in the *Hearings on National Science Foundation Legislation, 1975*.

CHAPTER 15

1. A brief cultural history of the environmental ethic can be found in Robert Cahn, *Footprints on the Planet, A Search for an Environmental Ethic*, Universe Books, 1978.

2. Carter Goodrich, *Government Promotion of American Canals and Railroads, 1800-1890*, Greenwood Press, 1974. Louis Hartz, *Economic Policy and Democratic Thought: Pennsylvania 1776-1860*, Harvard Univ. Press, 1948, Quadrangle Books, Chicago, 1968. Gerald D. Nash, *State Government and Economic Development: A History of Administrative Policies in California, 1849-1933, Institute of Governmental Studies, Univ. of California, Berkeley, 1964*.

3. Samuel P. Hays, *Conservation and the Gospel of Efficiency, The Progressive Conservation Movement, 1890-1920*, Harvard Univ. Press 1959.

4. L. A. Chambers, "Classification and Extent of Air Pollution Problems", in *Air Pollution, Vol. I: Air Pollution and Its Effects*, A. C. Stern, ed., second edition, Academic Press, New York, 1968.

5. Ronald Inglehart, *The Silent Revolution, Changing Values and Political Styles Among Western Publics*, Princeton Univ. Press, 1977.

6. Murray L. Weidenbaum and Robert De Fine, "The Cost of Federal Regulation of Economic Activity", American Enterprise Institute, Washington, DC, May 1978.

7. Interview of Walter W. Rostow by Leonard Silk, New York Times, May 21, 1978, Section 3.

8. J. K. Galbraith, *Economics and the Public Purpose*, Houghton Mifflin Company, Boston, 1973.

9. Ref. 8, p. 317.

10. Charles E. Lindblom, *Politics and Markets, The World's Political-Economic Systems*, Basic Books, 1977, p. 67.

11. Augustine Cournot, *Revue Sommaire des Doctrines Economiques*, 1877.

12. Adam Smith, *An Inquiry into the Nature and Causes of the Wealth of Nations*, edited by Edwin Cannan, The Modern Library, New York, p. 423, 1937.

13. B. W. Cone *et al.*, "An Analysis of Federal Incentives Used to Stimulate Energy Production," revised report by Battelle Pacific Northwest Laboratory, PNL-2410 REV, Dec. 1978. See also Gerard M. Brannon, *Energy Taxes and Subsidies*, a report to the Energy Policy Project of the Ford Foundation, Ballinger, 1974; "Federal Subsidy Programs," a staff study for the Joint Economic Committee of the Congress, U.S. Gov't. Printing Office, October 18, 1974; and General Accounting Office, "Nuclear Power Costs and Subsidies," June 13, 1979.

14. Since the actions of government are manifold it is arguable what constitutes subsidy. We have selected items which in our opinion are direct forms of tax and other assistance. The most important categories are: (a) for investor owned electric utilities investment tax credits and liberalized depreciation for tax purposes; (b) for publicly owned utilities, federal ownership of hydropower facilities, tax exemption, tax free bonds, low interest loans; (c) for both investor and publically owned utilities, federal R&D and federal nuclear fuel cycle activities; (d) for the oil and gas industry, intangible drilling expensing and what remains of the depletion allowance and the foreign tax credit.

15. Charles A. Berg, "Energy Conservation in Industry: the Present Approach, The Future Opportunities," report prepared for the Council on Environmental Quality of the Executive Office of the President (May 1979).

16. Talbot Page, *Conservation and Economic Efficiency, An Approach to Materials Policy*, Resources for the Future, Johns Hopkins University Press, 1977.

17. Clark W. Bullard III, "Energy Conservation Through Taxation," Center for Advanced Computation Document 95, University of Illinois (February 1974).

18. Bruce Hannon, Robert A. Herendeen, and Peter Penner, "An Energy Conservation Tax: Impacts and Policy Implications," report to the Council on Environmental Quality of the Executive Office of the President (July 1979).

19. Jay W. Forrester, "A Self-Regulating Energy Policy," Keynote Address of the AIAA Annual Meeting, *Astronautics and Aeronautics*, p. 40 (July/August 1979).

20. John B. Anderson (Ranking Republican on the House Ad Hoc Energy Committee in the 1977–78 Congress), "For a Tax of 50 Cents on Gas," Op Ed column in *The New York Times* (August 28, 1979).

21. A. H. Rosenfeld and A. C. Fisher, "Marginal Cost Pricing with Refunds per Capita," submitted to the 1978 California Energy Commission Hearings on Fuel Management (July 1978).

22. Robert M. Solow, "The Economics of Resources or the Resources of Economics," *Proceedings of the American Economic Association*, May 1974.

23. Colin Clark, "The Economics of Over Exploitation," *Science 181*, p. 630, Aug. 17, 1973.

24. M. A. Adelman, "Is the Oil Shortage Real?" *Foreign Policy* 9, p. 69, 1973.

25. Marc Roberts, "On Reforming Economic Growth" in *The No Growth Society*, Mancur Olson, ed., *Daedalus*, Fall, 1973.

26. The gasoline tax in several countries was as follows in 1978:

COUNTRY	TAX (¢/GALLON)	COUNTRY	TAX (¢/GALLON)
U.S.	12	Italy	154
Belgium	118	Netherlands	112
France	110	Sweden	69
Germany	102	United Kingdom	75

SOURCE: *Statistical Abstract of the United States*, p. 655, 1978.

27. U.S. gasoline consumption in 1978 was 114 billion gallons (*Monthly Energy Review*, Energy Information Administration, March 1979). A tax increase of $1.20/gallon would have generated additional revenues of $137 billion at the 1978 level of consumption. For comparison social security revenues in 1978 were about $124 (*Statistical Abstract*, p. 259, 1978). Since about 10% of all workers are not covered by social security, perhaps only 90% of the energy tax would be used to reduce the social security tax.

28. The revenue requirements of $137 billion (see note 27) could have been covered by a tax of $1.75 per million Btu at the 1978 level of energy consumption (78 quadrillion Btu).

29. The quantities for Table 15.2 were calculated as follows: The data on energy use by different income groups is from 1960–61. The essential assumption is then that the ratio of energy use to energy use by the average household in 1978 is the same as in 1960–61 for households with the same income ratio in 1960–61 and 1978.

SELECTED HOUSEHOLDS INCOME LEVELS 1960–61 (DOLLARS)	1978 HOUSEHOLD INCOME (DOLLARS)	ENERGY CONSUMPTION PER HOUSEHOLD (RELATIVE TO THE AVERAGE)*		
		RESIDENTIAL ENERGY	GASOLINE	INDIRECT ENERGY
2980	8,000	0.81	0.40	0.40
6580 (average)	17,700	1.0	1.0	1.0
9970	26,900	1.16	1.57	1.30

*Denton E. Morrison, "Equity Impacts of Some Major Energy Alternatives," Michigan State University, May 1977. These numbers are based on Robert Herendeen, "Affluence and Energy Demand," *Mechanical Engineering*, p. 18, October 1974.

The net impact of the energy tax for different income levels is calculated as follows:

a) The number of economically defined households was 72.9 million in 1975 and is estimated by us to be 77.7 in 1978 (from *Statistical Abstract of the United States*, 1978).

b) Residential energy consumption in 1975 was 16.22 quads or 222 million Btu per household (see ref. 1, Chapter 8). We assume that the energy use per household was the same in 1978. Non-renewable energy use per household is 208 million Btu, obtained by subtracting from total energy use per household hydroelectricity consumption, which amounted to 14 million Btu. [Households accounted for 34.5% of electricity demand and hydroelectricity production amounted to 3.15 quads in 1978.]

c) Gasoline consumption for autos and light trucks was 10.2 quads in 1975 (see Table 9.1), and we estimate that it was 11.3 quads in 1978 (assuming consumption increases in the same ratio as *total* motor gasoline consumption, as obtained from *Monthly Energy Review*, March 1979), or 145 million Btu per household.

d) Indirect non-renewable energy consumption in 1975 was 610 million Btu per household—the difference between total non-renewable energy consumption (963 million Btu, obtained from *Monthly Energy Review*, March 1979) and household consumption of non-renewable fuels, electricity and gasoline (353 million Btu).

e) The net increase in the cost of non-energy goods and services when a $1.75 per million Btu energy tax is levied is estimated as the difference between the impact of the fuels tax on indirect energy use ($429, $1068, and $1388 for the three households) and the impact of the reduction in industrial taxes on the cost of non-energy production. We assume that half the total tax reduction ($69 billion) accrues to industry, so that the total cost of non-energy production [$1524 billion = GNP ($2108 billion)—gross private domestic investment ($346 billion)—expenditures for energy ($238 billion)] is reduced by 100 × (69/1524) = 4.5%. We assume that the non-energy expenditures ($7140, $16,300, and $24,900 for the three households) are reduced by this percentage ($321, $734 and $1120 for the three households).

f) The social security tax is ordinarily 6.05% of income up to an income level of $17,700. We assume savings in this amount.

30. In 1978 the corporate income tax plus half the individual income tax amounted to $148 billion (see *Statistical Abstract of the United States: 1978,* p. 261). Thus the average level of the energy tax in 1978 would have been

$$\$(148 \times 10^9)/(78 \times 10^{15}\ \text{Btu}) = \$1.90/10^6\ \text{Btu}$$

or about 24¢ per gallon of gasoline equivalent energy. This tax should be compared to the actual average price of energy in 1978, about $3.06/10⁶ Btu (see note 31).

31. The following is an estimate of the quantities, prices, and total expenditures for energy in the U.S. in 1978[a]:

SECTOR	FUEL	QUANTITY (QUADS)	PRICE ($/10⁶ BTU)	EXPENDITURES ($10⁹)
Residential[b]	Electricity[c]	8.07 (2.32)	3.40	27.4
	Gas	6.00	2.58[d]	15.5
	Oil	2.98	3.53	10.5
	TOTAL	17.1	$3.13	53.4
Commercial	Electricity[c]	6.09 (1.75)	3.45	21.0
	Gas	1.68	2.29[d]	3.8
	Oil	4.17	2.66[e]	11.1
	Coal	0.28	1.50	0.4
	TOTAL	12.2	$2.98	36.3
Industrial	Electricity[c]	9.50 (2.73)	2.28	21.7
	Gas	8.27	1.88[d]	15.5
	Oil	6.74	2.40[e]	16.2
	Coal	3.57	1.10	3.9
	TOTAL	28.1	$2.04	57.3
Transportation	Electricity[c]	0.05 (0.015)	2.28	0.1
	Gas	0.54	2.29	1.1
	Gasoline	14.2	5.20	73.8
	Diesel Fuel	3.7	2.88	10.7
	Jet Fuel	2.1	2.88	6.0
	TOTAL	20.6	$4.45	91.7
GRAND TOTAL		78.0	$3.06	239

NOTES:

[a] Except where otherwise noted the data here were obtained from *Monthly Energy Review,* Energy Information Administration, U.S. Department of Energy, March 1979.

[b] *Monthly Energy Review* (see note a) gives residential/commercial totals only. Gas and oil consumption for the residential sector in 1977 is given in Office of Technology Assessment, *Residential Energy Conservation,* March 1979. We assume that the residential sector's share of the residential/commercial totals was the same in 1978 as in 1977 (42% for oil, 76% for gas).

[c] "1979 Annual Statistical Report," *Electrical World,* March 15, 1979. For electricity primary fuel consumption is listed, with electricity consumption given in parentheses. The electricity price shown is the total revenues divided by the primary fuel consumption.

[d] In 1977 average U.S. residential, commercial, and industrial natural gas prices were in the ratio 1/0.89/0.73 (*Statistical Abstract of the United States,* 1978). We assume this same ratio applies to 1978. The 1978 residential gas price is obtained from ref. (a).

[e] Estimate

32. In this book we have assumed that in the absence of an energy tax the price of energy would increase (in real terms) at an average rate of 2%/year (see Chapter 3). Thus the average price of energy in 2010 without the tax would be about $5.80/10⁶ Btu.

Assuming that the required tax revenues increase with GNP (to a level in 2010 that is 2.15 times the 1978 level) the tax on energy in 2010 in the negative energy growth scenario described in Chapter 13 if applied to all energy, would be

$$\frac{2.15 \times (\$148 \times 10^9)}{64 \times 10^{15} \text{ Btu}} = \$5.00/10^6 \text{ Btu}$$

33. With the tax shift in place the fraction of GNP spent on energy in 1978 would have been

$$\frac{\$(3.06 + 1.90)/10^6 \text{ Btu} \times (78 \times 10^{15} \text{ Btu})}{\$2128 \times 10^9} = 0.18$$

In 2010 this fraction would be reduced to

$$\frac{\$(5.80 + 5.00)/10^6 \text{ Btu} \times (64 \times 10^{15} \text{ Btu})}{2.15 \times (\$2128 \times 10^9)} = 0.15$$

For comparison about 11% of GNP was actually spent on energy in 1978.

34. Robert J. Samuelson, "Politics and a Gasoline Tax," *National Journal*, p. 1493 (September 8, 1979).

35. Elizabeth Drew, "A Reporter at Large: Charlie," *The New Yorker*, Jan. 9, 1978.

36. See Chapter 13, ref. 20.

37. Clark W. Bullard III, "Energy and Employment Impacts of Policy Alternatives," in *Energy Analysis: A New Public Policy Tool*, Martha W. Gilliland, Ed., AAAS Selected Symposium 9, Westview Press, Boulder, Colorado, 1978.

38. E. R. Berndt, "Aggregate Energy, Efficiency, and Productivity Measurement," *Annual Review of Energy*, Vol. 3, 1978.

39. Bruce Hannon, "System Energy and Recycling: A Study of the Beverage Industry," Center for Advanced Computation, Univ. of Illinois, Urbana, March 1973.

40. Bruce Hannon, "Energy Conservation and the Consumer," *Science 189*, p. 95, July 11, 1975.

41. *Federal Register*, FERC Application, p. 25459, June 13, 1978.

42. *Federal Register*, FERC petition for special relief, p. 25462, June 13, 1978.

43. *Federal Register*, ERA Interpretations of the General Counsel, p. 25080 and 25081, June 9, 1978.

44. *Federal Register*, FERC Application, p. 24122, June 2, 1978.

INDEX